D1175441

# Mechanical Properties of Metals

## Atomistic and Fractal Continuum Approaches

# Mechanical Properties of Metals

## Atomistic and Fractal Continuum Approaches


**C W Lung**

*International Centre for Materials Physics*
*Institute of Metal Research*
*Chinese Academy of Sciences*
*Shenyang, China*

**N H March**

*University of Oxford*
*Oxford, England*

*Published by*

World Scientific Publishing Co. Pte. Ltd.

P O Box 128, Farrer Road, Singapore 912805

*USA office:* Suite 1B, 1060 Main Street, River Edge, NJ 07661

*UK office:* 57 Shelton Street, Covent Garden, London WC2H 9HE

**British Library Cataloguing-in-Publication Data**
A catalogue record for this book is available from the British Library.

**MECHANICAL PROPERTIES OF METALS**
**Atomistic and Fractal Continuum Approaches**



ISBN 981-02-2622-5

Printed in Singapore.

# Preface

This book aims to bring together continuum elasticity theory, electronic structure and the concept of fractals as applied to roughness and toughness of metals.

Of course, a major contribution to fracture was made by Griffith, in which, though incomplete for reasons that are now largely understood, he derived an expression for the limiting strength of a material. This involved intimately the surface energy, which subsequently, at least in a simple metal like Al with $s$ and $p$ electrons, has been related to the energy of formation of a vacancy. However, a tremendous step forward came with the concept of a dislocation.

It was shown that if a limited area of one plane slips by one atomic distance over the neighbouring plane, the boundary of this area is a closed loop of dislocation. Once this loop is formed, glide can propagate across the plane by the spreading of the area, which is a motion of the dislocation line across its glide plane. General interest in dislocation theory was aroused by its success in providing atomistic theories for plastic deformation and crystal growth. Progress has been made in this field for more than one decade on the treatment of elastic anisotropy and dislocation mobility. For more than two decades, interests have concentrated on deepening our understanding of the structure and role of the dislocation core the behaviour of a pile-up of dislocations (simulating a crack). Interatomic forces (electronic structure) play a decisive role in the structure of the dislocation core and even in fracture. In particular, the temperature dependence of fracture toughness of materials has a close relationship to the interatomic forces. Molecular dynamics has been applied to understand the dislocation motion and the emission of dislocations at the crack tip under loading. Dynamics of crack propagation has become an area of considerable current interest for theoretical physicists and material scientists.

As to electronic structure, electron density theory based on a one-body potential $V(\mathbf{r})$ including electron-electron exchange and correlation interactions has transformed what can be done on electronic structure of both perfect and defective crystals. Nevertheless, it is still important to subsume ideas involved there into interatomic force fields, which can then be used to study extended defects such as surfaces, grain boundaries, dislocations and cracks. There is still, however, much to do in understanding, in metals, the role of collective effects (including plasmons) in determining mechanical properties and tribology of conducting materials. If our book proves to make a contribution to furthering the progress in relating and enriching ideas from continuum theory, from electronic structure, and from concepts of fractal structure, then that will be more than ample justification for the effort involved in the present project.

We are conscious that in some areas embraced in our book, there is rapid movement at the time of writing. If authors in electron theory or in very practical aspects of materials science and engineering see where we ought to do better, we shall count it a privilege if they write to us with positive suggestions for improvement.

Over a decade or more, diverse scientists have recognized that many of the structures common in their experiments have a special kind of geometrical complexity. Mandelbrot in his pioneering work introduced the concept of fractals and used the idea of a fractal dimension which often is not an integer to characterize the complex structure quantitatively. Fractals may be considered as systems which obey the law of self-similarity, or are self-affine.

Since Mandelbrot *et al.* (1984) showed that fractured surfaces are fractals in nature and that the fractal dimensions of the surfaces correlate well with the toughness of the material, many authors have found that the fractal dimension depends on the fracture properties of materials, but the values of it seem in a narrow range for measurements with a resolution down to the micron scale. This has led to much discussion on the universality and specificity of the fractal dimension of fractured surfaces. However, the roughness index (or local fractal dimension) is found to display wide differences depending on materials on a small length scale by means of scanning electron microscopy (SEM). Another problem is that the negative correlation of the fractal dimension of fractured surfaces with toughness of ductile materials is quite difficult to understand. These basic problems remain open at the time of writing and much remains to be done.

We could not end this Preface without acknowledging our indebtedness to other workers. C. W. Lung wishes to thank Professor P. L. Zhang for leading him into active research in materials physics, Professors T. S. Ge(Kê) and K. X. Guo for their advice, collaboration and much practical support, and Professors H. Wu, X. Li, J. Z. Gao and C. X. Shi for their continuous encouragement.

N. H. March is greatly indebted to Professor J. A. Alonso for many invaluable discussions on electrons in metals and alloys. Professor A. B. Lidiard played a prime role in interesting N. H. March in the problems of materials science and he is thanked here for the great influence he has thereby had on the present project. Dr. P. Schmidt has given N. H. March much encouragement and stimulation in his work in this field and support from the Office of Naval Research (ONR) is gratefully acknowledged. We are especially grateful to Professor V. Vitek for his permission to draw heavily on his lectures at the International Centre for Theoretical Physics (ICTP) in Trieste and from his contributions to a NATO Advanced Study Institute. The authors first met at ICTP Trieste, and much of their collaboration has resulted from frequent summer visits there. They thank Professors A. Salam and M. A. Virasoro for much hospitality. We thank also the many researchers who have sent us papers and have thereby helped us in preparing the book. Last, but not least, Ms G. Su has skilfully typed most of the manuscript and we are very grateful to her.

[illegible] discussions [illegible]
[illegible] in the problems of [illegible]
[illegible] great influence [illegible]
[illegible]
Research (NFR) is gratefully acknowledged. [illegible]
[illegible] International Centre for Theoretical Physics (ICTP) [illegible]
[illegible] NATO Advanced Study Institute [illegible]
[illegible]
[illegible] the book. [illegible]
[illegible] of the manuscript and [illegible]

## CONTENTS

# Chapter 1

# Background and Some Concepts

## Introduction

We assume knowledge of general solid state physics as in Rosenberg (1992). However, we shall begin by briefly summarizing a few concepts* that are basic to an understanding of later chapters below.

## 1.1. *Elastic and Plastic Regimes*

It is helpful to classify the discussion of mechanical properties by defining two regimes (i) elastic and (ii) plastic.

### 1.1.1. *Elastic Deformation*

The mechanical properties of materials are of vital importance in determining their fabrication and practical applications. Initially as a load is applied on the material, the nominal stress is defined as the load divided by the original cross section area, and the nominal strain as the extension divided by the original length. As the stress is increased, the strain increases uniformly and the deformation produced is completely reversible. This is so-called the elastic region. The stress and resulting strain are proportional to one another and obey Hooke's law.

From an atomistic point of view, if we pull two atoms apart or push them together by a force, the atoms can find a new equilibrium position in which the atomic and applied forces are balanced. The force in the bond is a function of the displacement. The deformation of the bond being reversible means that, when the displacement returns to the initial value, so does the force return

*Readers may skip this Chapter if they are familiar with this background material.

simultaneously to its corresponding value. The bulk elastic behaviour of large solid bodies is the aggregate effect of the individual deformations of the bonds which are the building blocks.

When the applied forces are sufficiently small, the elastic displacement is always proportional to force. This is Hooke's law. The elastic constant is a key parameter, to express the coefficient of proportionality between force and displacement. When the applied forces are large, the elastic displacement deviates from Hooke's law. The relation between force and displacement is nonlinear. This is then called nonlinear elasticity.

### 1.1.2. *Atomic Forces and Elastic Properties*

Taking NaCl type ionic crystals as an example, Cottrell (1964a) discussed the interaction energy of a pair of univalent ions at a distance $r$ as

$$U(r) = \pm\frac{e^2}{r} + \frac{B}{r^s} \tag{1.1.1}$$

where, $s \approx 9$, and where $+$ and $-$ refer to like and unlike ions respectively. Having summed the repulsive and attractive interactions with nearest neighbours, the total interaction energy of an ion can be written as

$$U_z = -A\frac{e^2}{r} + 6\frac{B}{r^s} \tag{1.1.2}$$

where $A$ is called the Madelung constant, equal to 1.7476 for the NaCl type crystals. At the equilibrium condition, $\frac{dU_z}{dr} = 0$, at $r = r_0$. Thus,

$$B = \frac{Ae^2 r_0^{s-1}}{6s} \tag{1.1.3}$$

and

$$U_z = -\frac{Ae^2}{r}\left[1 - \left(\frac{1}{s}\right)\left(\frac{r_0}{r}\right)^{s-1}\right]. \tag{1.1.4}$$

This is the work required to dissociate the crystal into 2N separate ions (N positive and N negative).

The elastic constant $E$,

$$E = \frac{f}{u} = \left(\frac{1}{6}\right)\left(\frac{\partial^2 U_z}{\partial r^2}\right)_{r=r_0} = \frac{(s-1)Ae^2}{6r_0^3} \tag{1.1.5}$$

where $\frac{U_z}{6}$ is the energy per each nearest-neighbour bond, and $u = r - r_0$, is the elastic displacement.

The bulk modulus of elasticity of the material is defined by

$$p = -K\frac{\Delta V}{V} \tag{1.1.6}$$

where $p$ is a hydrostatic pressure, $\frac{\Delta V}{V}$ is the volume change.

$$K = \frac{-p}{\left(\frac{\Delta V}{V}\right)} = -\frac{pr_0^2}{r_0^2\left(\frac{\Delta V}{V}\right)} \cong \frac{f}{r_0^2\left(\frac{3u}{r_0}\right)} = \frac{1}{3r_0}\left(\frac{f}{u}\right) = \frac{(s-1)Ae^2}{18r_0^4}\,. \tag{1.1.7}$$

In KCl, it gives $K^T = 1.88 \times 10^{11}$ dyn cm$^2$, whereas the observed value (extrapolated to OK) is $2 \times 10^{11}$ dyn cm$^2$. The corresponding calculations of elastic constants of metallic crystals are much more difficult for the laws of force are much more complicated. We shall discuss this in Chap. 5.

### 1.1.3. *Plastic Deformation*

Plastic deformation is characterized by a permanent deformation of the material. Unlike elastic deformation, it does not reverse on unloading but leaves the material with a permanent shape. This is called the plastic region. Between these two regions, there is a limiting stress, called the yield stress of the material, or the critical resolved shear stress for a single crystal.

The crystallinity of the structure is the prime cause of this behaviour, for it enables whole slabs of crystal to glide past one another. Each slip is a displacement, in certain glide direction, generally the crystal direction of closest atomic packing on certain crystal planes which is called the slip plane. In fcc and hcp metals, these are mainly close-packed planes, but in bcc metals, the situation is complicated. It will be discussed later. Slip begins on some small area of the surface.* The slip-front line between the slipped and unslipped areas is by definition a *dislocation* line. The glide motion of a dislocation is a property of a periodic crystal. The transition from the slipped to the unslipped region is spread over several atomic distances which is the width of the dislocation. Every atom in this transition region is pushed only a little further out of its original equilibrium site when it moves forward. This is the reason why dislocations can move easily in the crystal. Thus, the yield stress is much lower than the theoretical strength of crystals. Dislocation theory plays important role in understanding the microscopic processes in plastic deformation. Even the elastic theory of dislocations may explain many phenomena, such as yielding,

*See, for instance, the early paper of Chen and Pond (1952).

work hardening, etc. etc. It provides not only a deeper qualitative physical picture of plastic deformation but also to a certain degree a quantitative analysis of it.

Plastic deformation can also occur by twinning. The atoms slide, layer by layer to bring each deformed slab into mirror-image lattice orientation relative to the undeformed material. The critical stress of twinning is usually higher. Twins form at low temperature and under rapid deformation, e.g. bcc iron strained quickly at room temperature and slowly at 100K.

## 1.2. *Griffith Criterion: Role of Surfaces*

Griffith (1924) derived an expression for the elastic crack propagation on the basis of thermodynamic considerations. He reasoned that a crack would advance when the incremental release of stored elastic strain energy $dW_E$ in a body became greater than the incremental increase of surface energy $dW_s$ as new crack surface was created. For the two-dimensional case in plane stress

$$\begin{aligned} W_E &= \frac{\pi\sigma^2 c^2}{E} \\ W_s &= 4c\gamma_s \end{aligned} \tag{1.2.1}$$

where, $\sigma$ is the nominal stress; $E$, the elastic modulus; $2c$, the length of the crack, and $\gamma_s$ the specific surface energy.

The Griffith criterion can then be written as with $\sigma_F$, the fracture stress,

$$\sigma_F = \sqrt{\frac{2E\gamma_s}{\pi c}} \tag{1.2.2}$$

by the condition that

$$\frac{\partial}{\partial c} W_E \geq \frac{\partial W_s}{\partial c} .$$

Subsequent analysis* in fracture mechanics defines a parameter, crack extension force, $G = K^2/E$ (in plane strain) being equal to a critical value, $G_{Ic}$,

*This is associated with the names of Irwin (1957) and Inglis (1913). The analysis given by Inglis has been generalized by R. Löfstedt (Phys. Rev. *E55*, 6726, 1997) who has proposed an inequality involving a ratio of time scales to determine whether a material is brittle or ductile. One time scale is 'ductile', and is associated with the rate of decrease of the tensile stress at the tip of a narrow crack. The above ductile time scale is to be compared with a characteristic 'phonon time' $a/v_s$, where $v_s$ is the velocity of sound, and $a$ measures the lattice spacing. See also R. W. Armstrong (Mat. Sci. Eng. *1*, 251, 1996) also A. Kelly, W. R. Tyson and A. H. Cottrell, Phil. Mag. *15*, 567, 1967 and J. R. Rice and R. Thomson, Phil. Mag. *29*, 73, 1974).

the crack-resistance force of the material. For the elastic crack in an infinitely wide plate

$$G_{Ic} = \frac{K_{Ic}^2}{E} = \frac{\sigma_F^2 \pi c}{E} \tag{1.2.3}$$

where $K$ is the stress intensity factor: $K_{Ic}$ is called fracture toughness.

Comparing to Eq. (1.2.2), $G_{Ic} = 2\gamma_s$ — the two approaches lead to the same result although their methods are different. The specific surface energy plays a very important role in brittle fracture.

Engineering materials do not fracture in a completely elastic manner. The localized plastic deformation near crack tip gives the materials some toughness, or resistance to crack propagation.* Orowan (1948) proposed to add a term $\gamma_p$, the plastic work expanded during crack propagation to the elastic work $\gamma_s$ as an effective specific surface energy in Eq. (1.2.2). The Griffith equation is modified to read (in plane stress)

$$\sigma_F = \sqrt{\frac{2E}{\pi c}(\gamma_s + \gamma_p)} \sim \sqrt{\frac{2E\gamma_p}{\pi c}}\,. \tag{1.2.4}$$

(Some authors wrote $2(\gamma_s+\gamma_p)$ as $2\gamma_s+\gamma_p'$; then, $\gamma_p' = 2\gamma_p$.) From Eq. (1.2.4), it seems $\gamma_s$ is no longer an important factor in this process. However, Tetelman *et al.* (1967) showed that for the case of Fe–3% Si, by Frank-Read source multiplication (Cottrell 1964a):

$$\gamma_m = \text{const.}\, \gamma_s N_0^{\frac{3}{2}} \left(\frac{v_s}{v_c}\right)^2 T^{\frac{5}{2}} \tag{1.2.5}$$

where $\gamma_m$ is defined as the product of the work done in a unit volume element of material when the crack advances and the distance perpendicular to the crack in which the deformation is extensive. $N_0$ is the density of mobile dislocation sources, $v_s$ and $v_c$ are velocities of sound and the crack respectively. $\gamma_m$, like $\gamma_p$ is a measure of the intrinsic toughness of a solid. In Eq. (1.2.5), $\gamma_s$ is a multiplying factor and not an addition term. The change of $\gamma_s$ directly influences the change of $\gamma_m$.

Moreover, Lung and Gao (1985) calculated the relative $K_{ic}$ value of metals with a simplified dislocation motion model and BCS dislocation distribution function at the crack tip

$$G_{ic}^p \cong 2\gamma_p \propto W_i = E_0 (K_{ic}^0)^2 F_i(\theta_0) r_i^*(\theta_0)^{\frac{1}{2}} \tag{1.2.6}$$

---

*A general article on failure of solids is that of M. Marder and J. Fineberg (Phys. Today, September 1996, p. 24).

where $K_{ic}^0$ is the fracture toughness in linear elastic case ($= \sqrt{EG_{ic}^0}$ or $\sqrt{2\gamma_s E}$); $r_i^*$, the plastic zone size; and $E_0 \propto E^{-1}$. The $E_0$ in Eq. (1.2.6) is proportional to the inverse of the elastic modulus of materials, and $F_i(\theta)$, the angular dependent function respectively. $\theta_0$ is the direction of $r_{\max}^*$ of the plastic zone.

Comparing Eqs. (1.2.6) with (1.2.5), the two approaches lead to the same conclusion that $\gamma_s$ plays the role of a multiplying factor in the expression of critical crack extension forces. For a multiplying factor,

$$\frac{\Delta(\gamma_s f)}{(\gamma_s f)} = \frac{\Delta\gamma_s}{\gamma_s} + \frac{\Delta f}{f}\,. \qquad (1.2.6)'$$

The relative change of $\gamma_s$ is as important as that of $f$. If we consider the underlying role of atomic forces in the structure of dislocation core and dynamics, the role of interatomic forces is not only in the surface energy term but also in the dislocation core structure.

## 1.3. *Peierls Stress and Barrier*

A dislocation experiences an oscillating potential energy as it glides in a crystal. In the Peierls model (Peierls, 1940), the bonds across the glide plane were considered to interact via an interatomic potential, while the remainder of the lattice was linearly elastic. Nabarro (1957) gave an analytical expression for the dislocation core model. One can approximately estimate the ideal lattice resistance to dislocation motion by means of the Peierls model. The resolved applied stress necessary to move the dislocation over the Peierls barrier is called the Peierls stress, $\sigma_p$. The Peierls stress comes from the expression for the Peierls energy which changes for a translation of the dislocation by a distance smaller than the Burgers vector.

Figure 1.1, reproduced from Nabarro (1967), shows the Peierls model of a dislocation. The material above A and below B is regarded as forming an elastic continuum. The force between the rows A and B is a periodic function of the displacement.

As the dislocation moves through the lattice, it passes through an unsymmetrical configuration to a different symmetrical configuration in which one half plane of atoms on the expanded side of the glide plane lies midway between two half planes on the compressed side. Further motion passes through unsymmetrical configurations back to a state equivalent to the original. The dislocation moves if a finite force acts on it. The critical stress is the Peierls stress. After a lengthy calculation, the approximate energy of misfit

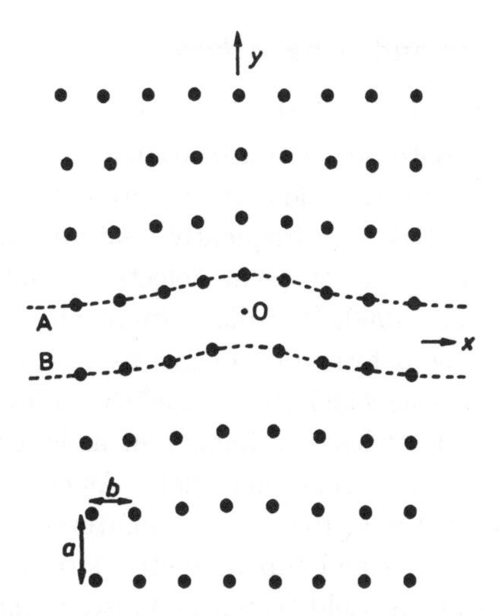


Fig. 1.1. Peierls's model of a dislocation. The material above A and below B is regarded as forming an elastic continuum. The force between the rows A and B is a periodic function of the displacement.

is given by (Nabarro, 1967)

$$E = \left[\frac{b^2\mu}{4\pi(1-\nu)}\right]\left\{1 + 2\cos 4\pi\alpha \exp\left(\frac{-4\pi\zeta}{b}\right)\right\} . \tag{1.3.1}$$

The force acting on unit length of the edge dislocation is,

$$F = -\left(\frac{1}{b}\right)\frac{dE}{d\alpha} = \frac{2b\mu}{(1-\nu)}\sin 4\pi\alpha \exp\left(\frac{-4\pi\zeta}{b}\right) \tag{1.3.2}$$

where $\zeta = \frac{a}{2(1-\nu)}$ is a parameter measuring the width of dislocation, $\alpha b$, the displacement of the centre of the dislocation from the original equilibrium position, $\mu$, the shear modulus and $\nu$ Poisson's ratio.

The maximum value of Eq. (1.3.2) is the critical shear strength; the Peierls stress is given by

$$\sigma_p = \frac{2\mu}{4\pi(1-\nu)}\exp\left(\frac{-4\pi\zeta}{b}\right) . \tag{1.3.3}$$

Considering the spirit of this model, and extending the displacement of the centre of the dislocation to include the thermal vibration amplitude, the temperature dependence of the crss can be obtained (see Lung *et al.*, 1966; or later Sec. 10.3).

## 1.4. *Dislocation Core and Atomic Force*

Early development of dislocation theory, and related theoretical treatment of metallic properties controlled by dislocations, focussed most attention on the effects of long-range elastic fields. In the present context (see also Vitek, 1995) mechanical properties were frequently analyzed in terms of long-range dislocation-dislocation, dislocation-point defects, etc interactions. The attitude prevailing in the late 1960s that dislocation cores were of but secondary importance in the plastic deformation of metals was radically altered in the next two decades. It became widely recognized then that dislocation core phenomena could play a role at least as important as long-range interactions in the deformation behaviour of many materials. As emphasized in the studies of Vitek and co-workers (Vitek, 1985) clear signatures of core effects are to be found in deformation modes and slip geometry, strong orientation and temperature dependences of the yield stress, and also in anomalous temperature dependence of the yield and flow stresses (see also Duesbery and Richardson, 1991).

Significant impetus for such atomistic modeling has been the marked improvement in experimental techniques (see Appendix 2.1), such as high resolution electron microscopy (HREM), that are capable of atomic resolution.

## 1.5. *Stacking Faults*

Rosenberg (1992) discussed how close-packed planes of hard spheres can be stacked to form, say, an fcc structure, Fig. 1.2 being reproduced from his account.

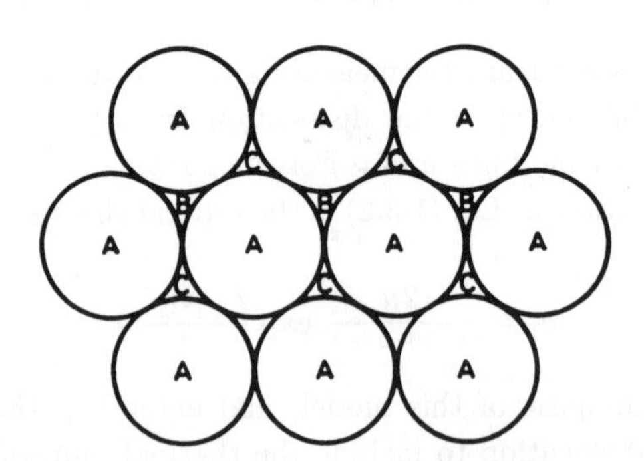


Fig. 1.2. The close-packed array of spheres. Note the three different possible positions. A, B and C for the successive layers.

The first and second layers can be in positions labelled A and B while the third layer can be placed above the C positions. The pattern continues as ABCABC..., the pattern repeating at every third layer.

A stacking fault* in such an fcc structure occurs if this sequence gets disturbed, as in ABCBCABC.... Here a layer A is missing, while in the sequence ABCABACABC... an extra A layer has been introduced. While stacking faults can, at least in principle, extend through the entire crystal, they usually occupy only a part of the plane. In this last case, of a stacking fault which terminates within the crystal, the configuration at the termination is referred to as a partial dislocation.

## 1.6. *Glissile and Sessile Dislocations*

Dislocations that can move by pure slip are called glissile. Dislocations which cannot glide, but have to move by some form of mass transport are called sessile (Read, 1953).

In crystals, the dislocation core spreads to certain crystallographic planes containing the dislocation line. If the core spreads into one of such planes, the core is planar and is glissile. If the core spreads into several non-parallel planes of the zone of the dislocation line, it is non-planar and is sessile. In the former case the dislocation moves easily in the plane of the core spreading, while in the latter case, it moves only with difficulty (Vitek, 1992). A Shockley partial is a partial dislocation, the Burgers vector of which lies in the plane of the fault. Then, Shockley partials are glissile. A Frank partial is a partial dislocation, the Burgers vector of which is not parallel to the fault. Then, Frank partials are sessile.

## 1.7. *Concept of Fractals*

Over a decade or more, diverse scientists have recognized that many of the structures common in their experiments have a quite special kind of geometrical complexity. The pioneering work was that of Mandelbrot (1977, 1979, 1982, 1988) who drew attention to the particular geometrical properties of such things as the shore of continents, tree branches, or the surface of clouds.

---

*Stacking faults remain a challenge for interatomic force fields. A novel system which might test N-body force lows discussed in Chapter 8 has arisen from the study of S. A. de Vries *et al.*, (Phys. Rev. Lett. *81*, 381, 1998). These authors have studied the influence of Sb on the formation of stacking faults due to Ag(111) growth using X-ray scattering (see also related theoretical studies of S. Oppo *et al.*, Phys. Rev. Lett. *71*, 2437, 1993).

Mandelbrot used the word 'fractal' for these complex shapes, in order to emphasize that they are to be characterized by a non-integer (fractal) dimensionality.

Our interest here is the fractal aspects of fractured surfaces. Mandelbrot *et al.* (1984) gave an elegant route for determining the fractal dimension $D$ of the fractured surface. Their work pointed to a correlation between toughness and $D$. Further studies were performed by Lung (1986), Pande (1987), Lung and Mu (1988) and Xie and Chen (1988). Bouchaud *et al.* (1990) later reported their findings that for a variety of rupture modes and materials the observed fractal dimensions were the same to within the error bars. Dauskardt *et al.* (1990) reported a fractal dimension $D \cong 2.2$, which, when combined with the studies of Bouchaud *et al.* (1990) may turn out to be a universal value (but also see below).

Though it is known that cracks in nature can have fractal character, at the time of writing it is still difficult to specify just how this fractal nature arises. For it is certainly true that the mechanisms leading to fracture are highly material dependent (see Liebowitz, 1984). This, we will discuss further in Sec. 4.17.

Progress has resulted from modelling the growth of a single, connected crack. With the assumption of central forces numerical simulations of media with a breaking probability proportional to the elongation of springs revealed that the cracks resulting are fractal (Louis *et al.*, 1986; Hinrichen *et al.*, 1989). The fractal dimension of such cracks appears to be sensitive to the type of external force (e.g. uniaxial tension, shear, uniform dilatation) but since only rather small cracks can be grown, more precision is lacking. Herrmann (1989) has considered therefore deterministic models.

## 1.8. *'Glue' and Related Models of Interatomic Force Fields*

*Ab initio* simulation of complex processes is fairly commonplace at the time of writing. But it is still eminently worth while (compare Heine, 1994) to ask whether empirical models for accounting for interatomic bonding can be refined so as to make them, even if not completely satisfactory, at least widely useful.

In the field of interest of our book, namely metals and metallic alloys, the various types of 'glue' models (Finnis and Sinclair, effective medium, embedded atom, etc.) developed since the early 1980s embody metallic many-atom bondings and therefore are major advances over earlier models.

One can certainly anticipate continuing refinements of such empirical potentials. So far, they have been mainly, though not exclusively, fitted to some experimental quantities relating to perfect or nearly perfect metallic crystals, e.g. equilibrium lattice spacing, coheseive energy, elastic moduli, plus some particular phonon characteristics. But one would envisage, in refinements (see also Heine, 1994) information about unusual bonding geometries (e.g. at reconstructed metal surfaces) or perhaps an interstitial atom at the maximum of its migration barrier, also being embodied.

## 1.9. *Pair Potentials*

Simulations using empirical interatomic potentials can often supply efficient and usually inexpensive routes for studying ionic structure and dynamics in metallic systems. For a long time, pair potentials were used very extensively in such simulation studies. They can reproduce usefully total energies for many systems. But when one turns to elastic properties, deficiencies begin to emerge (e.g. their inability to reproduce the so-called Cauchy discrepancy: see for instance Johnson, 1972). This situation can be remedied by the addition to the pair potential contribution of a volume-dependent, structure independent energy (the reasons being set out in Chaps. 6 and 7). But in specific examples, such as fracture of surfaces, where the volume is ambiguous, pair potential models need transcending. A further difficulty in the (simplest) pair potential scheme comes up in the determination to the vacancy formation energy $E_v$ (compare Johnson, 1987). It is found empirically that this energy $E_v$ is typically about 1/3 of the cohesive energy. In contrast, the straightforward pair potential models predict that, excluding the contribution from relaxation which is modest in close-packed metals, these two energies are equal. These limitations of the simple pair potential approximation have been addressed by the development of empirical many-body potentials which is the major theme of Chap. 8.

## 1.10. *Grain and Twin Boundaries*

Most solids do not occur as single crystals. Usually, they are assemblies of small crystallites randomly oriented with respect to one another. The boundaries between them are referred to as grain boundaries (Mclean, 1957; Kê, 1947, 1990). Rosenberg (1978) gives a figure (Fig. 3.7, p. 42) of crystallites and grain boundaries in $\alpha$-brass: the specimen having been first cold-rolled and then annealed. Generally the structure of a grain boundary is complicated, but

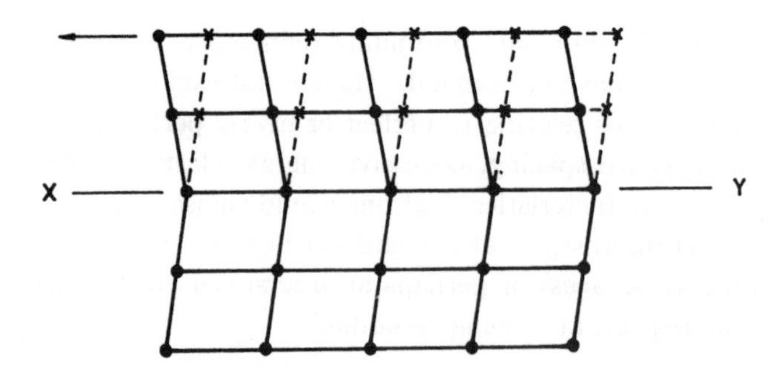


Fig. 1.3. Twinning. The crystal structure is reflected in the plane XY which forms the twin boundary. The vertical sides of the crystal as drawn are no longer smooth.

a special case of importance is that for which the orientation of neighbouring grains is very similar: such cases are termed low-angle boundaries. Their geometry is relatively simple and can be expressed in terms of dislocations.

We shall return to grain boundaries in Chap. 8 when we report atomistic structures with some realistic force laws. However, let us summarize here a few basic facts on twin boundaries.

Crystals are frequently produced with a fault which is such that one region of the crystal is a mirror image of the other part. The atoms in one region are in positions produced by reflecting the atoms in the second part at some symmetry plane of the crystal. Figure 1.3, reproduced from Rosenberg (1992), is an example of twinning. The crystal structure is reflected in the plane XY which forms the twin boundary.

Twinning frequently occurs in metals which have a small stacking fault energy as this fact then implies that the additional energy needed for any small atomic mismatch is small. Twinning can also happen during deformation. Twinning planes can often been seen by optical microscopy, and the presence of twins can be detected by X-ray diffraction. This is due to extra sets of spots which are produced from the twinned regions.

## 1.11. *Alloy Formation: Rules and Models*

### 1.11.1. *Solid Solubility: Hume-Rothery Factors*

Hume-Rothery *et al.* (1934, 1969) in very early work proposed several factors controlling the extent of solid solubility. Even at the time of writing, these factors form a useful basis for discussing the formation of extensive or restricted solid solutions (see also Alonso and March, 1989).

(a) Size effects

The 'size rule' asserts that solid solutions should not be anticipated if the atomic sizes of solute and solvent differ by more than 15%. Provided that 'other factors are favourable' solid solutions may form if the size difference is less than this value. Waber *et al.* (1963) applied this rule to some 1400 solid solutions. 90% of the systems predicted to be insoluble by the above 'size rule' were indeed found to have limited solid solubility, the distinction between a limited and an extensive solid solution being taken at 5 atomic %. However, of the systems predicted to form extensive solid solutions, only 50% were found to occur. In other words, it would appear to be the case that a favourable size factor is a necessary but not sufficient condition for the formation of solid solutions with extensive solubility.

(b) Electrochemical factor

The second rule of Hume-Rothery states that the electrochemical nature of the two elements involved must be similar for solid solutions to be expected. On the other hand, if their electrochemical characters are very different, compound formation is likely to occur. A measure of the electrochemical natures of the elements is afforded by their electronegativity. Introduced into chemistry by Pauling and by Mulliken, electron density theory, to be summarized in Chap. 6 holds promise that eventually a fully quantitative measure of this important chemical concept will be possible via the chemical potential of the inhomogeneous electronic charge cloud in an atomic (or molecular) system. Difference in electronegativity between two atoms, A and B say, drives the redistribution of charge as these approach one another to form, say, a stable AB molecule. Such charge transfer turns out to contribute to the enthalpy of formation.

Darken and Gurry (1963) made important progress in the prediction of solid solubility when they made simultaneous use of size and of electrochemical factors. In particular, these workers constructed a plot in which the coordinates are the electronegativity and the atomic radius. The values of these coordinates characterize the position of each chemical element in the plot.

Figure 1.4 reproduced from Alonso and March (1989), shows such a plot for various solutes in a Ag host. The solid circles indicate alloys in which extensive solution are found. The solid squares refer to alloys in which limited or zero solid solubility is obtained. The ellipse drawn in Fig. 1.4 approximately acts as a 'boundary' between soluble and insoluble impurities. To summarize the above very briefly, one can say that only chemically similar elements are mutually soluble (see also Alonso and March, 1989).

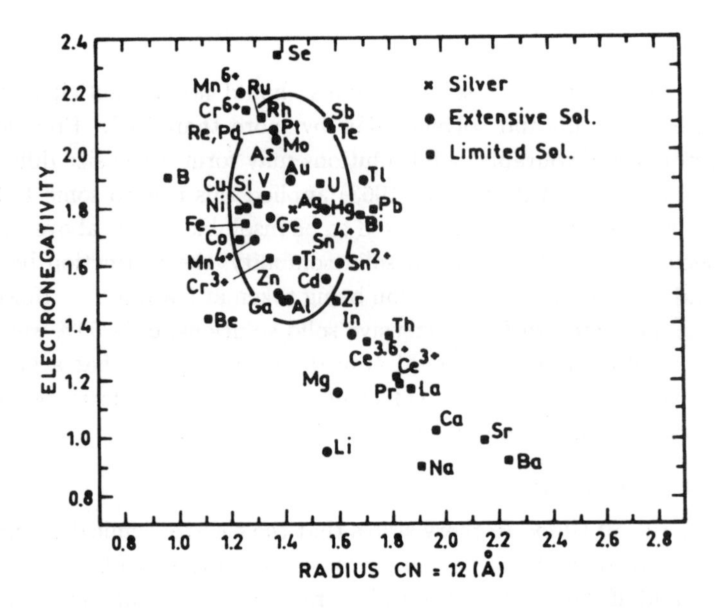


Fig. 1.4. Darken–Gurry plot for various solutes dissolved in Ag. Circles indicate alloys in which extensive solutions are found: squares indicate alloys in which limited or no solid solubility is found. The ellipse approximately separates soluble and insoluble impurities. (Redrawn after Waber *et al.*)

## (c) Valence-difference effect

Hume-Rothery formulated a third rule which states that a higher-valent metal is more soluble in a lower-valent host than vice versa. Darken and Gurry (1963) have re-formulated this rule in a slightly broader framework by asserting (see also Alonso and March, 1989) 'a disparity in valence is conducive to low solubility and this disparity has an especially pronounced effect when the solute valence is lower than the solvent valence'

Alonso and March (1989) separate two aspects of the above rule:

(i) The statement concerning absolute solubility: a difference in valence leads to low solubility and
(ii) The assertion concerning the relative solubilities of A in B and B in A.

The work of Gschneidner (1980) and of Goodman *et al.* (1983), see also Watson *et al.* (1983), has clarified the situation regarding relative solubilities

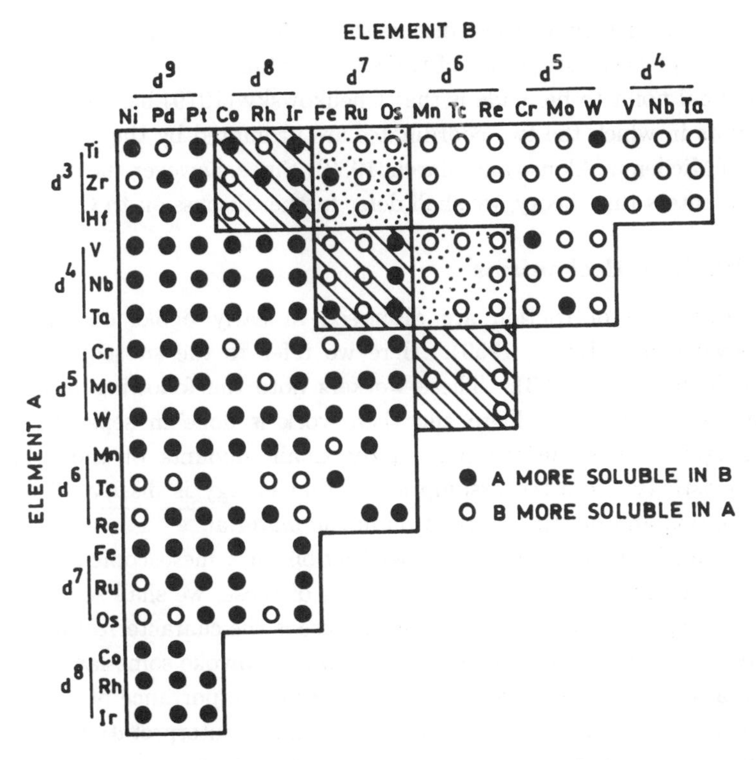


Fig. 1.5. Experimental relative solubilities in transition-metal alloys. The dotted region indicates systems where the d-bands of the constituents are equally far from half-filled: the hatched region indicates a diagonal boundary region in which no clear bias appears in the relative solubilities. (Redrawn after Watson *et al.*)

(see also Alonso and March, 1989). Gschneidner studied 300 systems formed by two metals having different valence and for which the terminal solid solubilities are known at both ends of the phase diagram. The result was that 55% of the systems do not confirm to the rule. The work of Goodman and co-workers deals with the relative solubilities in transition-metal alloys. Figure 1.5, reproduced from Alonso and March (1989) serves to illustrate the findings of these authors. The chart can be divided into two regions separated by a diagonal boundary (the cross-hatched region). Only in one of the regions is the relative-valence rule obeyed. In contrast, in the other region the rule obeyed is that 'the lower-valent metal is more soluble in the higher-valent one than vice-versa' (see also Alonso and March, 1989). We shall return to this point briefly in Chap. 6

on electronic structure, when we discuss the role of d-band filling in determining some of the physical properties of transition metals.

The final point to be made is that Chelikowsky (1979) has presented solubility plots in which the two coordinates are prompted by the semi-empirical theory of Miedema (Alonso and March, 1989). Since, however, electronic structure also enters this theory, we shall defer further discussion to Chap. 6.

## 1.12. *Friction Mechanisms*[*]

The mechanisms of friction are discussed in early books on the subject (e.g. Bowden and Tabor, 1950). Here we refer to the subsequent account of Stoneham *et al.* (1993). These workers note the following mechanisms: (i) Adhesion: surfaces adhere and then work is done in separating them. (ii) Ploughing: one surface pulls away small amounts of the other and (iii) Anelasticity: here the assumption is that energy is dissipated by dislocation motion and plastic deformation in the material.

We shall, in later chapters, discuss friction on a mesoscopic scale as well as specific atomistic studies. As to the first of these, we shall see below that two main steps are involved. The first of these is the characterization of rough surfaces and their contact. The second step is to invoke some law of friction. In such a law, we want to stress here the central importance of atomic force microscope (AFM) data (see Appendix 2.5) and its interpretation.

Tribology, the study of surfaces in moving contact, is an important area for technology. In spite of this, friction, at the time of writing, is not well understood at an atomistic level. Persson (1994) has posed some fundamental questions as follows:

(1) What is structure of sliding interface: both geometric and electronic?
(2) Where does the sliding take place?
(3) What is the physical origin of the sliding force?

Persson follows these somewhat general points with some more specific questions:

(i) Why is the frictional force $F$ usually proportional to the load $N$?
(ii) What is the microscopic origin of 'stick-and-slip' motion? (see following page).

[*]The reader who requires an advanced account should refer to the book *Physics of Sliding Friction*, Eds. B. N. J. Persson and E. Tosatti, NATO ASI series E: Applied Sciences, Vol 311 (1996): (Kluwer: Dordrecht).

Persson (1994) has discussed the theory of friction. He asserts that, during sliding of a metal block on a metal substrate (in the absence of any lubricant), the frictional force is due mainly to the shearing of cold-welded contact 'points' junctions (see also Bowden *et al.*, 1967). After a junction has been formed, it is first elastically deformed, a slow process if the sliding velocity $v$ is small, followed by plastic deformation involving rapid motion of dislocations, and other fast, nonadiabatic, rearrangement processes. As the block slides across the substrate, junctions are continually 'broken' and 'formed' at a rate proportional to the sliding velocity $v$. Persson argues that this is the reason why the frictional force is velocity independent. He asserts that the fundamental problem in sliding friction is to understand the microscopic features of the above rapid processes and then to relate these to the macroscopic movement of the metal block over the metal substrate.

Landman (1995) emphasized the importance of atomic-scale simulation in this problem. Simulations often reveal that the physical behaviour of materials at interfaces can be very different from that in the bulk. For example, the traditional picture of stick-slip motion ascribed it to a negative slope in the friction-velocity function but subsequent work has put this in reverse: in many cases stick-slip behaviour is found to be the primary process — arising from successive freezing and melting transitions of the shearing film — and it is this which gives rise to the negative friction-velocity dependence.

[illegible] the theory of friction. He asserts that [illegible] the dependence [illegible] to the [illegible] of [illegible] junctions (see also Bowden et al., 1967). [illegible] a good [illegible] of the sliding velocity [illegible] [illegible] constituting the basis [illegible] continual [illegible] consider the sliding velocity. Persson argues that this is the reason why [illegible] He asserts that the [illegible]

[illegible] the importance of [illegible] that the [illegible] behaviour of material [illegible] different from that in [illegible]. For example, the [illegible] behaviour assumed to be [illegible] [illegible] transitions of [illegible] [illegible] gives rise to the negative friction-velocity dependence.

# Chapter 2

# Phenomenology and Experiments

It is well known that the elastic properties of metals are determined primarily by the atomic forces, and the lattice imperfections only introduce small deviations from the behaviour of a perfect crystal of an amount of not more than order of a few per cent. The plastic properties are determined primarily by these imperfections. Dislocations, grain boundaries, vacancies, and interstitial atoms affect the plastic properties greatly. There was abundant direct experimental evidence for the existence of dislocations. Many prominent phenomena of plastic properties could be described in the framework of the continuum theory of dislocations. Books (Indenbom and Lothe, 1992; Hirth and Lothe, 1982; Nabarro, 1967; Friedel, 1964; Seeger, 1955; van Bueren, 1960; Read, 1953; Cottrell, 1953) and reviews have been published from time to time. This book is not intended to be an exhaustive attempt to cover all aspects in this field; instead it will try to cover rather fully, the atomistic* and fractal approaches. The motivation of this intention will be developed in this chapter.

Earlier developments of dislocation theory, and related theoretical analyses of mechanical properties controlled by dislocations, concentrated on the effects of long range elastic fields of dislocation-dislocation, dislocation-point defects and dislocation-particles interactions. In the seventies and eighties when dislocation core phenomena were recognized to play at least as important a role as long-range interactions in the deformation behaviour of most materials, atomistic studies of dislocations started and became common for investigation of structures of dislocation cores in the eighties. The progress of computer science and recognition of the role of the dislocation core made such studies feasible from the mid-eighties onwards (Vitek, 1994, 1995).

---

*This embraces interatomic force fields and electronic structure.

## 2.1. *Plastic Deformation of bcc Metals*

### 2.1.1. *Deviation from the Schmid Law*

We first summarize the Schmid law about the mechanism of plastic flow in crystals: The plastic flow always occurs on close-packed planes in the direction of the densest atomic packing, and begins when the resolved shear stress on this slip system reaches a critical value. For close-packed metals, experiments in Zn (Jillson, 1950); Mg (Barke and Hibbard Jr, 1952), $\alpha$-brass (Fenn Jr *et al.*, 1950) and Al (Rosi and Mathewson, 1950) showed no significant deviation from the predictions of the Schmid law. For bcc metals, experiments in Fe-Si alloys (Sĕsták and Libovický 1963a; 1963b) showed very different behaviour. The crss* is strongly dependent on temperature and on the orientation of the applied stress. Surface slip traces are sometimes diffuse and wavy, tending to lie parallel to non-crystallographic planes. Curiously, in the twenties, Taylor and his coworkers (1926; 1928) found that in the cases of $\alpha$-iron and $\beta$-brass slips were not confined to low index crystal planes and thereby established evidence of deficiencies in the Schmid law. However, few scientists paid attention to it. Extensive investigation of the mechanical behaviour of bcc metals and alloys started in J. W. Christian's group at Oxford in the mid-sixties. Excellent reviews on experimental and theoretical studies have been published (Vitek, 1992; Duesbery, 1989; Christian, 1983; Kubin, 1982; Vitek, 1975; Taylor, 1992). The principal finding of the deformation behaviour of bcc materials is that the effects observed are due to intrinsic characteristics of the bcc lattice; especially to the properties of the cores of screw dislocations in this structure. They found the reason why bcc metals, different from close-packed metals, deviate from the Schmid law. The slip was not confined to well-defined low index crystal planes and the flow stress was found to be dependent on the orientation of the slip plane. We shall not repeat the details of these papers, here. Rather, we will briefly report some experiments which have not been well known. Zhou (1963) found in his Mo single crystal experiment that the slip could be any apparent plane between $\{110\}$ and $\{112\}$, but the direction is always $\langle 111 \rangle$. The apparent slip plane could change at different temperatures and it depends on the orientation of the crystal. The apparent slip traces on $\{112\}, \{123\}, \{145\} \ldots$ observed are possibly combined in two groups (or four groups) of $\{110\}$ plane along the same $\langle 111 \rangle$ direction. These he termed "confined slips". Lung and He (1964) found that in molybdenum single crystals,

*CRSS refers throughout to critical resolved shear stress.

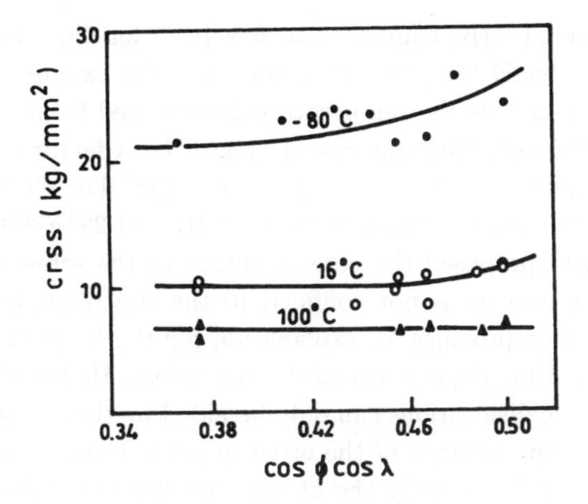


Fig. 2.1. (a) The relationship of crss with $\cos\phi\cos\lambda$.

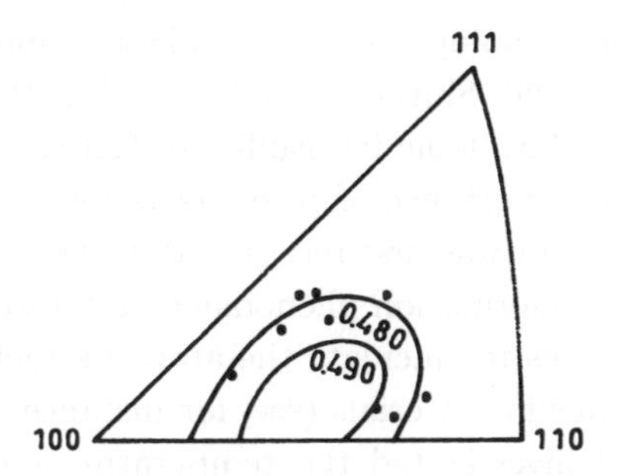


Fig. 2.1. (b) The orientation of Mo single crystals.

the critical resolved shear stress (CRSS) depends upon the $\cos\phi\cos\lambda$ value,* and that this effect becomes increasingly important at lower temperatures (Fig. 2.1).

### 2.1.2. *Temperature Dependence of Critical Resolved Shear Stress of bcc Metals*

In bcc metals, the crss is very large. For example, the crss for $\langle 110\rangle$ loading in $W$ at 77K is much larger than the crss in Cu at 4.2K. Normalised to the shear modulus, the crss values for bcc crystals range from 0.003 $\mu$ for $K$

*$\cos\phi\cos\lambda$ is called Schmid factor to characterize the orientation of the slip plane and direction of cryetals under loading.

to 0.005 $\mu$ for $W$ at 77K, compared to $5 \times 10^{-5}$ for Cu. The large crss in bcc crystals is strongly temperature-dependent. For example, the crss in $W$ drops by a factor of 4 as the temperature is increased from 77K to 450K; in Mo, between 4.2K and 420K the crss decreases by a factor of 20. In $K$, the crss drops by a factor of 10 between 1.5K and 25K. For Cu in contrast, the crss varies by a factor of two only between 4.2K and 295K (Duesbery, 1989). Hirsch (1960) first proposed the core structure of the screw dislocation and pointed out that its core is not confined to the slip plane but extends into several planes. This provides an explanation for the large Peierls stress and the strong temperature dependence of the flow stress. He found that the movement of such sessile dislocations can only be aided by thermal activation if the width of the core spreading is of the order of a few lattice spacings. Then, it became very desirable to study the atomic structure of dislocations directly. Observation of atomic configurations is feasible now with the development of high resolution electron microscopy. Studies with positron annihilation effects* have found dislocation trapping useful for understanding this problem (Shen *et al.*, 1986; Shi *et al.*, 1990; Shirai *et al.*, 1992). The theoretical descriptions of interatomic forces are available for studies of lattice defects at present (see e.g. Vitek, 1989). The importance of core effects for understanding the basic feature of plastic behaviour was first recognized in the case of bcc metals. The core structures control deformation phenomena. Interatomic forces have been used in dislocation studies to elucidate the atomic structure and atomic level properties of dislocations in materials (see, for instance, Chapter 3).

Lung *et al.* (1964) investigated the temperature dependence of the crss of molybdenum single crystals. The orientation of the crystals were chosen according to two requirements: (1) Considering that the slip system of Mo single crystals with the same orientation may vary with the testing temperature, they chose the orientation in such a way that (110)[111] remained to be the only slip system throughout the testing temperature range. (2) The variation of $\cos\phi\cos\lambda$ value was kept within 0.01 in order to minimize its effect upon the crss. The experimental results showed the logarithm value of crss varies linearly with the absolute temperature. The absolute value of crss and the slope of the line are influenced by the purity of specimens. The experimental results did not agree with Fisher's simplified treatment of Cottrell-Bilby's theory ($\sigma \alpha T^{-1}$), neither with Cottrell's simplified theory ($\sigma(T) = \sigma(0) - kT^{1/3}$), nor with Seeger's dislocation-forest model ($\sigma = A - BT$). For comparison,

*See Chapter 9 for full details.

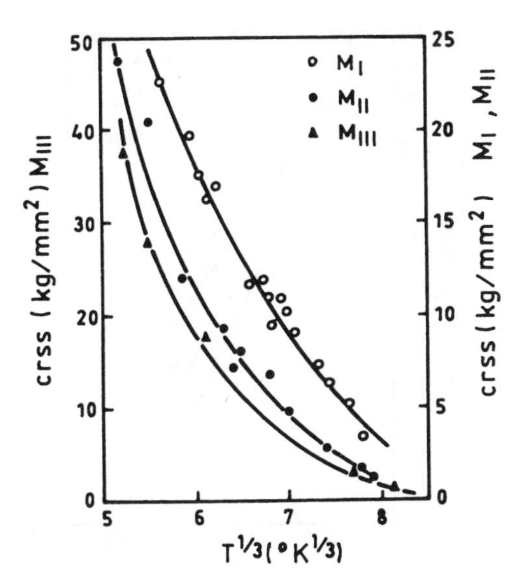


Fig. 2.2. (a) CRSS$-T^{1/3}$ relationship.

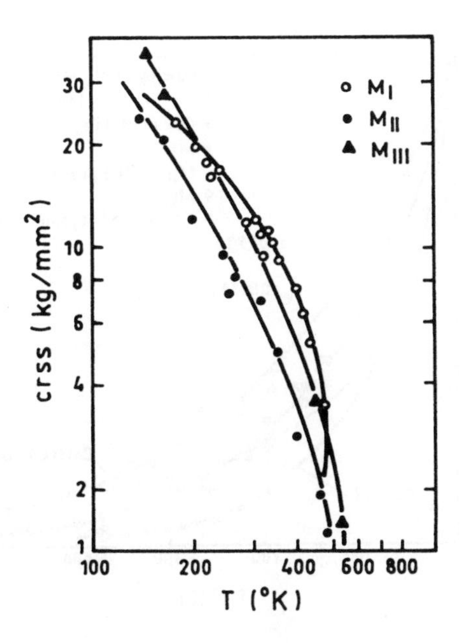


Fig. 2.2. (b) ln CRSS$-$ln $T$ relationship.

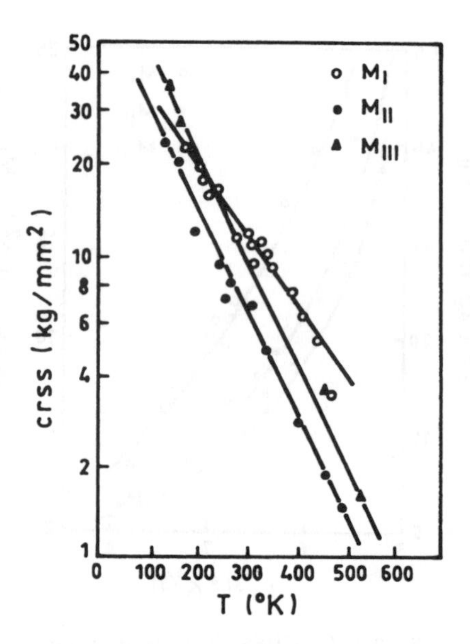


Fig. 2.2. (c) ln CRSS$-T$ relationship.

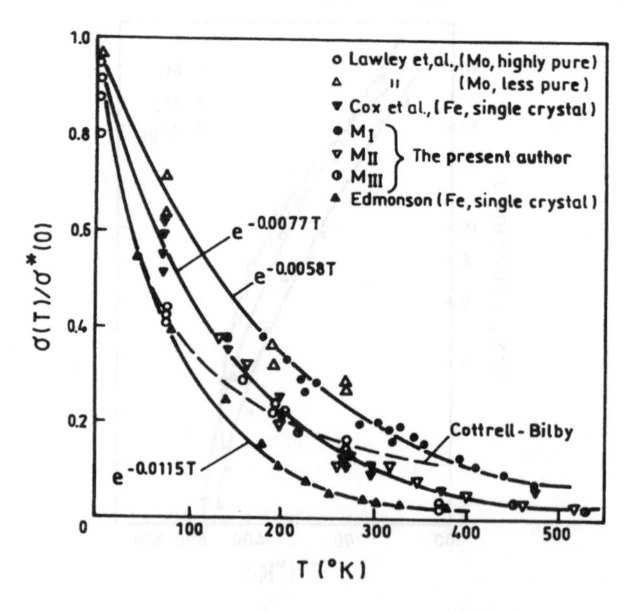


Fig. 2.2. (d) $\sigma(T)/\sigma(0) - T$ relationship.

in Fig. 2.2, it is of interest to note that the yield data obtained by Lawley *et al.* (1962) can be well represented by exponential relationship given by them except that the result at 77K appears to depart from their curve.

The absolute value of crss can be expressed by the following equation approximately:

$$\sigma(T) = \sigma(0)\, e^{-BT}\,.$$

This relationship is consistent with that proposed by Petch *et al.* (1958) based on P-N model* and also with that by Castaing *et al.* (1981), Lagerlof *et al.* (1994) and Suzuki *et al.* (1995) on the experimental data of temperature dependence of the plastic flow stress of covalent crystals.†

He *et al.* (1966) found that the $B$ value is lowered when more carbon atoms are dissolved in $\alpha$-iron. The concentration of carbon in $\alpha$-iron was measured by internal friction method in that work. In order to explain this, they assumed that there exists a strong exchange bond between the impurity atoms and the dangling atoms with unpaired electrons along the dislocation axes. When a dislocation line anchored at both ends is subjected to move, vigorous vibrations accur with a rise in temperature. The distance between the impurity and dangling atoms will be lengthened with the consequence of weakening the chemical exchange bond between them, and this will, in turn, lower the frictional stress of motion of dislocations. Lung *et al.* (1964) was one group of the earliest authors who recognized the important role of the dislocation core and used a simple interatomic force model to explain the temperature dependence of the crss of Mo-single crystals.

## 2.2. *Phonons, Electrons and Plasticity*

When a dislocation moves through a medium it produces a varying strain field. Dissipative processes within the medium relax these strains, generating entropy and retarding the motion of the dislocation. Shear stresses can be relaxed by several mechanisms, each of which can be represented by a coefficient of viscosity $\eta$. Here, we consider specifically the effects of phonon viscosity and of electron viscosity.

---

*The Peierls-Nabarro model has been referred to in Section 1.3.

†After analysis of experimental results of Mo, Fe, Al and Mg single crystals, Lung *et al.* (1997) pointed out that the approximate exponential relationship between crss and the absolute temperature is common, even for bcc single crystals, provided that the slip plane and direction are kept the same.

### 2.2.1. *Phonon Drag of Dislocations in Metals*

The method of treating the contribution of viscosity to the drag on a dislocation developed by Mason (1955) for phonon viscosity $\eta$ was modified by Nabarro (1967). For a screw dislocation with displacement

$$s = \frac{b}{2\pi} \tan^{-1} \frac{y}{x - vt}, \tag{2.2.1}$$

the velocity is

$$\dot{s} = \frac{bv}{2\pi} \frac{y}{(x - vt)^2 + y^2}. \tag{2.2.2}$$

The rate of dissipation of energy in unit volume is

$$\frac{\eta}{2} \left\{ \left( \frac{\partial \dot{s}}{\partial x} \right)^2 + \left( \frac{\partial \dot{s}}{\partial y} \right)^2 \right\} \tag{2.2.3}$$

where $\eta$ is the viscosity of the medium. In cylindrical coordinates, this become $b^2 v^2 \eta/(8\pi^2 r^4)$. The total dissipation from unit length of the dislocation is

$$W = \frac{b^2 v^2 \eta}{4\pi} \int_{r_0}^{\infty} \frac{dr}{r^3} \tag{2.2.4}$$

where $r_0$ is the inner cut-off radius. Equating the power supplied by a shearing stress $\sigma$ to the power dissipated by unit length of dislocation, $W = b^2 v^2 \eta / 8\pi r_0^2$.

$$\sigma b v = b^2 v^2 \eta/(8\pi r_0^2) = B_1 v^2 \tag{2.2.5}$$

Then

$$B_1 = \left( \frac{b^2}{8\pi r_0^2} \right) \eta. \tag{2.2.6}$$

### 2.2.2. *Electron Drag of Dislocations in Metals*

The electron contribution to energy losses during plastic deformation is in principle determined by the same dissipative processes in the electron system of the metal which determine the electrical conductivity.

The dislocation viscous drag coefficient will be denoted by $B_2$ and is defined by the equation (Hirth and Lothe (1982)):

$$\frac{F}{L} = \sigma b = B_2 v_D \tag{2.2.7}$$

Here, $F/L$ or $\sigma b$ is the force per unit length that must be applied to the dislocation to keep it in uniform motion with velocity $v_D$. By the definition of viscosity

$$\eta = \sigma/(v_D/c_t) \tag{2.2.8}$$

where $c_t$ is the transverse-sound-wave velocity. Comparing the above equations:

$$\eta = \frac{B_2 c_t}{b}\,. \tag{2.2.9}$$

The method of treating the contribution of viscosity to the drag on a dislocation was developed by Mason (1955):

$$B_2 = b^2\eta/(8\pi\lambda^2) \tag{2.2.10}$$

where:

$$\begin{aligned} \eta &= Nm\lambda\bar{v}/3 \\ \lambda &= \sigma_e m\bar{v}/(Ne^2) \end{aligned} \tag{2.2.11}$$

where $\lambda$ is the electron mean free path, $N$ is the number of electrons in unit volume, $e$ is the charge of each, $m$ their mass, $\bar{v}$ their mean speed and $\sigma_e$ the electrical conductivity. Then (Nabarro, 1967):

$$B_2 = (bNe)^2/(24\pi\sigma_e)\,. \tag{2.2.12}$$

It is seen from Eq. (2.2.12) that $B_2$ is directly proportional to the electrical resistivity. For copper at room temperature a dislocation speed of $2.5 \times 10^5$ cm sec$^{-1}$ would be attained under a stress of $3 \times 10^5$ dyn cm$^{-1}$, so this drag is very small, but is essential at low temperature, at which phonons are 'frozen out'.

However, many discussions on this relationship arose. Some authors thought that the basic contribution to the energy dissipation takes place by processes of electron scattering in the short-wavelength part of the packet where a macroscopic description is not allowed and where it is required that the spatial dispersion of electrical conductivity is taken into account (Alshits, 1992).

### 2.2.3. *Superconductivity and Plasticity*

An increased plasticity of materials entering the super-conducting state was found at the end of 1960's by Pustovalov *et al.* (Pb) (1969) and by Kojima and Suzuki (Pb) (1968). The stress required to maintain the deformation in constant strain-rate tests drops. For creep measurements at constant stress, Soldatov *et al.* (1971) found that the strain rate increases by factors of up to 250. For stress-relaxation experiments at constant strain, Suenage and Galligan (1970) demonstrate that the stress drops suddenly. The stress-change effects observed are typically of the order of 0.1% to 10%, but effects as large as 50% have been reported. Pustovalov *et al.* (1969) have reported an average change of 30% for the critical resolved shear stress of lead single crystals with tension axes close to ⟨110⟩ and of 20% for crystals of some different orientations.

For polycrystalline samples of comparable microstructure, the yield stress was found to decrease by 35 to 40%. Ni *et al.* found that the critical current density ($J_c$) of YBCO superconductors increases synchronously with the improvement of mechanical properties (Vickers hardness) (Ni *et al.*, 1993). From these experiments it was suggested that the strain variation depends on the influence of the normal to superconducting state.

Different models have been proposed to explain the experimental results (Granato, 1971; Suenaga and Galligan, 1971). In the superconducting metals the electron drag in the superconducting state reduces drastically. Discussions can be found in review papers given by Startsev (1983) and Nabarro (1980).

The resistance to the motion of a dislocation, expressed by the drag coefficient $B_D$, is as follows:

$$B_D = F_v/v_D = \sigma b/v_D\,. \tag{2.2.13}$$

The coefficient $B_D$ has the dimensions of viscosity. $v_D$ is the velocity of the dislocation, $\sigma$, the stress and $b$ the Burgers vector. At ordinary temperatures, the drag coefficient is due principally to the scattering of phonons. At low temperatures, the phonon scattering tends to a low value, and $B_D$ is determined principally by the interaction between the dislocation and the conduction electrons. One of the mechanisms assumes that the Fermi surface is adiabatically distorted to conform to instantaneous strain, and then energy is dissipated as the Fermi surface relaxes to its equilibrium form.

The theories of Granato (1970) and of Natisk (1972) both start from the vibrating-string model of a dislocation. However, the idea of Hutchison and McBride (1972) is different. These workers assumed that the dislocation breaks away from an obstacle under the applied stress, and impinges on the next obstacle with an appreciable kinetic energy. The kinetic energy is turned to heat when the dislocation strikes the second obstacle, and raises the local temperature. Since the thermal conductivity in a superconductor is low, the effective temperature is raised for an appreciable time, and the stress $\sigma$ required to maintain a prescribed strain rate $\dot{\varepsilon}$ is decreased. This model depends on the role of obstacles. It cannot explain why these effects still exist in high purity single crystalline metals.

Ultrasonic measurements have particular advantages for this problem (Granato, 1990).* They are particularly well suited for the study of quantum

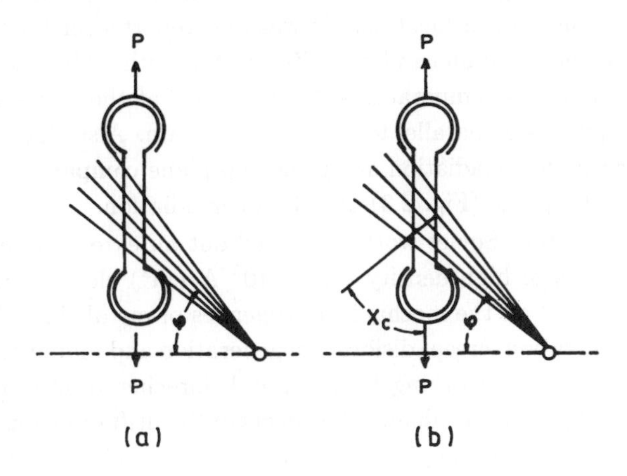


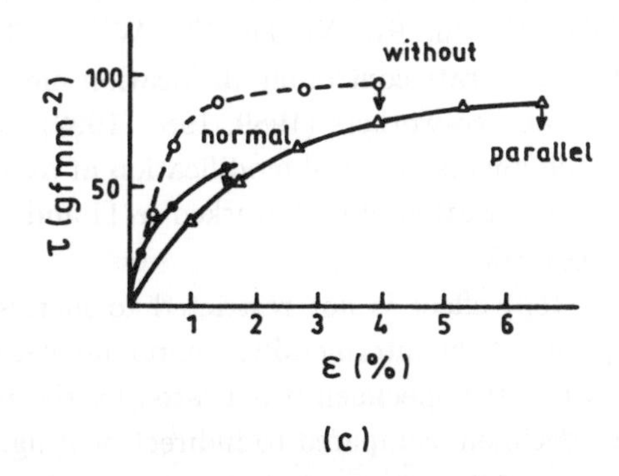


Fig. 2.3. Zn crystals irradiated with 1 MeV electrons during plastic deformation at 77K: (a) electron beam parallel to the (0001) glide plane: (b) beam normal to glide plane: (c) resulting stress-elongation curves for the beams parallel and normal to the glide plane and without irradiation. From Troitskii and Likhtman (1963).

---

*See also Appendix 2.4.

system. Information about energy gaps and stress coupling strengths are readily obtained from ultrasonic attenuation peaks and velocity changes at low temperature for defects in concentrations of a few ppm.

### 2.2.4. *The Electroplastic Effects in Metals*

It has long been recognized that an electric current (drift of electrons) can enhance the mobility of dislocations. It was first reported in 1963 by Troitskii and Likhtman (1963). Upon irradiating Zn crystals with 1 MeV electrons under uniaxial tension at low temperatures, they found that the flow stress was less when the slip plane was parallel to the electron beam. Also, the elongation at fracture increased for irradiation along the slip plane compared to irradiation normal to the slip plane (Fig. 2.3) or without irradiation.

Troitskii and other Soviet workers carried out an extensive series of studies into the effects of high density ($10^3 - 10^5$ A/cm$^2$) electric current pulses ($\sim 100$ $\mu$s duration) on the mechanical properties of metals including the flow stress, stress relaxation, creep, dislocation generation and mobility, brittle fracture, fatigue and metal working (Conrad and Sprecher, 1989). High current densities were employed to enhance the effects of the drift electrons, short pulse times to reduce Joule heating. Conrad and co-workers (1989) conducted a series of investigations into the effects of single, high density ($10^5 - 10^6$ A/cm$^2$) current pulses ($\sim 60$ $\mu$s) on the tensile flow stress at 300K of a number of polycrystalline metals (Al, Cu, Pb, Ni, Fe, Nb, W, Sn, Ti) representing a range in crystal structures, rate-controlling dislocation mechanisms and electronic properties. Lai and co-workers (1989, 1988, 1992) found that electric current pulses affect the microstructural modification at atomic level in amorphous alloys, the recrystallization of cold worked $\alpha$-Ti and cyclic softening in a secondary hardening peak.

The electroplastic (ep) effect is not restricted to metals. Martin (1980) found that the creep rate of the intermetallic compound $V_3Si$ at elevated temperatures increased when the specimen was heated by the passage of electric current through the specimen, compared to indirect heating.

Direct measurements of the effects of an electric current on dislocation mobility have been made by Zuev *et al.* (1978). Employing the etch pit technique, the effect of a single current pulse applied simultaneously with a mechanical stress pulse of $10^{-2} - 10^{-1}$ s duration on the velocity of dislocations in Zn crystals at 77–300K was studied. The increase in dislocation velocity for the parallel direction is larger than for the antiparallel directions. This showed a

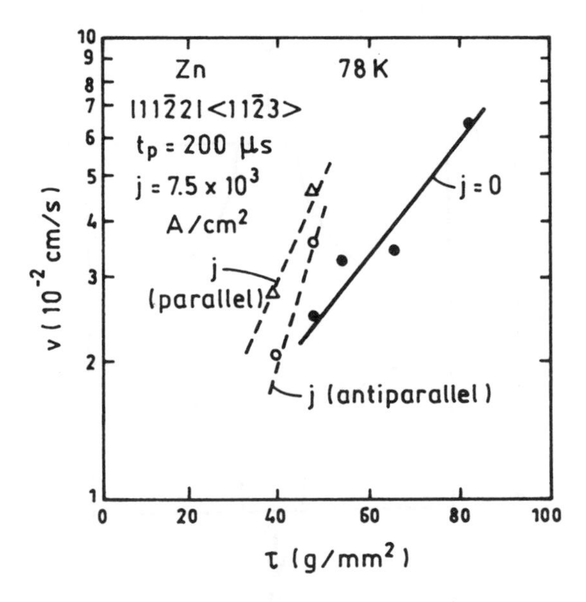


Fig. 2.4. Effect of a single current pulse ($j = 0.75 \times 10^3$ A/cm$^2$, $t_\text{p} = 200$ $\mu$s) on the velocity of $\{11\bar{2}2\}$ $\langle 11\bar{2}3 \rangle$ dislocations in Zn parallel and antiparallel to the electron current as a function of the mechanically applied stress $\tau$. Data from Zuev *et al.* (1978).

vectorial effect of the current, i.e. an electron wind effect (Fig. 2.4) (Fig. 19 of Conrad and Sprecher, 1989). The fact that an increase in velocity over that for $j = 0$ still occurred when the current was antiparallel to the dislocation velocity suggests a scalar effect of the current in addition to an electron wind.

## 2.3. *High Temperature Strength of Alloys*

It has been found empirically that the strength of metals and alloys is closely related to interatomic bonding. Diffusion coefficient, Debye temperature, elastic modulus, melting temperature and sublimation energy are important parameters for characterization of the high temperature strength of metals and alloys (Ocipov, 1960), especially for long time services of mechanical components.

### 2.3.1. *Diffusion Creep*

Nabarro (1948) and Herring (1950) proposed a diffusion creep model to explain the relationship between creep and diffusion. If a tensile stress is applied, vacancies flow in the directions indicated in Fig. 2.5.

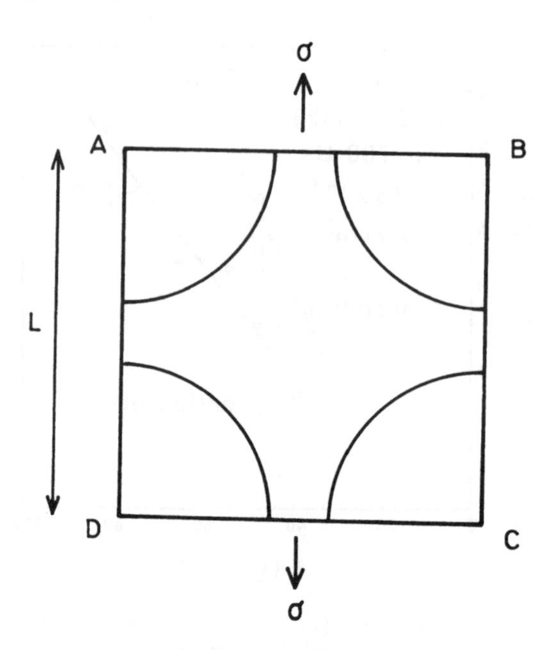


Fig. 2.5. Stress-directed flow of vacancies inside a grain ABCD.

Along AB and CD the vacancy excess is:

$$\frac{C - C_0}{C_0} \cong \ln \frac{C}{C_0} = \frac{\sigma b^3}{kT} . \quad (2.3.1)$$

If $L$ is the grain size, the rate at which vacancies migrate from the ends to the sides is:

$$\frac{\partial v}{\partial t} \cong LD\sigma b^3/(kT) \quad (2.3.2)$$

and the creep rate is:

$$\dot{\varepsilon} = \frac{1}{L^3}\frac{\partial v}{\partial t} = \alpha \frac{D\sigma b^3}{L^2 kT} . \quad (2.3.3)$$

According to Eq. (2.3.3), diffusion creep is linearly dependent on the diffusion coefficient, which is in relation to interatomic forces between atoms in metals and alloys. Equation (2.3.3) has been supported by experiments on wires or foils of Ag (Greenough, 1952), Au (Buttner *et al.*, 1952), Cu (Pranatis and Pound, 1955) and Cu-Sb alloys (Tipler, see also Mclean, 1962) at a creep rate, to within an order of magnitude.

Löfstedt (1997) has suggest that the 'diffusion constant' $d$ of dislocations is experimentally accessible via the creep of a bar under its own weight. The proposal is that the time rate of change $\dot{l}$ of the height of the bar is $\dot{l} = -d(4/3)(\mu\rho\, g/E)(1+\nu)$ where $E$ is Young's modulus and $\nu$ is Poisson's ratio. However, atomic diffusion may perhaps enter such a measurement as an additional creep mechanism.

### 2.3.2. *Effect of Solute Atoms*

Solute atoms in high melting point metals have stronger effects on strengthening the metal at high temperatures. This was explained by the climbing mechanism of dislocations in which the strong interatomic force increases the activation energy for diffusion.

In metals, the electric dipole produced in a metal by an edge dislocation creates a field which can act on the nuclear charge of an impurity atom. In a metal the cloud of conduction electrons takes energies which range between two limits $E_0$ and $E_M$. These vary with the dilatation $\delta$ of the lattice (Friedel, 1964):

$$\Delta E_M = -\frac{4}{15}(E_M - E_0)\delta\,. \tag{2.3.4}$$

Near to an edge dislocation, the dilation $\delta$ is not uniform. The electrons in the compressed regions ($\Delta E_M > 0$) tend to flow towards the dilated regions ($\Delta E_M < 0$) to compensate for the differences $\Delta E_M$ of the Fermi level. Peutz (1963) calculated this kind of interaction, and demonstrated that the binding energy is small. For example, for Cu-Zn, $W_e = -0.005$ eV. It is not important in most cases. It is one order of magnitude smaller than the elastic interaction. It may however become important for large differences of valency. The description for the dislocation core structure seems too simple.

de Hosson (1980) and Wang *et al.* (1993) studied the electronic structure of dislocation core which was not calculated in details previously. From the modified atomic coordinates and by use of the recursion method, Wang *et al.* (1993) calculated the electronic structure of edge dislocation in iron. According to their calculations, it seems that the binding energy is large and cannot be ignored compared to the elastic interaction.

The electrostatic interaction is the main term in an ionic solid. A straight edge dislocation introduces locally an excess of positive (or negative) charge. Similar excess charges of alternating signs, appear on the successive lattice planes along an edge dislocation line. Such a dislocation attracts electrostatically charged impurities of both signs. The binding energy can be appreciable and probably of the same order as that of the elastic binding energy.

Most impurity atoms can form a sort of chemical exchange bond with the unpaired electrons of the atoms at the dislocation core. The binding energy is then of the order of a few eV; much larger than that corresponding to purely electrostatic interaction (Van Bueren, 1960).

## 2.4. *The Crack and Fracture*

Thomson (1986) proposed three prototype cracks: cracks which cleave, cracks which emit dislocations, and plastically blunted notches which merely activate external plastic flow in the surrounding medium. The physical mechanism in each case are quite distinct, and each type tends to correspond to a particular geometry.

### 2.4.1. *Cleavage Cracks*

The crack is capable of cleavage advance without intrinsic generation of dislocation. (By intrinsic is meant generation from the crack tip itself, independent of external sources). The form of the crack in the cleavage case is atomically shaped at its tip.

### 2.4.2. *Emitting Cracks*

In this case, dislocations are emitted from the tip without cleavage when the crack is stressed. Crack advance takes place by ledge formation or slide off the crack tip.

### 2.4.3. *Plastically Blunted Cracks*

The effect of plasticity external to the crack is to blunt the crack by means of the steps formed on the cleavage surfaces when dislocations are annihilated there. The final macroscopic shape of the crack blunted by dislocation absorption will be determined by the combined effect of the available slip systems, stress distribution around the crack, the characteristics of the sources such as operating stress, etc., and the mobility and mean free path of the dislocations. In general the shape of the crack will not be dominated by crystallographic features as in the case for the cleaving or emitting crack. The crack advance takes place by accretion of the voids to the main crack.

Qualitatively, the brittle form is associated with the cleavage crack, while the emitting crack and the plastically blunted crack are forms of ductile fracture. The atomic structure of the tip will dictate whether the crack is a cleaver or an emitter, and the properties of the dislocations and their sources in the medium and their interactions with the crack will determine whether the core of the crack can ever be stressed sufficiently so that cleavage or emission is possible.

### 2.4.4. *The Condition for Intrinsic Brittleness*

Rice and Thomson (1974) suggested that the condition for intrinsic brittleness in a material is the activation energy for dislocation emission from cracks.

The combined criterion for cleavage/emission in pure Mode I is:

$$\begin{aligned} K_{IE} &< K_{IC} \quad \text{emission} \\ K_{IE} &> K_{IC} \quad \text{cleavage}. \end{aligned} \tag{2.4.1}$$

The added subscript $E$ on $K$ refers to the critical values for emission. This form of the cleavage-emission criterion was first given by Mason (1979). Later, Rice (1992) introduced a parameter, the unstable stacking energy $\gamma_{us}$, which characterizes the resistance to displacement along the slip plane, and thus to dislocation nucleation. In terms of $\gamma_{us}$, the emission criterion is as follows (see Zhou *et al.*, 1993):

$$K_{IIe} = \sqrt{2\gamma_{us}\mu/(1-\nu)} \tag{2.4.2}$$

where $\mu$ is the shear modulus, $\nu(\doteq 0.25)$ is Poisson's ratio. In a systematic atomic analysis, Zhou *et al.* (1993) have found that Eq. (2.4.2) is very accurate for a two-dimensional hexagonal model, using a "Mode II" geometry in which an edge dislocation is emitted along the cleavage plane ahead of the crack. Simulations in the 2D hexagonal lattice by Zhou *et al.* (1994) have shown that for a variety of force laws, the ductile-brittle crossover is independent of $\gamma_s$, and is determined by a critical value of the $\gamma_{us}$ only. They found the dependence of $K_{1e}$ on $\gamma_s$ to be the same as that of $K_{1e}$ in the form:

$$K_{1e} = 2\sqrt{\gamma_s\mu(1+\nu)}f(\gamma_{us}). \tag{2.4.3}$$

The brittle-to-ductile crossover is determined by the ratio of $K_{1e}$ to $K_{1c}$: not $K_{1e}$ itself. The new finding is striking in that all previous criteria express the crossover as a competition between the values of $\gamma_s$ and $\gamma_{us}$. That is, a low ratio of $\gamma_s/\gamma_{us}$ would imply a brittle material. The new criterion replaces this competition with a simple critical value for $\gamma_{us}$. It is interesting that the new criterion for the ductile-brittle crossover is equivalent to structure-dependent critical value for the Peierls energy of the dislocation in the material. Thus, the interatomic force plays an important role in crack propagation processes. Using a 35-hundred-atom simulation cell, Zhou *et al.* (1997), with an EAM-type potential* for copper, calculated the dislocation emission at a crack tip in a ductile material.

### 2.4.5. *Dislocation Shielding, Antishielding and Annihilation at the Crack Tip*

The discussion on intrinsic brittleness in a material is the ideal case. The term "intrinsic" assumed an intrinsic homogeneous property of the crack in the given material (Zhou *et al.*, 1993). In the material, there is no dislocation formation at sources outside the core region of the crack. Actually, in the vicinity of a crack, pairs of dislocations with opposite Burgers vectors may be created by various multiplication mechanisms at dislocation sources. Lin and Thomson (1983) have analyzed the stress on the source and the shielding of the crack tip as a function of the number of dislocations produced. The local stress intensity factor $k$ can be written as

$$k = K^C + K^D \tag{2.4.4}$$

where,

$$K^D = -\sum_{j\neq 1}' \frac{\mu b}{(2\pi x_j)^{1/2}}$$

and $K^C$ is the stress intensity factor due to the crack tip, $x_j$ is the distance of the $j$ dislocations piling up from the crack tip and $k$ is termed the effective stress intensity factor which modifies the original $K^C$ with the effects of dislocations near the crack tip, if a dislocation source exists in front of a sharp crack from which a pair of dislocations with opposite Burgers vectors has been emitted. It seems possible if the local stress near the crack tip exceeds the maximum shear stress to form a semi-circle of a Frank-Read source. Positive ones, with shielding Burgers vectors, are repelled by the crack tip and negative ones, with anti-shielding Burgers vectors, are attracted towards the crack tip (Zhou *et al.*, 1989). Negative ones may be absorbed and annihilated by the open cleavage surface by producing steps on this surface (Lung, 1990).

*See Chapter 8.

The entry of dislocations into the tip blunts the cracks (Weertman, 1980). A dislocation free zone is formed (Lung, 1990; Ha *et al.*, 1900).

The elementary processes may be Bardeen-Herring climb sources, the double-cross-slip mechanism, vacancy disc nucleation and nucleation of glide loops (Hirth and Lothe, 1982). The central physical quantities are the line-tension force and the energy of the dislocation line. The influence of crystal structure on the atomic configurations of dislocations and with the stresses and energies associated with them must be considered. The treatment of non-linear region as a thin strip in the glide plane in the Peierls-Nabarro model has successfully allowed analytical results. However, this model is somewhat physically unrealistic in view of the cylindrical symmetry of dislocation fields. There were a number of atomic calculations with the use of pair potentials to describe atomic interactions (Puls, 1981; Gehlen, 1972). Thomson *et al.* (1992) studied imperfections by use of lattice Green's functions. This method built on earlier work by Kanzaki (1957) on lattice statics and Tewary (1973) on Green's-function methods.

2.4.6. *Dislocation Dynamics and Fracture*

Roberts (1991, 1993, 1994) proposed a dynamical model for dislocation motion near crack tips. When dislocations are at stresses well above any friction stress and if the stress is kept constant, the array would continue to grow, position of the dislocations depending on stress history of the test, and their stress, temperature and velocity relation. Effective $k$ at the crack tip is reduced by dislocation stress field (Eq. 2.4.4).

The extent of shielding depends on the number of dislocations and on their positions. Using the dislocation motion model to monitor $K_{\text{eff}}$ with time, the material will fracture as $K_{\text{eff}}$ reaches $K_{Ic}$.

The dislocation velocity is:

$$v_D = A\left(\frac{\sigma}{\sigma_0}\right)^m e^{-\frac{U_d}{kT}} \tag{2.4.5}$$

and the values of $A$, $\tau_0$, m and $U_d$ are known accurately from the literature. Thus knowing the velocity of each dislocation, its position change over a small time interval can be calculated. Then, knowing the position of each dislocation, the $K_{\text{eff}}$ change over a small time can be calculated. The material fractures at $K_{\text{eff}} = K_{Ic}$.

Relating to $U_d$, many calculations have been performed. Bullough and Perrin (1968) constructed an atomic model to study the core configurations of certain dislocations and vacancies in iron. The core structures controlling deformation phenomena have later been reviewed by a number of authors (Duesbery, 1989; Vitek, 1992).

## 2.5. *Power Law Relation between the Plastic Strain and the Number of Cycles to Fatigue Failure*

During fatigue,* plastic strain occurs in every cycle. It is well known that Manson-Coffin relationship holds in fatigue processes of many metals (Raraty and Suhr, 1960). The relationship is given by

$$\Delta\varepsilon_p \propto N_f^{\frac{-1}{2}} \tag{2.5.1}$$

where $\Delta\varepsilon_p$ is the plastic strain and $N_f$ is the number of cycles to fatigue failure. Suppose that the plastic strain occurring in every cycle is $\varepsilon_{pf} \propto N_f^{-1}$. Then,

$$\Delta\varepsilon_p \propto \varepsilon_{pf}^{\frac{1}{2}}\,. \tag{2.5.2}$$

It seems that this process possesses a scaling property in this dynamical system.

## 2.6. *Statistical Behaviour for the Fracture of Disordered Media*

Most metals and alloys are not homogeneous and their heterogeneity strongly influences their mechanical behaviour. We should obtain information at the macroscopic level starting from a microscopic one and taking into account local fluctuations due to disorder. Fracture is much more sensitive to disorder and mean field theory is expected to give rather poor results in either two or three dimensions. Fracture naturally enhances the effect of the preexisting heterogeneities. Some models indicate that the final stage of rupture can be interpreted as a critical point (Arcangelis *et al.*, 1989). We should emphasize the relevance of the collective behaviour of microscopic or mesoscopic-elements in the presence of disorder (i.e. random variations of the local properties from one element to the next, or randomness in the initial state or during the time evolution, etc.).

*See ASM Handbook, Vol. 19: *Fatigue and Fracture* (ASM International: Materials Park, OH, USA) 1996.

There are three basic experimental facts which show that disorder plays an important role in real materials.

### 2.6.1. *Statistical Fluctuations in the Rupture Stress of Materials*

It is well known that two pieces made of the same material under the same condition, will not break at the same stress or at the same time under same stress loading. The statistical fluctuations are very often large (Volkov, 1960). In high strength materials, the stress concentration is well localized. The fracture behaviour is highly sensitive to defects in materials. A micro crack of the order of several $\mu$m's would lead the material to fracture. These sample to sample fluctuations are present in most brittle materials, and on all scales.*

### 2.6.2. *The Dependence of the Mean Fracture Stress on the Sample Size*

In case of usual sample size, say several cms, the dependence of the mean fracture stress on the sample size is not important. However, it was well known that the strength of thin glass wire depends on its diameter. The strength of a thin glass wire of 22 $\mu$m diameter is 22 kg/mm$^2$; and of 16 $\mu$m diameter, 107 kg/mm$^2$; of 12.5 $\mu$m diameter, 146 kg/mm$^2$ and of 8 m diameter, 207 kg/mm$^2$. It was explained that the probability of existence of dangerous cracks is high in large size materials (Frenkel, 1950).

Similarly, if the system size remains constant, that if the characteristic size of the microstructure changes, the mean fracture stress will be affected. This has been observed in many materials, where the microstructure is controlled by the grain size. It is explained by the Hall-Petch relationship based on dislocation theory. These size effects tell us that we cannot neglect the small scale structure of the medium even if the interaction between atoms has been chosen to be the same.

### 2.6.3. *Local Damage Zones*

The shape of the stress-strain curve is shown schematically in Fig. 2.6. This shape is quite frequently encountered in many materials such as concrete.

*A. Buchel and J. P. Sethna (Phys. Rev. E55, 7669, 1997) have studied extensively the statistical mechanics of cracks. More specifically, these workers studied a class of models for brittle fracture, namely elastic theory models that have cracks but do not allow for plastic flow. Their conclusion is that such models display a transition to fracture under applied load that has similarities to the first-order liquid-gas transition.

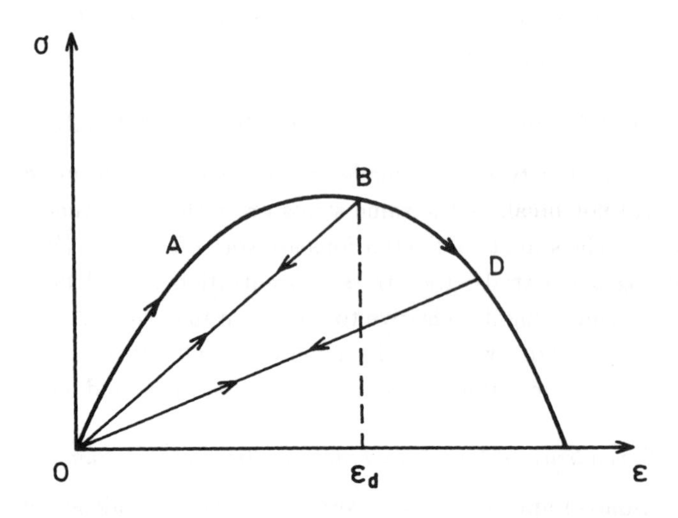


Fig. 2.6. Typical behaviour of a simple elastic solid with damage.

In the decreasing part of the curve, damage has concentrated in local zones. Treating this model numerically with a finite element code, we will see that the localization of damage is very dependent on the mesh size used. The finer the mesh, the more localized the damage. Its behaviour is not scale invariant.

Facing these experimental facts, Herrmann and Roux (1990) proposed mesoscopic lattice models for fracture alternative to molecular dynamics and continuum mechanics. These methods do not pretend to describe nature on an atomistic level as molecular dynamics but have their validity at much larger length scales where the medium can be described by continuous vector fields. The big advantage of the lattice models is that they allow very naturally for the introduction of disorder. Small cracks are sharply defined and it is possible to simulate simultaneously many cracks within rather moderate lattice sizes.

Cracks can be grown on a lattice by deterministic rules and their patterns can be fractal* due only to the interplay of anisotropy and memory. Herrmann and de Arcangelis (1990) showed cracks of various shapes generated by different breaking criteria. They discussed the universal scaling behaviour. They found that the distribution of local strain has multifractal scaling properties just before the system breaks fully apart.

---

*See Section 1.7.

## 2.7. *The Roughness of the Crack Surface*

In a common fracture sequence, an initially smooth and mirror-like fracture surface begins to appear misty and then evolves into a rough, hackled region with a limiting velocity of about six-tenths the Rayleigh speed. In some brittle materials, the crack pattern can also exhibit a wiggle of a characteristic wavelength. Some experiments have shown that violent crack velocity oscillations occur beyond a speed of about one-third the Rayleigh speed and are correlated with the roughness of the crack surface. These may be universal behaviour in fracture (Abraham, 1994). On the other hand, materials are not homogeneous in microstructures. If the stress is uniform, the crack would propagate along the cleavage plane. The nonuniform stress distribution combined with the anisotropic properties of crystals would make many complicated cases for the crack to change cleavage plane abruptly (Thomson, 1986). Then, the crack propagates along zigzag path even in a single crystal. This is the other origin of the roughness of the crack surface. For intergranular crack, the material fractures along the grain boundaries. This is also an origin of the roughness of crack surface (Lung, 1986). In Chapter 4 we shall see that the fractal dimension can describe the roughness of crystal surface as a continuous variable parameter for fracture properties of materials.

## 2.8. *Dynamic Instabilities of Fracture*

Ching *et al.* (1996) found that moving cracks are strongly unstable against deflection in essentially all conventional cohesive-zone models of fracture dynamics. In the ideal central force model, cracks are stable in the limit of the crack speed $v$ approaching zero, and are unstable at nonzero speeds. The stability appears only when the cohesive shear stress is larger than its central-force value. They found that the instability is governed by detailed mechanisms of deformation and decohesion at crack tips; it cannot be detected by quasistatic far-field theories that consider only energy balance and neglect relevant dynamic degrees of freedom. They stressed that the challenge for future research is to understand why dynamic fracture sometimes seems to be stable. What mechanisms might cause the cohesive shear stress to be larger than its central-force value? What is the role of dissipative forces?

The picture of intrinsic instability found by them shows the necessity to study more complete dynamic models of deformation and decohesion at crack tips.

There are some other physical sources proposed to form fractal structures of many fractured surfaces. They will be discussed in Chap. 4.

# Chapter 3

# Introduction to Extended Defects and Mechanical Strength

Dislocation theory relates the plastic properties of crystals to atomic structure. Dislocations are more complicated than most other lattice defects. It is not easy to work with real laws of interatomic force, derived from quantum mechanics (see Chaps. 6–8). The strain field of a dislocation has a long range part, and this part can be discussed rigorously from simple elasticity theory. The amount of work on applications of the theory to the understanding of the structure-sensitive properties of crystals greatly exceeds that on the pure theory itself. The topic is vast, and it will not be possible to give a comprehensive description in one chapter. We will limit ourselves to an overview for those who already have some textbook knowledge of dislocation theory (see, for example, Rosenberg, 1992). We will concentrate our attention on dislocation mobility and fracture. Much of what follows is based on the writings by Lothe (1992) and Cottrell (1964): see also Hull and Bacon (1984).

## 3.1. *Some Basic Theory of Crystal Dislocations*

### 3.1.1. *Dislocations and Slip*

— Slip line. One part of the crystal slides as a unit across a neighbouring part along the slip direction lying in the surface of slip. The line of intersection of this surface with the outer surface of the crystal is called a slip line.

— Schmid's law. Slip begins on a given plane and direction when shear stress resolved on that plane and direction reaches a critical value.

— Dislocation. The boundary line between a slipped and an unslipped area is called a dislocation line. A dislocation line can never end within a crystal; it must form a closed ring or end at a free surface or be joined to other dislocation lines (Fig. 3.1).

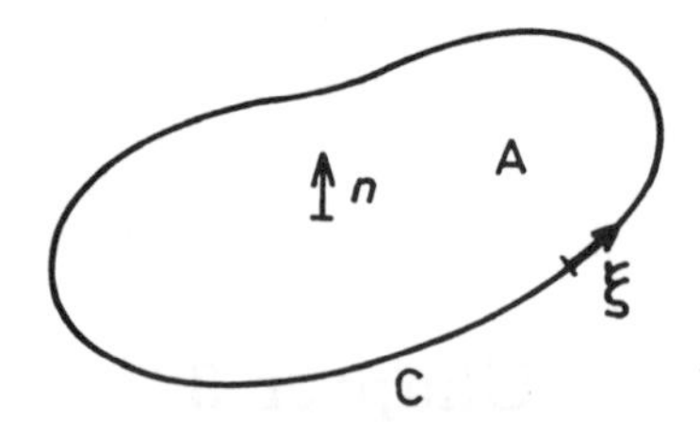


Fig. 3.1. A loop $C$ enclosing a surface $A$.

— Slip direction. The slip usually occurs along the crystallo-graphic direction of closed package in the slip plane, even when the applied stress is not precisely oriented along this direction, and it would occur instead along the line of greatest stress.

To create a dislocation in a perfect crystal, slip should occur over part of a slip plane. The shear strength of a perfect lattice may be calculated. The limiting shear strength of the lattice is given by:

$$\sigma_m = \frac{\mu}{2\pi}\frac{b}{a} \sim \frac{\mu}{10} \tag{3.1.1}$$

where, $b$ and $a$ are lattice parameters in the slip plane and perpendicular direction respectively. Using better approximations to the law of interatomic force in the slip plane reduces the value of $\sigma_m$ to about $\mu/30$. It can be seen that high stress is needed to create a dislocation. This is the theoretical shear strength or the upper limiting yield strength of the lattice (Cottrell, 1964). Calculations on this problem will be discussed later (Zhou *et al.*, 1994; Rice, 1992).

Lattice defects distort the electron and phonon structure of the crystal; the electronic levels are displaced (see Chap. 6), and local vibrations can arise (Indenbom, 1992).

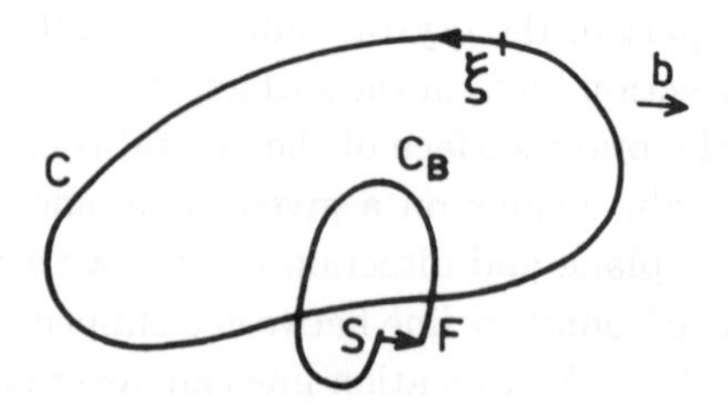


Fig. 3.2. The construction of a Burgers circuit.

### 3.1.2. *Burgers Vector*

Before giving the definition of the Burgers vector, we use the slip definition to classify dislocations.

(i) Super dislocations are those with Burgers vectors larger than the unit lattice vectors. These may be either perfect or imperfect.
(ii) Unit dislocations, which have a Burgers vector equal to a unit lattice vector.
(iii) Partial dislocations, which have a Burgers vector smaller than a unit lattice vector.

The Burgers vector **b** for a given loop can be found by the following procedure: construct a Burgers circuit $C$ encircling the dislocation in a right-handed way relative to the dislocation line (Fig. 3.2). The circuit would be closed if the crystal was perfect. The circuit will not be closed; then the vector from the starting point S to the finishing point F is the Burgers vector **b**.

According to this definition, the local Burgers vector depends on the lattice strains. If we define b more precisely, meaning the corresponding translational vector in an unstrained reference lattice, **b** is well defined and independent of the Burgers circuit and is a characteristic vector for the whole dislocation line (Lothe, 1992).

In real crystals, the cut surface would prefer to have the Burgers vector producing a low energy stacking fault or twin boundary. Thus, only the one or two shortest perfect lattice Burgers vectors are stable in high symmetry crystals: $\frac{1}{2}\langle 110\rangle$ and $\langle 100\rangle$ in fcc, $\frac{1}{2}\langle 111\rangle$ and $\langle 100\rangle$ in bcc, and $\langle 0001\rangle$ and $\frac{1}{6}\langle 11\bar{2}0\rangle$ in hcp (Hirth, 1992).

The sign of the direction of a dislocation $\xi$ was arbitrarily chosen. If the sign of $\xi$ is reversed, the sign of **b** also changes. The Burgers vector **b** as well as $\xi$ must be specified for a complete characterization of a dislocation loop $C$. A straight dislocation with $\mathbf{b} \parallel \xi$ is a screw dislocation. A straight dislocation with $\mathbf{b} \perp \xi$ is an edge dislocation.

If the dislocation branches, the sum of the Burgers vector of the dislocations after branching must be equal to the Burgers vector of the initial dislocation (Fig. 3.3). If we assume that all the dislocations are directed towards (or outwards from) the branching point (node) the sum of their Burgers vectors must be equal to zero (Fig. 3.4). This is the condition for conservation of the Burgers vector along the dislocation.

$$\sum_i b_i = 0 \tag{3.1.2}$$

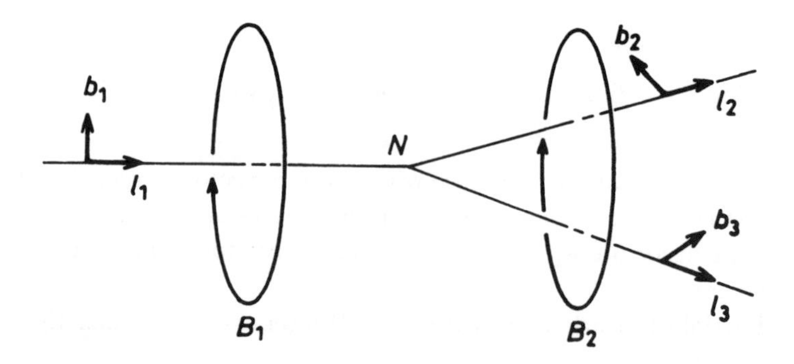


Fig. 3.3. Branch of a dislocations line.

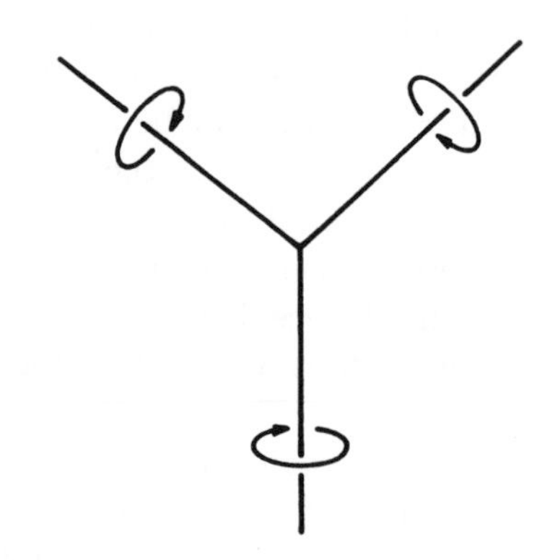

Fig. 3.4. Burgers vectors at a node.

### 3.1.3. *Glide and Climb*

Suppose that the element $d\mathbf{l} = \xi dl$ of the dislocation loop in Fig. 3.1 moves by $\delta r$ (Fig. 3.5). The motion of $d\mathbf{l}$ contributes to the change in the enclosed area A with

$$\delta\mathbf{A} = \delta\mathbf{r} \times d\mathbf{l} \tag{3.1.3}$$

and material in the amount

$$\delta V = \mathbf{b} \cdot d\mathbf{A} = (d\mathbf{l} \times \mathbf{b}) \cdot \delta\mathbf{r} \tag{3.1.4}$$

must be removed over the cut surface A in the process ($\delta V < 0$, added) in order that material continuity be preserved.

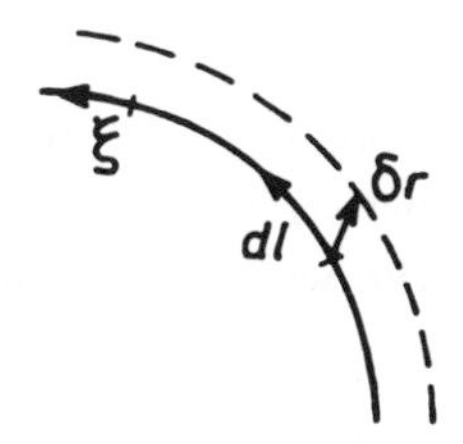


Fig. 3.5. Motion $\delta\mathbf{r}$ of an element $d\mathbf{l}$.

If $\delta r$ lies in the plane containing both $b$ and $dl$, then $\delta V = 0$. In this case, no material transport to or from the dislocation element is required. This is a conservative motion of the dislocation. Such motion is called glide. The glide plane is $(d\mathbf{l} \times \mathbf{b})$.

Motion out of the glide plane requires material transport to or from the dislocation, and is called climb. These are non-conservative motions. Since the dislocation "climb" out of its true glide plane (defined by $\mathbf{b} \cdot d\mathbf{A} = 0$) as it moves. The net amount of material removed is given by

$$V = \mathbf{b} \cdot \int_A d\mathbf{A} \,. \tag{3.1.5}$$

It is proportional to the area projected onto a plane perpendicular to $\mathbf{b}$, no matter how $\mathbf{A}$ is chosen, provided the loop is given. Voids ($\delta V > 0$) or interstitial atoms are produced in the track of the moving dislocation.

### 3.1.4. *Jogs*

When non-parallel dislocations with non-parallel Burgers vectors cut through one another, jogs can be produced. The type of jog produced by such an intersection can be found by establishing whether the $\xi$ of one dislocation crosses the slip plane of the other from the negative to the positive face, or vice versa.

In the case of an edge cutting an edge, only the stationary dislocation acquires a jog; in the case of an edge cutting a screw, a jog appears only in the moving dislocation. The reason is that a jog must be normal to the slip plane of the dislocation; a jog in the slip plane would immediately be eliminated since the dislocation would straighten out by pure glide.

In general, a jog in a moving dislocation produces a drag that slows down the dislocation. The details are shown in the book by Read (1953).

### 3.1.5. *Forces on Dislocations*

If we have formed a dislocation loop C in a crystal with Burgers vector **b** which has elastic self-energy $E$, another stress $\sigma_2$ is present (due to another loop $C_2$ or externally applied forces), the work needed for the creation of loop $C_1$ is reduced by

$$E_{1-2} = -\int_{A_1} \mathbf{b}_1 \cdot \sigma_2 \cdot d\mathbf{A}_1 \,. \tag{3.1.6}$$

Then, the total energy will be

$$E = E_1 + E_2 + E_{1-2} \,. \tag{3.1.7}$$

Let us now deform the loop $C_1$ by moving the elements $d\mathbf{l}_1$ by $\delta r_1$, while the field $\sigma_2$ is kept constant and the configuration of the loop $C_2$ is not changed. Then, $\delta\sigma_2 = 0$ and $\delta E_2 = 0$. From Eq. (3.1.7),

$$\delta E = \delta E_1 + \delta E_{1-2} \tag{3.1.8}$$

where, by Eqs. (3.1.6) and (3.1.3)

$$\delta E_{1-2} = -\oint_{C_1} [(\mathbf{b}_1 \cdot \mathbf{d}_2) \times d\mathbf{l}_1] \cdot \delta\mathbf{r}_1 \tag{3.1.9}$$

or:

$$\delta E_{1-2} = -\int_{C_1} d\mathbf{f}_{1-2} \cdot \delta\mathbf{r}_1 \tag{3.1.10}$$

where $df_{1-2}$ is the force on element $d\mathbf{l}_1$ due to the stress $\sigma_2$ and is given by (Peach and Koehler, 1950)

$$d\mathbf{f}_{1-2} = (\mathbf{b}_1 \cdot \sigma) \times d\mathbf{l}_1 \,. \tag{3.1.11}$$

If the shear stress $\sigma_2$ is on the slip plane $(\mathbf{b}_1 \times d\mathbf{l}_1)$, then, the force per unit length acting on the dislocation line is simplified as

$$F = \sigma b \,. \tag{3.1.12}$$

## 3.2. *Elastic Field of Straight Dislocation*

The stress field of a straight dislocation line is long-range, falling off as $r^{-1}$. At a distance $10^4 b$ the stress is of order $10^{-5}\ \mu$, i.e. as large as the yield stress of a soft crystal. Hence dislocations strongly interact with one another elastically.

3.2.1. *Summary of Isotropic Elasticity Theory*

In isotropic solids

$$C_{ijkl} = \mu(\delta_{ik}\delta_{jl} + \delta_{il}\delta_{jk}) + \lambda\delta_{ij}\delta_{kl} \tag{3.2.1}$$

where $\lambda$ is the Lamé constant and $\mu$ the shear modulus. The relation between stress, $\sigma_{ij}$, and strain, is given by

$$\sigma_{ij} = \lambda(\epsilon_{11} + \epsilon_{22} + \epsilon_{33})\delta_{ij} + \mu\epsilon_{ij} \tag{3.2.2}$$

where

$$\epsilon_{ij} = \frac{1}{2}\left(\frac{\partial u_i}{\partial x_j} + \frac{\partial u_j}{\partial x_i}\right) \tag{3.2.3}$$

and $u_i$ and $u_j$ are displacements along $i$ and $j$ directions respectively.

The inverse of Eq. (3.2.2) is

$$\epsilon_{ij} = \frac{1}{2\mu}\sigma_{ij} - \frac{\nu}{E}(\sigma_{11} + \sigma_{22} + \sigma_{33})\delta_{ij} \tag{3.2.4}$$

where $\nu$ is the Poisson ratio

$$\nu = \frac{\lambda}{2(\mu + \lambda)} \tag{3.2.5}$$

and $E$ is Young's modulus

$$E = \frac{\mu(3\lambda + 2\mu)}{\mu + \lambda}\,. \tag{3.2.6}$$

It follows simply from Eqs. (3.2.5) and (3.2.6) that

$$\mu = \frac{E}{2(1 + \nu)}\,. \tag{3.2.7}$$

From Eq. (3.2.4) $\epsilon_{11}$ can be written as

$$\epsilon_{11} = \frac{1}{E}[\sigma_{11} - \nu(\sigma_{22} + \sigma_{33})]\,. \tag{3.2.8}$$

Navier's equation expressed by displacement (3.2.10) can be obtained by substituting (3.2.2) to stress equilibrium Eq. (3.2.9) (Peach and Koehler, 1950)

$$\sigma_{ij,j} + F_i = 0 \tag{3.2.9}$$

$$\mu\nabla^2 u_i + (\lambda + \mu)e_{,i} + F_i = 0 \tag{3.2.10}$$

where,

$$\nabla^2 = \frac{\partial^2}{\partial x_i \partial x_i}$$

$$e = u_{j,j} \,.$$

Beltrami–Michell's compatibility equation expressed by stress can be obtained by substituting (3.2.4) to strain compatibility equation,

$$\epsilon_{ij,kl} + \epsilon_{kl,ij} - \epsilon_{ik,jl} - \epsilon_{jl,ik} = 0 \tag{3.2.11}$$

$$\nabla^2 \sigma_{ij} + \frac{1}{1+\nu}\Theta_{,ij} = -\frac{\nu}{1-\nu}\delta_{ij}F_{k,k} - (F_{i,j} + F_{j,i}) \tag{3.2.12}$$

where,

$$\Theta = \sigma_{k,k} \,.$$

In the special case when the body force is constant, then

$$\nabla^2 \Theta = -\frac{1+\nu}{1+\nu}F_{i,i} = 0$$

is a harmonic function. Then, from Eq. (3.2.12),

$$\nabla^2 \sigma_{ij} = -\frac{1}{1+\nu}\Theta_{,ij} \tag{3.2.13}$$

$$\nabla^2 \nabla^2 \sigma_{ij} = \nabla^4 \sigma_{ij} = 0 \,. \tag{3.2.14}$$

From the relation between stress and strain, we also have

$$\nabla^4 \epsilon_{ij} = 0 \,.$$

Both $\sigma_{ij}$ and $\epsilon_{ij}$ satisfy biharmonic equations.

For plane deformation problems, it is useful to introduce Airy's stress function, $\Phi$, which is given by

$$\sigma_{11} - V = \Phi_{,22} \,, \quad \sigma_{12} = -\Phi_{,12} \,, \quad \sigma_{22} - V = \Phi_{,11} \,. \tag{3.2.15}$$

It can be shown that the elastic equation for plane deformation is satisfied by any solution of the biharmonic equation

$$\nabla^4 \Phi = 0 \,, \tag{3.2.16}$$

where the body force has been ignored. The solution of Eq. (3.2.16) is

$$\Phi(x,y,z) = \Phi_0(x,y) + z\Phi_1(x,y) - \frac{\nu}{2(1+\nu)} z^2 \nabla^2 \Phi_0 \tag{3.2.17}$$

where $\Phi_0$ and $\Phi_1$ have the forms

$$\begin{aligned} \Phi_0 &= \frac{1}{2} x\phi(x,y) + f_0(x,y) \\ \Phi_1 &= \frac{1}{4} k(x^2+y^2) + f_1(x,y) \end{aligned} \tag{3.2.18}$$

and $f_0$, $f_1$ are harmonic functions. $\sigma_{ij}$ can be obtained by Eq. (3.2.15) (the body force has been ignored $V \approx 0$). Using Eq. (3.2.4), $\epsilon_{ij}$ could be calculated explicitly. Here, $\nabla^2\Phi_0$ in Eq. (3.2.17) is given by

$$\nabla^2 \Phi_0 = \Theta_0(x,y) \tag{3.2.19}$$

$$\Theta_0 = \Theta - kz \tag{3.2.20}$$

and $\Theta$ is defined by Eq. (3.2.12),

$$\Theta = \sigma_{kk} = \sigma_{11} + \sigma_{12} \,. \tag{3.2.21}$$

### 3.2.2. *Elastic Field of An Edge Dislocation*

An edge dislocation is a case of plane strain. Consider a long edge dislocation along the localized axis. The discontinuity in displacement, $b$, is normal to the line. Hence we seek solutions for $u_x$ and $u_y$, $u_z = 0$ and $\partial u_x/\partial z = \partial u_y/\partial z = 0$. Taking the separable form

$$\Phi = R(r)\Theta(\theta) \tag{3.2.22}$$

and solving the resulting ordinary differential equations, the solution for a positive edge dislocation is obtained of the form

$$\Phi = -Dr \ln r \sin\theta = -Dy \ln(x^2+y^2)^{1/2} \tag{3.2.23}$$

where

$$D = \mu b/(2\pi(1-\nu)) \,.$$

The stresses are given by

$$\begin{aligned}
\sigma_{xx} &= -\frac{\mu b}{2\pi(1-\nu)} \frac{y(3x^2+y^2)}{(x^2+y^2)^2} \\
\sigma_{yy} &= \frac{\mu b}{2\pi(1-\nu)} \frac{y(x^2-y^2)}{(x^2+y^2)^2} \\
\sigma_{xy} &= \frac{\mu b}{2\pi(1-\nu)} \frac{x(x^2-y^2)}{(x^2+y^2)^2} \\
\sigma_{zz} &= \nu(\sigma_{xx}+\sigma_{yy}) = -\frac{\mu b \nu}{\pi(1-\nu)} \frac{y}{x^2+y^2} \\
\sigma_{xz} &= \sigma_{yz} = 0\,.
\end{aligned} \tag{3.2.24}$$

The displacements are

$$\begin{aligned}
u_x &= \frac{b}{2\pi}\left[\tan^{-1}\frac{y}{x} + \frac{xy}{2(1-\nu)(x^2+y^2)}\right] \\
u_y &= -\frac{b}{2\pi}\left[\frac{1-2\nu}{4(1-\nu)}\ln(x^2+y^2) + \frac{x^2-y^2}{4(1-\nu)(x^2+y^2)}\right].
\end{aligned} \tag{3.2.25}$$

In cylindrical coordinates, the formulae for the stresses are

$$\begin{aligned}
\sigma_{rr} &= \sigma_{\theta\theta} = -\frac{\mu b \sin\theta}{2\pi(1-\nu)r} \\
\sigma_{r\theta} &= \frac{\mu b \cos\theta}{2\pi(1-\nu)r} \\
\sigma_{zz} &= \nu(\sigma_{rr}+\sigma_{\theta\theta}) = -\frac{\mu b \nu \sin\theta}{\pi(1-\nu)r}\,.
\end{aligned} \tag{3.2.26}$$

The displacements in Eq. (3.2.5) take the form

$$\begin{aligned}
u_r &= \frac{b}{2\pi}\left[-\frac{(1-2\nu)}{2(1-\nu)}\sin\theta\ln r + \frac{\sin\theta}{4(1-\nu)} + \theta\cos\theta\right] \\
u_\theta &= \frac{b}{2\pi}\left[-\frac{(1-2\nu)}{2(1-\nu)}\cos\theta\ln r - \frac{\cos\theta}{4(1-\nu)} - \theta\sin\theta\right].
\end{aligned} \tag{3.2.27}$$

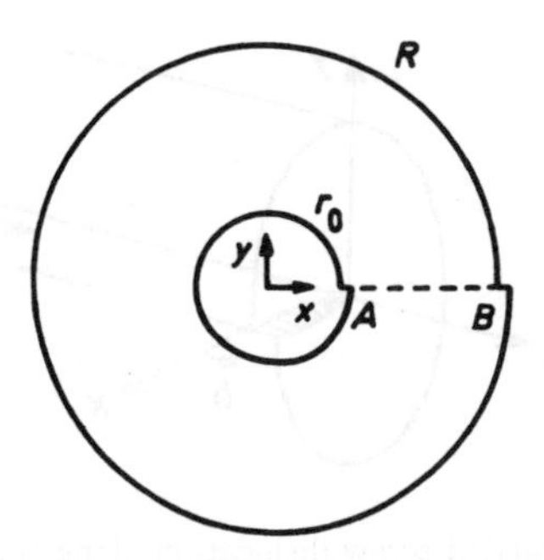


Fig. 3.6. The formation of an edge dislocation by slip.

The energy per unit length $W_e/L$ in the region between two cylinders of radii $r_0$ and $R$ (Fig. 3.6) can be calculated as the work done on the plane $AB$,

$$\frac{W_e}{L} = \frac{\mu b^2}{4\pi(1-\nu)} \ln\left(\frac{R}{r_0}\right) . \tag{3.2.28}$$

The work done on the outer cylinder of radius $R$ just compensates the same amount of work extracted at the inner cylinder of radius $r_0$.

### 3.2.3. *Elastic Field of Screw Dislocation*

The field is not one of plane deformation and so cannot be found from the biharmonic equation. For a right-handed screw dislocation along the localized axis in an infinite medium, the displacement is:

$$u_Z = \frac{b\theta}{2\pi} = \frac{b}{2\pi} \tan^{-1}\left(\frac{y}{x}\right) . \tag{3.2.29}$$

The discontinuity surface is chosen on $y = 0$, $x > 0$. The stresses can be obtained from Eq. (3.2.29) as

$$\begin{aligned} \sigma_{xz} &= -\frac{\mu b}{2\pi} \frac{y}{x^2 + y^2} \\ \sigma_{yz} &= \frac{\mu b}{2\pi} \frac{x}{x^2 + y^2} \\ \sigma_{xy} &= \sigma_{xx} = \sigma_{yy} = \sigma_{zz} = 0 . \end{aligned} \tag{3.2.30}$$

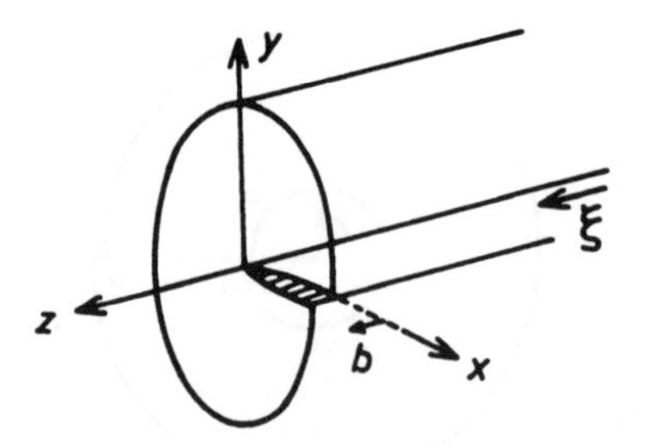


Fig. 3.7. A right-handed screw dislocation along the axis of a cylinder.

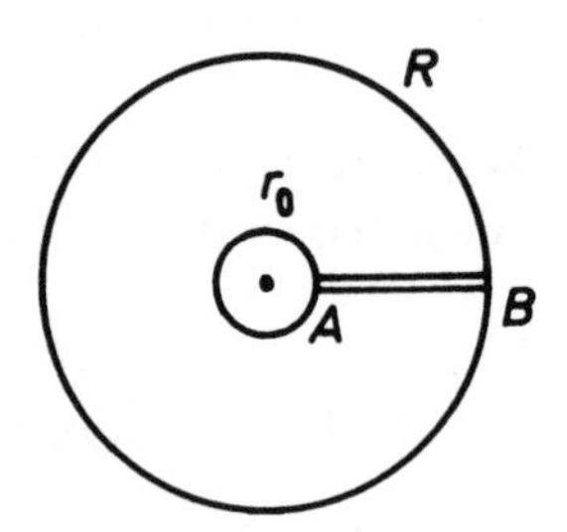


Fig. 3.8. A region defined by two coaxial cylinders.

In cylindrical coordinates (Fig. 3.7)

$$\begin{aligned}\sigma_{\theta z} &= \frac{\mu b}{2\pi r} \\ \sigma_{xz} &= \sigma_{r\theta} = \sigma_{rr} = \sigma_{\theta\theta} = \sigma_{zz} = 0\,.\end{aligned} \tag{3.2.31}$$

The energy per unit length $W_s/L$ in the region between two cylinders of radii $r_0$ and $R$ (Fig. 3.8) is

$$\frac{W_s}{L} = \frac{1}{2}\int_{r_0}^{R} \sigma_{yz}(y=0)\, b dx = \frac{\mu b^2}{4\pi} \ln\left(\frac{R}{r_0}\right)\,. \tag{3.2.32}$$

Equations (3.2.28) and (3.2.32) diverge as $r_0 \to 0$, indicating the failure of the continuum description near the core.

### 3.2.4. *Uniform Dissociation Model of Dislocation Core*

Lothe (1992) considered the divergencies of dislocation theory and proposed a spread-out dislocation core instead of the ideal line dislocation to remove this

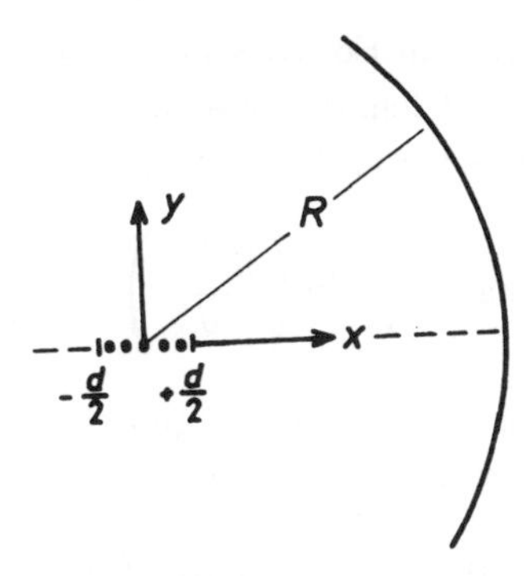


Fig. 3.9. The dislocation of the standard core.

divergence. His mathematically simple model for the core for the screw may be used consistently in all calculations, as a standard reference core. A uniform dissociation of the core within a strip of width $d$, as indicated in Fig. 3.9 was assumed.

A strip of width $dx$ within the interval $[-d/2, d/2]$ contributes with $db$,

$$db = bdx/d \tag{3.2.33}$$

to the Burgers vector, and the relative displacement over the discontinuity surface is then

$$\begin{aligned} \Delta u_z &= 0 & x &< -\frac{d}{2} \\ \Delta u_z &= b\left(\frac{1}{2} + \frac{x}{d}\right), & -\frac{d}{2} &< x < \frac{d}{2} \\ \Delta u_z &= b & x &> \frac{d}{2}. \end{aligned} \tag{3.2.34}$$

The stress given by this continuous distribution of dislocations is written as

$$\sigma_{yz} = \frac{\mu b}{2\pi d} P \int_{\frac{-d}{2}}^{+\frac{d}{2}} \frac{1}{x - x'} dx' .$$

That is

$$\begin{aligned} \sigma_{yz} &= \frac{\mu b}{2\pi d} \ln\left(\frac{\frac{x+d}{2}}{\frac{x-d}{2}}\right) & x &> \frac{d}{2} \\ \sigma_{yz} &= \frac{\mu b}{2\pi d} \ln\left(\frac{\frac{d}{2} + x}{\frac{d}{2} - x}\right) & -\frac{d}{2} &< x < \frac{d}{2}. \end{aligned} \tag{3.2.35}$$

The stress-distribution, given in Eq. (3.2.35) is logarithmically divergent, but integrable. The work per unit length done on the discontinuity surface $y = 0$, $-d/2 < x < R$ can be derived from Eqs. (3.2.34) and (3.2.35)

$$\frac{W}{L} = \frac{1}{2}\int_{\frac{d}{2}}^{R} \sigma_{yz}\Delta u_z dx = \frac{\mu b^2}{4\pi}\ln\left(\frac{\mathrm{Re}^{3/2}}{d}\right), \quad \text{when } R \gg d. \tag{3.2.36}$$

The work done on the outer cylinder can be ignored compared with Eq. (3.2.36). The total elastic energy per unit length within a cylinder of radius $R$ is

$$\frac{W_s}{L} = \frac{\mu b^2}{4\pi}\ln\left(\frac{\mathrm{Re}^{3/2}}{d}\right). \tag{3.2.37}$$

For brevity, Lothe (1992) defined a parameter $\rho$ which is

$$\rho = de^{-3/2}/2. \tag{3.2.38}$$

Equation (3.2.37) can be more simply written as

$$\frac{W_s}{L} = \frac{\mu b^2}{4\pi}\ln\left(\frac{R}{2\rho}\right). \tag{3.2.39}$$

The $W_s/L$ value of a single dislocation in infinite media remains divergent at outer radius $R$. If we consider the dislocation dipole, it is a well-defined value. Let $r$ be the separation of dislocations of opposite sign in the plane of core dissociation,

$$\frac{W}{L} = 2\frac{\mu b^2}{4\pi}\ln\left(\frac{r}{2\rho}\right), \quad r \gg 2\rho e^{3/2}. \tag{3.2.40}$$

Analogous to the treatment of the screw dislocation, except that now there is a non-zero energy input at the cylindrical surface, the elastic energy per unit length edge dislocation can be written as

$$\frac{W_e}{L} = \frac{\mu b^2}{4\pi(1-\nu)}\ln\left(\frac{R}{2\rho}\right) + \frac{\mu b^2(1-2\nu)}{16\pi(1-\nu)^2}. \tag{3.2.41}$$

For a dipole consisting of two opposite sign standard-core edge dislocations in the same glide plane and separated by a distance $r$,

$$\frac{W}{L} = 2\frac{\mu b^2}{4\pi(1-\nu)}\ln\left(\frac{r}{2\rho}\right), \quad (r \gg 2\rho e^{3/2}). \tag{3.2.42}$$

Beyond the slip plane, other core dissociations, say in the climb plane, could be imagined. For the same separation, the radial dislocation-dislocation interaction force is the same in the climb plane as in the glide plane, the energy per unit length of a dipole in the climb plane and with climb core dissociation is the same form as Eq. (3.2.42). The total energy of the dislocation core contains both an elastic part and a misfit energy part. The real width and form of the dissociation is determined by the balance between elastic terms and misfit terms. Planes with a misfit energy function allowing wide dissociation will thus be the favored planes for core dissociation. Anyhow, the model of uniform glide plane dissociation is considered as a standard reference.

### 3.2.5. *Mixed Dislocations*

A mixed dislocation can be considered as a superposition of a screw dislocation and an edge dislocation. Strains and stresses are obtained by the superposition of screw and edge solutions. From Eqs. (3.2.39) and (3.2.41), it follows that,

$$\frac{W}{L} = \frac{\mu b^2}{4\pi}\left(\cos^2\alpha + \frac{\sin^2\alpha}{1-\nu}\right)\ln\left(\frac{R}{2\rho}\right) + \frac{\mu b^2(1-2\nu)\sin^2\alpha}{16\pi(1-\nu)^2} \tag{3.2.43}$$

for the standard core mixed dislocation, and that for the standard core dipole lying in the glide plane

$$\frac{W}{L} = 2\frac{\mu b^2}{4\pi}\left(\cos^2\alpha + \frac{\sin^2\alpha}{1-\nu}\right)\ln\left(\frac{r}{2\rho}\right), \quad (r \gg 2\rho e^{3/2}). \tag{3.2.44}$$

### 3.2.6. *Dislocations in Anisotropic Media*

The calculation of the self-stress of dislocations is quite complicated for anisotropic crystals. For many purposes, a calculation assuming isotropy is adequate in view of the inaccuracy in the theories other than the neglect of anisotropy. However, reactions and interactions between dislocations in low-index crystallographic directions are necessarily to be treated with theories including anisotropy.

## 3.3. *Interactions of Dislocation with Other Defects*

### 3.3.1. *Interactions with Point Defects*

With the dislocation axis lying at a distance $d > a$ from the point defect, where $a$ is the distance between neighbouring glide planes, the dislocation interaction with the defect can be described in an elastic approximation. With $d \leq a$, an analysis in terms of interatomic interactions is needed, to compute the behaviour of the impurity atom in the dislocation core. In this section, we shall consider the first case.

The elastic interaction of dislocations with a point defect is by the interference term in the elastic energy (Indenbom and Cheronov, 1992).

$$W = S_{ijkl} \int dV \sigma_{ij}^{(1)} \sigma_{ki}^{(2)} \tag{3.3.1}$$

where index (1) refers to the dislocation, and index (2) to the point defect. According to the Colonnetti theorem (see Indenbom, 1992), the work of internal stresses on total deformations of any origin vanishes

$$W = -\int dV \sigma_{ij}^{(1)} \epsilon_{ij}^{0} = -V_0 \sigma_{ij}^{(1)} \epsilon_{ij}^{0} \tag{3.3.2}$$

which means that the energy of elastic interaction between the dislocation and the point defect is equal (but opposite in sign) to the work of the internal stresses $\delta\nu_{ij}^{(1)}$ due to the dislocation of the self-deformations of the point defect $\epsilon_{ij}^{0}$. At distances of the order of the lattice parameter, various non-linear effects become significant, while modulus and non-linear effects far-from the dislocation diminish more rapidly than the first-order size effect and become insignificant.

Considering an edge dislocation interacting with a dilation center, we take into account only the pressure distribution in the dislocation elastic field (Lothe, 1992).

$$p = -\frac{1}{3}\sigma_{ii} = \frac{\mu b}{3\pi}\left(\frac{1+\nu}{1-\nu}\right)\frac{\sin\theta}{r} \tag{3.3.3}$$

Here, **b** is the Burgers vector, **r** is the distance from the dislocation axis to the point defect and $\theta$ is the azimuth angle taken from the Burgers vector towards the extra half-plane. If the dislocation is located along the $Z$-axis, the Burgers vector is directed along the $x$-axis and the dilation center, and the self-volume is $\delta v = \epsilon_{ij}^{0} V_0$, is at the point with coordinates $(x, y, z)$. Then, the interaction energy:

$$W = \frac{\mu b}{3\pi}\delta V \frac{(1+\nu)}{(1-\nu)}\frac{y}{x^2+y^2}\,. \tag{3.3.4}$$

Taking $y_0$ as the distance between the point defect and the glide plane, the energy $W$ assumes its extreme value when $x = 0$

$$W_0 = \frac{\mu b}{3\pi y_0}\delta V\left(\frac{1+\nu}{1-\nu}\right) \tag{3.3.5}$$

which is the binding energy. The interaction force can be calculated from Eq. (3.3.4)

$$F = \pm\frac{\partial W}{\partial x} = \mp 2W_0\frac{xyy_0}{(x^2+y^2)^2}\,.$$

The upper sign means directed from the point defect to the dislocation and vice versa.

### 3.3.2. *Interactions with Planar Free Surface*

Consider a right-handed screw dislocation $s$ parallel to a planar free surface, $B$, as shown in Fig. 3.10. The distance from $s$ to $B$ is $l$. If we superpose the self-stress of the screw and that of an imaginary screw of the same strength and opposite sign at the mirror position outside the solid, the boundary condition that at the free surface, $\nu_{xz} = 0$ is met. The image stress at the dislocation then is

$$\sigma_{yz} = \frac{\mu b}{4\pi l} \tag{3.3.6}$$

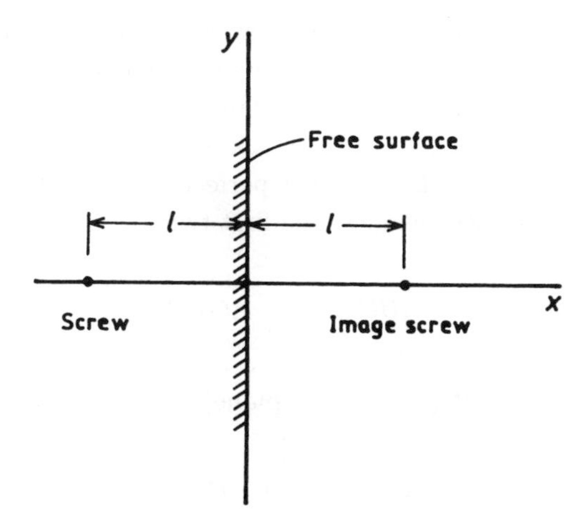


Fig. 3.10. A screw dislocation parallel to a free surface and an image dislocation.

The screw is drawn toward the surface by a force

$$\frac{F_x}{L} = \sigma_{yz} b = \frac{\mu b^2}{4\pi l} \tag{3.3.7}$$

per unit length.

As the image problem for the edge dislocation, even for the relatively simple problem as above the mathematics is somewhat lengthy and tedious.

More complicated image-force problems are discussed by Eshelby, (1979); Lothe, (1992).

## 3.4. *Crystal Lattice Effects*

In the continuum model of dislocations, the displacement of the cut and the magnitude of Burgers vector are arbitrary. In a crystal lattice, the cut surface would have an unacceptably large surface energy unless the Burgers vector were a translation vector of the perfect lattice or a vector producing a low energy stacking fault or twin boundary. Morever, if a dislocation splitting is energetically favorable, the dislocation can split into two product dislocations. Thus, only the one or two shortest perfect lattice Burgers vectors are stable in high symmetry crystals: $\frac{1}{2}\langle 110\rangle$ and $\langle 100\rangle$ in fcc, $\frac{1}{2}\langle 111\rangle$ and $\langle 110\rangle$ in bcc, and $\langle 0001\rangle$ and $\frac{1}{6}\langle 1120\rangle$ in hcp (Read, 1953; Hirth, 1992).

### 3.4.1. *Stacking Order in Closed-Packed Structures*

Closed-packed fcc and hcp are two simple examples of spheres stacked in close packing. First lay down a closed-packed layer; call it the A layer. The second layer can go either of the two sets of hollows on the first layer; call these the $B$ and $C$ positions, respectively. Every layer in the stack has to lie in one of the three positions — $A, B,$ or $C$ — if the stack is close packed. The normal f.c.c. sequence is $\ldots ABCABCABC\ldots$ in a perfect crystal. Three simple faults have relatively low surface energies: the first twin,

$$\ldots ABCA \quad \big| \quad B \quad \big| \quad ACBA \ldots$$

the intrinsic stacking fault (removing a plane),

$$\ldots ABCA \quad \big| \quad BA \quad \big| \quad BCA \ldots$$

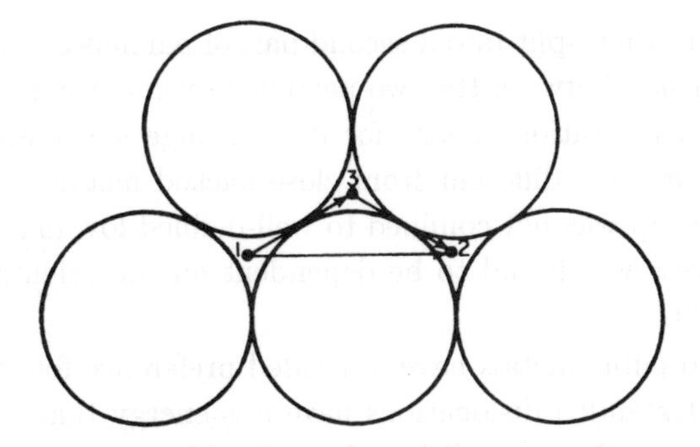


Fig. 3.11. Atomic scale view normal to a {111} glide plane in an fcc crystal.

and the extrinsic stacking fault (inserting a plane)

$$\ldots ABCAB \;|\; A \;|\; CABC \ldots .$$

The value of the extrinsic fault energy $\gamma_E$ is close to $\gamma_I$, and is the double value of the twin boundary energy $\gamma_T$, $2\gamma_T = \gamma_E = \gamma_I$, (Hirth, 1992). $\gamma_T = 24\ \mathrm{mJm^{-2}}$ and $15\ \mathrm{mJm^{-2}}$ for Cu and Au respectively.

### 3.4.2. *Partial and Extended Dislocations in fcc Crystals*

In Fig. 3.11, motion of a perfect $\frac{1}{2}[011]$ dislocation shears an atom from position 1 to position 2. Alternatively, the dislocation can dissociate into a partial $\frac{1}{6}[112]$, shearing the atom from 1 to 3 and creating an intrinsic stacking fault and a partial $\frac{1}{6}[\bar{1}21]$ completing the shear from 3 to 2 and annihilating the fault.

Suppose a stacking fault does not extend all the way through the crystal, it must terminate on a line imperfection. The boundary of a stacking fault is called a partial dislocation. There are various partial dislocations. Read, 1953 and Hirth, 1992, deal with partial dislocations and stacking faults in detail. Two parallel partials connected by a strip of stacking fault form an extended dislocation.

### 3.4.3. *Stacking Faults and Extended Dislocations in Plastic Deformation*

Extended dislocations affect the form of slip lines. If an extended screw dislocation wants to shift from one plane to another, it must first collapse into a single dislocation, then split into a second pair of partials on the second plane. The strong repulsion between the two partials prevents their coming together. There will be an activation energy for it to change slip planes. This is the reason why bcc metals, different from close-packed metals, deviate from the Schmid law, the slip was not confined to well-defined low index crystal planes and the flow stress was found to be dependent on the orientation of the slip plane (Vitek, 1992).

Face-centered-cubic metals have a decided preference for slip on the close-packed planes. Extended dislocations have less energy than others, and they lie on $\{111\}$ planes; therefore dislocations should form more easily on $\{111\}$ planes. The critical stress for the Frank-Read mechanism is proportional to dislocation energy; the mechanism should therefore operate preferentially on $\{111\}$ planes.

If the slip takes place on a plane that intersects the stacking fault, it cuts the fault into two halves, which are displaced parallel to the slip vector. These boundaries of the halves are partial dislocations. The slip that cuts the fault has to supply the energy to create the partials. Therefore a stacking fault is a barrier to slip on an intersecting plane. It hardens the crystal. Furthermore, the two new partials must have equal and opposite Burgers vectors, which are held close together by a strong attraction. This would resist the slip on the intersecting plane of the partials from being moved farther apart.

Dislocations can harden intersecting slip systems. An extended dislocation is hard to cut; where it jogs, the partials have to come together. The mutual repulsion of the partials which resists unsplitting before it can be cut hardens the system.

### 3.5. *Dislocation Motion over Peierls Barrier*

A dislocation experiences an oscillating potential energy as it glides in a crystal. In the Peierls model (1940), the bonds across the glide plane were considered to interact via an interatomic potential, while the remainder of the lattice was linearly elastic. Nabarro (1947) gave an analytical expression for the dislocation core model. Accurate theoretical evaluation of the Peierls barrier poses a very difficult problem, even with modern computers at hand. Another model is the Frenkel-Kontorova model (1938). This model has been widely used and has received much attention. It assumed that an infinite set of atoms were tied together in a chain by elastic bonds, and situated in the periodic potential of a substratum. When the number of atoms does not agree with the substrate potential valleys, defects will arise. These defects, in current terminology called solitons, can be considered to picture dislocation cores. For a review of current models and methods in numerical dislocation modelling, see Puls' paper (1981).

### 3.5.1. *Peierls Model*

One can approximately estimate the ideal lattice resistance to dislocation motion by means of the Peierls model. The resolved applied stress necessary to move the dislocation over the Peierls barrier is called Peierls stress, $\sigma_p$. It depends on crystal structure and elastic moduli. Nabarro (1947) derives

$$\sigma_p = \frac{2\mu}{1-\nu} \exp\left(-\frac{2\pi\xi}{b}\right) . \tag{3.5.1}$$

Here, $\xi = \frac{1}{2}a(1-\nu)$, is the dislocation core width, $a$ the lattice period, $\mu$ is the shear modulus, $\nu$ is Poisson's ratio and $b$ is the Burgers vector.

Equation (3.5.1) comes from the expression for the Peierls energy,

$$W(\alpha) = \frac{\mu b^2}{4\pi(1-\nu)} + \frac{W_p}{2} \cos(4\pi\alpha) \tag{3.5.2}$$

where $\alpha$ is a parameter for a translation of the dislocation by a distance $\alpha b$.

Hirth and Lothe (1982) pointed out that Eq. (3.5.2) predicts that, $\alpha = 0$ and $1/2$, correspond to maximum energy configuration and $\alpha = 1/4$ corresponds to the minimum energy. This point has led to some confusion in many literatures. In spite of the inadequacy of the model, the Peierls-Nabarro model can be used to demonstrate that a small finite stress is required to move a dislocation and that this stress decreases rapidly with increasing spread of misfit width across the glide plane.

### 3.5.2. *Frenkel-Kontorova Model*

In this model, the atoms above the slip plane are replaced by a series of mass points connected by identical springs, and the atoms on the bottom of the slip plane are replaced by a sinusoidal potential substrate. We may think of this configuration as a series of chain connected balls relaxing onto a corrugated trough under the force of gravity (Frenkel and Kontorova, 1938). With one more ball than at normal trough, there are potential minima, a one-dimensional dislocation is formed, having stable and unstable equilibrium configurations

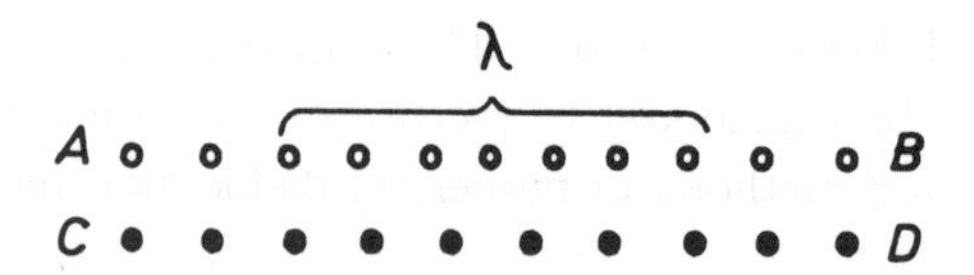


Fig. 3.12. The model of Frenkel and Kontorova.

just as for the Peierls energy (Fig. 3.12). The dislocation begins to move when its energy is larger than a certain minimum energy $W_0$ (see Frenkel, 1957).

$$W_0 = \frac{4}{\pi}\sqrt{mw_0^2 A}$$

where $w_0$ is the propagation velocity of a sound wave, $m$ is the mass, $A$ is the work done for an atom to move from one appropriate place to another appropriate place. The maximum effective length of the dislocation (or the width of the dislocation line) $\lambda_0$ is:

$$\lambda_0 = \frac{a}{2\pi}\sqrt{\frac{mw_0^2}{A}}\,. \qquad (3.5.3)$$

The Frenkel-Kontorova model also has the feature that the Peierls stress drops markedly with increasing dislocation width.

Atomic calculations and interpretation of experimental results have led to the general agreement that the Peierls barrier exists and that the magnitude of $\sigma_p$ varies from $\sim (10^{-3} - 10^{-4})\ \mu$ for closest packed metals and alkali-halide crystals to $\sim 10^{-2}\ \mu$ for elemental semiconductor with directed covalent bonds. The bcc metals are in an intermediate position. There, microscopic core models somewhat different from the original Peierls model apply (Hirth, 1960; Vitek and Kroupa, 1966; Vitek, 1974).

### 3.5.3. *Phase Transition: Analytical Mechanism Motion of Dislocations*

When the externally applied stress is less than the Peierls stress, simple mechanical motion will not take place. It can move over only by means of fluctuations helping it to overcome energy barriers. The original dislocation position corresponds to the metastable phase and the role of a "new phase" is played by dislocation segments which have overcome the barrier (Lothe and Hirth, 1959). The configuration that a dislocation segment in a new valley next to the original valley lies between two kinks which take the dislocation over the hill between the two valleys is called a double kink. The double kink plays the role of an embryo of "new phase". In the spirit of nucleation theory as applied to phase transitions, we assume an Arrhenius law:

$$n = n_0 \exp\left[-\frac{E(\sigma)}{kT}\right] \tag{3.5.4}$$

for the nucleation frequency of double kinks. This quantity, together with the kink mobility, will specify the time for complete 'transition' of the entire dislocation. Calculation of the energy barrier $E(\sigma)$ for double-kink nucleation involves a specification of the saddle point in the many-dimensional potential for double-kink formation on an originally unkinked dislocation. The saddle point is determined by application of equilibrium conditions to the dislocation in an interaction with the periodic potential and external stress $\sigma$. In order to arrive at a complete solution of the problem of determining $n$, including the preexponential factor $n_0$, we must resort to the kinetic equations for the distribution function describing fluctuational deviations of the dislocation out of the potential valley. Explicit solutions have been obtained in two limiting cases: The case of an applied stress which is small compared with the Peierls stress $\sigma_p$ and the case of an applied stress that is close to $\sigma_p$. Since overcoming the barrier proceeds by changes of the shape of the dislocation, one must start with a general geometrical study of the topography of the energy in the space of all possible dislocation configurations. Such an approach gives a rigorous foundation for the theory of kinks on dislocations and makes it possible to describe the transition from one valley to the next one in terms of fluctuations around the saddle point, and which can be described in kink terminology, relative to fluctuations about equilibrium on an otherwise straight dislocation in the potential valley (see Indenbom *et al.*, 1993).

For the cases $\sigma \ll \sigma_p$,

$$n = \frac{Da(U_0'')^{3/2}}{kT\sqrt{k}} L e^{\frac{-V_0}{kT}} \frac{e^{-\gamma}\gamma^{\gamma}}{\Gamma(\gamma)} \tag{3.5.5}$$

where we have introduced the abbreviation $\gamma = (\sigma ab/kT)(k/U_0'')^{1/2}$. For high values of $\gamma$, the activation energy $V_0$ is given by:

$$V_0 = 2V_1 - \sigma ab\sqrt{\frac{K}{U_0''}} \ln\left(\frac{\sigma_*}{\sigma}\right), \quad (\sigma_* \sim \sigma_p). \tag{3.5.6}$$

For the cases $\sigma \sim \sigma_p$, and $\sigma_p - \sigma \ll \sigma_p$,

$$n = 2^{15/8} \frac{15Dab^{11/8}|U_0'''|^{3/8}}{(2\pi kT)^{3/2}K^{1/4}} L(\sigma_p - \sigma)^{11/8} \exp\left(-\frac{V_0}{kT}\right) \tag{3.5.7}$$

where

$$V_0 = \frac{2^{1/4} 48\sqrt{k}(\sigma_p b)^{5/4}}{5|U_0'''|^{3/4}} \left(1 - \frac{\sigma}{\sigma_p}\right)^{5/4} \tag{3.5.8}$$

Here, $k$ is the line tension, $U(y)$ is a planar potential (Peierls potential) and $U_0''$ and $U_0'''$ are evaluated at the point $y = 0$ and the inflection point of the potential curve respectively, $a$ is the distance from a valley to the neighbouring valley along $y$-axis, $L$ is the length of the segment, and $D$ is the "atomic" diffusion coefficient. Equations (3.5.6) and (3.5.8) are convenient for comparison with experiments.

### 3.5.4. *Dislocation Velocity*

The quantity $n/L = J$ is interpreted as the probability of double-kink nucleation per unit time per unit length of dislocation. When the dislocation is short, the process would not involve the kink-kink annihilation. In this case, the dislocation velocity shows simple proportionality with the length of the dislocation,

$$v_D \sim an \sim aJL\,. \tag{3.5.9}$$

For sufficiently long segments, nucleation will take place in parallel at different places along the dislocation and the process will then also involve the rate of annihilation by kink-kink collisons. For a straight, originally unkinked dislocation, the mean distance between kinks after a time $t$ will be $\sim (1/Jt)$. The transition time for motion into the next valley is estimated by requiring that the distance travelled by a kink $v_n t$ equals the distance it must go to annihilation with another kink, that is the mean kink separation. This balance gives approximately

$$t \sim \frac{1}{\sqrt{Jv_n}} \tag{3.5.10}$$

for the transition time. Equation (3.5.10) gives a dislocation velocity

$$v_D \sim \frac{a}{t_n} \sim a\sqrt{Jv_n} \tag{3.5.11}$$

where, $v_n$ is the velocity of the kink which is determined by the kink mobility $\mu$ (the Einstein relation $\mu = D_1/kT$), the discreteness parameter $a$ and the force per unit length $\sigma b$.

### 3.5.5. *Quantum Motion of Dislocations*

At sufficiently low temperatures, quantum effects should be important and change a sharp decrease of dislocation mobility into a gradual transition to an athermal regime.

Weertman (1958) considered the dislocation in a sort of 'particle' picture and used the expression

$$W \sim \exp\left[-\frac{2}{\hbar}\int \sqrt{2mU(x)}dx\right] \tag{3.5.12}$$

for the tunneling transition probability. In this expression, $m$ was taken to the mass of one atom of the material, and the barrier was described to be of height $2V_1$ and width $\sim b$.

Indenbom *et al.* (1992) pointed out that for dislocation process, it is quasi-particle tunneling which corresponds to double-kink widening. The potential barrier for the process is

$$V(1) = 2V_1 - \sigma abl\,. \tag{3.5.13}$$

The mass related to double-kink widening can be deduced from the kinetic energy in the string as function of the velocity of widening $l^o$. Calculating the integral in Eq. (3.5.12), they obtained:

$$w \sim \exp\left(-\frac{16}{3}M^{1/2}\frac{V_1^{3/2}}{\hbar\sigma ab}\right)\,. \tag{3.5.14}$$

The difference between Eq. (3.5.14) and Weertman's result consists in the stress dependence of the transition probalility.

The most likely candidates for appreciable quantum effects will be crystals where the zero point motion has a large relative amplitude, such as inert-gas crystals and also solid hydrogen, solid deuterium and solid methane.

### 3.5.6. *Flow Limit*

In the simplest case, the macroscopic deformation velocity $\epsilon$ can be written as:

$$\dot{\epsilon} = \rho_D b v_D \tag{3.5.15}$$

where $\rho_D$ is the density of mobile dislocations. Equation (3.5.15) can be employed to interpret experiments on the temperature dependence of the flow limit $\sigma(T)$. The $\sigma(T)$ must be such that the condition

$$\dot{\epsilon} = \rho_D b v_D = \text{const.} \tag{3.5.16}$$

is fulfilled, where the constant is the velocity imposed by the machine.

It is interesting to discuss $\sigma(T)$ of a composite system of edge and screw dislocations (Lung, 1995; Jiang and Lung, 1996).

$$v_e = v_0 \left(\frac{\sigma}{\sigma_e}\right)^m$$

$$v_s = v_0 \left(\frac{\sigma}{\sigma_s}\right)^m \tag{3.5.17}$$

$$\bar{v} = (\rho_e v_e + \rho_s v_s)/\rho_T$$

$$\dot{\epsilon} = b v_0 \sigma^m \left(\frac{\rho_e}{\sigma_e^m} + \frac{\rho_s}{\sigma_s^m}\right). \tag{3.5.18}$$

Let:

$$x = \frac{b v_0 \sigma^m}{\dot{\epsilon}}, \quad x_1 = \frac{\sigma_e^m}{\rho_e}, \quad x_2 = \frac{\sigma_s^m}{\rho_s}$$

$$\frac{1}{x} = \frac{1}{x_1} + \frac{1}{x_2}. \tag{3.5.19}$$

Unlike the configuration of resistors in series, these processes seem like the configuration of resistors in parallel. We should note that the small one usually controls the change of the whole system. This conclusion seems in contradiction to the widely accepted concept, the difficult part, the screw, controls the change of $\sigma$ in bcc metals and alloys.

Moreover,

$$\frac{\Delta\sigma}{\sigma} = \frac{\sigma_s^m \rho_e}{\sigma_e^m \rho_s + \sigma_s^m \rho_e}\left(\frac{\Delta\sigma_e}{\sigma_e}\right) + \frac{\sigma_e^m \rho_s}{\sigma_e^m \rho_s + \sigma_s^m \rho_e}\left(\frac{\Delta\sigma_s}{\sigma_s}\right). \tag{3.5.20}$$

Let:

$$\alpha = \frac{\sigma_s^m}{\sigma_e^m}, \quad \beta = \left(\frac{\Delta\sigma_s}{\sigma_s}\right) / \left(\frac{\Delta\sigma_e}{\sigma_e}\right), \quad \rho = \frac{\rho_s}{\rho_e}$$

and then

$$\frac{\Delta\sigma}{\sigma} = \frac{\alpha + \rho\beta}{\rho + \alpha}\left(\frac{\Delta\sigma_e}{\sigma_e}\right). \tag{3.5.21}$$

The relative values of $\alpha$ and $\rho\beta$ determine which kind of dislocations contributes more in $\Delta\sigma/\sigma$. If $\rho\beta > \alpha$, the screw contributes more in the deformation. Figure 3.13 (Fig. 2(c) of Jiang and Lung, 1996) shows the relationship of total stress change with $(\frac{\Delta\sigma_e}{\sigma_e})$ and $s$, where $s = \rho_s/(\rho_s + \rho_e)$, $\alpha = 10$ and

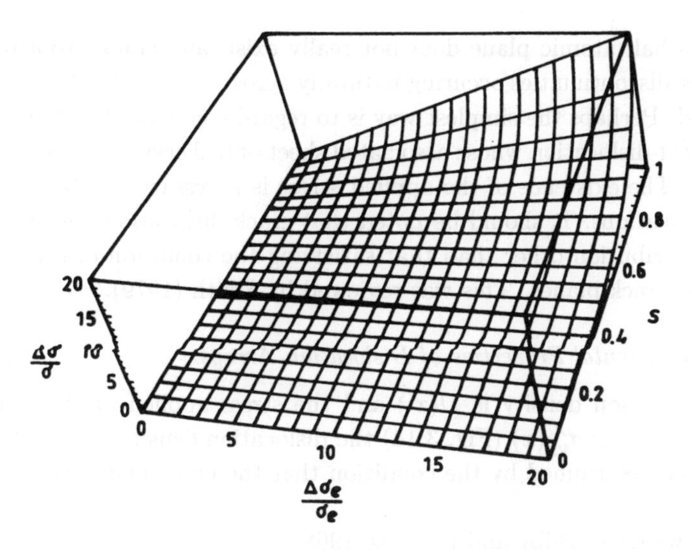


Fig. 3.13. Total stress changes vs $\Delta\sigma_e/\sigma_e$ for $\alpha = 10$ and $\beta = 1$: contribution from the screw. (Jiang and Lung, 1996)

$\beta = 1$. This figure shows that the screw contributes mainly when $s > \frac{10}{11}$. We believe that in many bcc metals, this condition is satisfied. Therefore, the above may not contradict with the previous experimental results and concrete conclusions on bcc metals. However, Lung *et al.* doubt the general concept that the difficult part controls the change of the whole composite system with the configuration in parallel. We noticed that Pharr and Nix (1979) proposed a model of strong dependence of the mobile dislocation density, $\rho$, on effective stress to account for some of the plastic flow behaviour observed in fcc metals (Pharr and Nix, 1979). Stein and Low (1960) have found that the temperature dependence of the applied stress on edge dislocation moving at constant velocity is also very strong in Fe-Si single crystals.

## 3.6. *Dislocations and Cracks*

There are many similarities between the mechanical behaviour of a macrocrack and dislocations piled up against a locked dislocation. Both have stress concentration effects. The Griffith formula for brittle fracture of solid may be derived from the fundamental properties of dislocations piled up against a locked dislocation (Hirth and Lothe, 1982). But the physical concept of a crack dislocation is not the same as an ordinary one (Bilby and Eshelby, 1968).

The extra half atomic plane does not really exist, and crack dislocations are defined by discontinuities occuring naturally across unwelded cuts when a body is stressed. Perhaps the simplest way is to regard a crack dislocation as a type of imperfect dislocation whose associated sheet of bad crystal is a missing plane of atoms. The existence of the applied stress is necessary for the existence of crack dislocations. It should be noted that crack dislocation densities cannot be any distribution other than that satisfying the condition of the stress-free surface on crack planes. This was reviewed by Smith (1979).

### 3.6.1. *Fundamental Properties of Dislocation Arrays*

If the dislocation density is $D(x')$ and the stress applied at the point $x$ on the crack plane is $\sigma_{yy}^A(x)$ (Fig. 3.14) the dislocation density distribution function can be determined by the condition that the crack planes should be free surfaces.

Then we have (Bilby and Eshelby, 1968)

$$\sigma_{yy}^A(x) + A \int_{-a}^{a} \frac{D(x')dx'}{x - x'} = 0 \,. \tag{3.6.1}$$

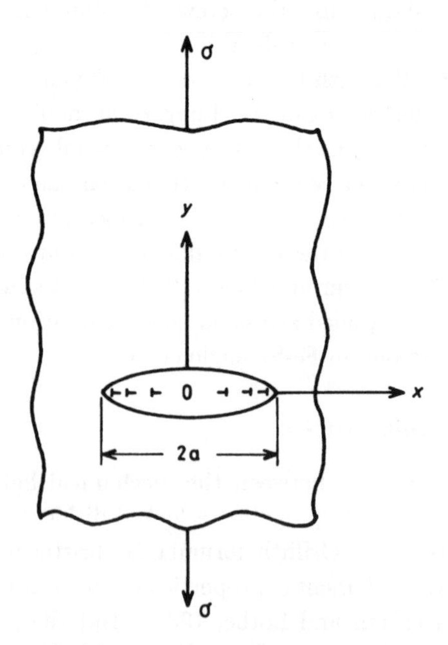


Fig. 3.14. The crack dislocation model.

$A = (\mu b)/(2\pi(1-\nu))$, $\mu$ is the shear modulus, $b$ is the Burgers vector. Solving Eq. (3.6.1), the dislocation density distribution function $D(x')$ can be determined.

The total stress $\sigma^A + \sigma^D$ can be determined at any point of the solid material and

$$\sigma_{ij}^D(\mathbf{r}) = b \int_{-a}^{a} D(x')\sigma_{ij}^y(x - x', y, z)dx' . \tag{3.6.2}$$

$\sigma_{ij}^D(r)$ is the stress at point $\mathbf{r}$ due to the dislocation, the core of which is at $x'$ and the Burgers vector of which is [010].

Since $\sigma^D$ is much larger than $\sigma^A$ in the neighbourhood of the crack tip, we need to consider only $\sigma^D$. Now we may displace the origin of coordinates to the crack tip. Let $s = x - a$, which is taken to be small. After some calculation, the expressions for $\sigma^D(s)$ and $D(s)$ may be obtained,

$$\sigma^D = \frac{K^D}{(2\pi)^{1/2}} s^{-1/2} \tag{3.6.3}$$

$$D(s) = \frac{K^D}{A\pi(2\pi)^{1/2}}(-s)^{-1/2} . \tag{3.6.4}$$

By comparing expression (3.6.3) with that in conventional fracture mechanics, we see that they are similar in mathematical form,

$$\sigma^c = \frac{K^c}{(2\pi)^{1/2}} s^{-1/2} . \tag{3.6.5}$$

Therefore,

$$\frac{\sigma^c(s)}{\sigma^D(s)} = \frac{K^c}{K^D} = \alpha . \tag{3.6.6}$$

We do not know whether they are similar or equivalent, as this depends on whether or not $\alpha$ is equal to unity. Although the general solution of $K^D$ from Eq. (3.6.1) is the same as that of $K^c$ in the central symmetrical crack case as has been done in symposium ASTM, STP 381, it is still necessary to compare the numerical results from Eqs. (3.6.3) and (3.6.4)with those from conventional fracture mechanics. According to the definition of the stress intensity factor,

$$K^D = (2\pi)^{1/2} \lim_{S\to 0} S^{1/2}\sigma^D(S) \tag{3.6.7}$$

and from Eq. (3.6.4),

$$(K^D)^2 = 2\pi A^2 \lim_{x\to a} D^2(x) . \tag{3.6.8}$$

Solving the integral equation, Eq. (3.6.1), $D(x)$ is obtained and the $K^D$ may be obtained from Eq. (3.6.8). Expressions (3.6.4) and (3.6.8) have been described by Bilby and Eshelby (1968); but no practical numerical results calculated with this method have been reported. It is not so easy to solve the integral equation for $D(x')$ analytically because of the variety of distribution of $\sigma_{yy}^A(x)$.

### 3.6.2. *Solution of Integral Equation with Chebyshev Polynomials*

$T_n(x)$ and $U_n(x)$ are a pair of normalized orthogonal polynomials between $(-1,+1)$, the weight function of which are $(1-x^2)^{-1/2}$, respectively. $T_n(x)$ and $U_n(x)$ are called Chebyshev Polynomials.

$$\begin{aligned} &T_0(x)=1, \quad T_1(x)=x, \quad T_2(x)=2x^2-1, \quad T_3(x)=4x^3-3x, \ \ldots \\ &U_0(x)=0, \quad U_1(x)=1, \quad U_2(x)=2x, \qquad\quad U_3(x)=4x^2-1, \ \ldots \end{aligned} \tag{3.6.9}$$

The relation between $T_n(x)$ and $U_n(x)$ is:

$$\frac{1}{\pi}\int_{-1}^{1}\frac{1}{y-x}\left[\frac{T_n(y)}{(1-y^2)^{1/2}}\right]dy = U_n(x)\,. \tag{3.6.10}$$

If the applied stress $\sigma^A(x)$ can be expanded to be polynomials of $U_n(\eta)$ terms,

$$\sigma^A(x)=\sum_{n=0}^{n}a_n x^n=\sum_{n=0}^{n}(a_n a^n)\eta^n=\sum_{n=1}^{n+1}C_n(a)U_n(\eta) \tag{3.6.11}$$

where $\eta = x/a$, $U_n(\eta)$ has the form of (3.6.9). Comparing (3.6.9) with (3.6.1), Lung (1980) obtained:

$$A\int_{-1}^{1}\frac{D(\eta)d\eta}{\eta-\frac{x}{a}}=\sigma_{yy}^A\left(\frac{x}{a}\right)\,. \tag{3.6.12}$$

Using the relations of (3.6.10) and (3.6.12), Lung obtained $D(\eta)$ and then the expression for $K^D$. After certain steps of calculation and check for the edge crack case, Lung (1980) applied this method to a rotary plate (a model for the steam turbine wheel) with an edge crack at the central hole. The result obtained is compared with that of the finite element method (Fig. 3.15). The difference in percentage are about 2–15%. However, the calculation based on the dislocation theory is simpler.

The Chebyshev polynomials may be applied to more general cases. Any stress distribution function $\sigma(x)$ can be expressed in terms of $U_n(x)$ in principle

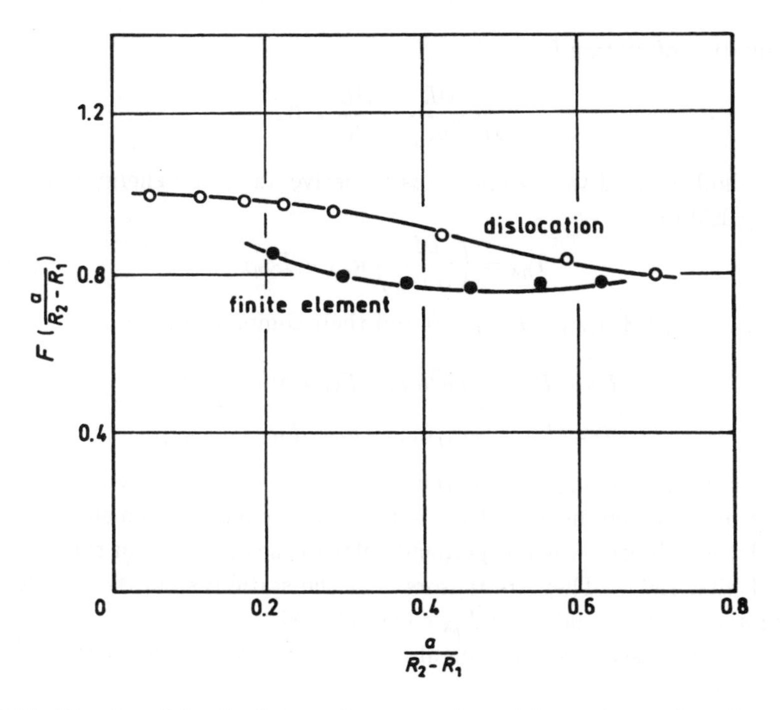


Fig. 3.15. The stress intensity factors of a rotary plate with an edge crack at the central hole. (Lung, 1980)

and $\sigma(x)$ can be obtained from design data. From the above, it was shown that the fractural behaviour of the dislocation model gives approximately the same results as that calculated by other methods in fracture mechanics.

### 3.6.3. *Elastic Energy Momentum Tensor in Defect Problem*

In general, point defects, dislocations, crack tips, etc. and all other defects may be considered as singularities in an elastic field. Their mechanical behaviour obeys the law of the general elastic field theory. Their differences reflect their specificities. Eshelby has discussed the interaction between the applied stress and the defects (point defects and dislocations) (Eshelby, 1956). The Lagrangian density for the free elastic field:

$$L = \frac{1}{2}\rho\dot{u}^2 - W(u_{i,j}) \tag{3.6.13}$$

with an external force density $f_i$, is not taken into account in the Lagrangian.

The equation of motion is:

$$\frac{\partial}{\partial x_i}\frac{\partial L}{\partial u_{i,j}} - \frac{\partial L}{\partial u} = 0\,. \tag{3.6.14}$$

The methods of field theory enable us to derive an elastic energy momentum tensor (EEMT),

$$T_{\eta\lambda} = \left(\frac{\partial L}{\partial u_{i,\eta}}\right) u_{i,\lambda} - L\delta_{\eta\lambda} \tag{3.6.15}$$

where $\eta, \lambda = 1, 2, 3, 4, x_4 = t,\ u_4 = 0$ and their components are

$$\begin{aligned} T_{jl} &= P_{jl} - \frac{1}{2}\rho\dot{u}^2\delta_{jl}\,, \quad T_{44} = W + \frac{1}{2}\rho\dot{u}^2 \\ S_j &= T_{j4} = -\sigma_{ij}\dot{u}_i\,, \quad\ \ g_i = T_{4l} = \rho\dot{u}_i^2 u_{i,l} \end{aligned} \tag{3.6.16}$$

and $P_{jl} = W\delta_{jl} - \sigma_{ij}u_{i,l}$ and the relation $\sigma_{ij} = \partial W/\partial u_{ij}$ is used.

The physical significance of $T_{jl}$ is the $l$ component of momentum flowing through the unit area which is perpendicular to the $l$ axis in unit time (Landau and Lifshitz, 1961). They are tensors. For the static case in the absence of a body force, only $P_{jl}$ and W of $T_{\eta\lambda}$ come into existence.

Eshelby has used an elastic singularity motion model to derive the following expression:

$$F_l = \int_{\Sigma} P_{jl} ds_j \tag{3.6.17}$$

where $F_l$ is the $l$-component of the forces acting on all the sources of internal stresses as well as elastic inhomogeneities in region I (Fig. 3.16). These forces are caused by sources of internal stresses and elastic inhomogeneities in region II and by the image effects associated with boundary conditions.

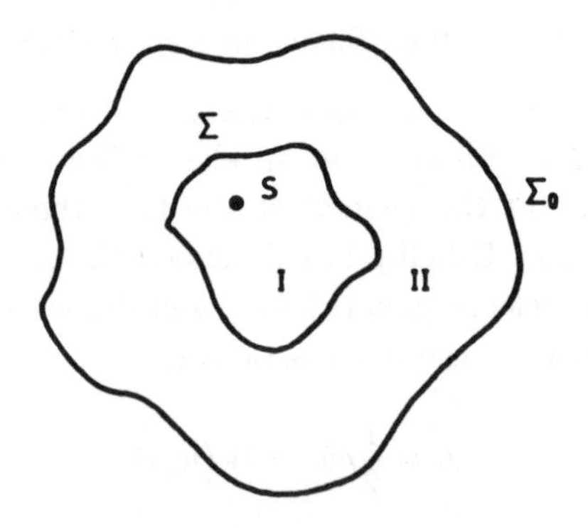


Fig. 3.16. The internal stress source in an elastic continuum medium.

Suppose there is only one defect in region I; $F_l$ may be taken as the force acting on the defect. Formula (3.6.17) is the form in three dimensions. If one uses the form in two dimensions, the components of $F_l$ are,

$$\begin{aligned} F_1 &= \oint \left[ W dx_2 - \mathbf{t} \cdot \left( \frac{\partial \mathbf{u}}{\partial x_1} \right) ds \right] \\ F_2 &= \oint \left[ W dx_1 - \mathbf{t} \cdot \left( \frac{\partial \mathbf{u}}{\partial x_2} \right) ds \right] \end{aligned} \tag{3.6.18}$$

where $\mathbf{t}$ is the traction vector on $s$ having unit outward normal $\mathbf{n}$, $\mathbf{u}$ is the displacement vector and $W$ is the energy density.

Assuming a straight dislocation line in infinite homogenous continuous media, its energy will not change due to its position change. Therefore, the first form for this dislocation in (3.6.18) is zero. The force acting on the dislocation is mainly the second term. Then,

$$\begin{aligned} F_l &= F_1 \cos\alpha + F_2 \sin\alpha \\ &= -\oint \mathbf{t}^0 \cdot \left[ \frac{\partial \mathbf{u}}{\partial x_1} \cos\alpha + \frac{\partial \mathbf{u}}{\partial x_2} \sin\alpha \right] ds \\ &= -\oint \mathbf{t}^0 \cdot \frac{\partial u}{\partial s} ds \\ &= -t^0 \oint du \\ &= -t^0 \cdot \mathbf{b} \end{aligned} \tag{3.6.19}$$

where $\alpha$ is the angle between $l$ and $x_1$, $\oint du = \mathbf{b}$, $\mathbf{t}^0$ the surface traction for the displacement of the dislocation. If one lets $t^0 = (\sigma_{11}, 0)$, then

$$F_2 = -\sigma_{11} b\,. \tag{3.6.20}$$

Similarly,

$$\begin{aligned} F_1 &= \sigma_{23} b\,, \quad F_2 = -\sigma_{13} b \quad \text{(screw)} \\ F_1 &= \sigma_{12} b\,, \quad F_2 = -\sigma_{11} b \quad \text{(edge)}\,. \end{aligned}$$

Here, the expressions are derived from the EEMT.

Suppose there is a crack tip; $F_1$ may be taken as the force acting on the crack tip. Expression (3.6.18) is similar to the $J$ integral in fracture mechanics (Lung, 1980).

If mode II deformation is mixed with mode I, the direction of $F_{lo}$, the maximum $F_1$, would not coincide with the $x_1$ direction. Suppose it has an angle $\alpha$

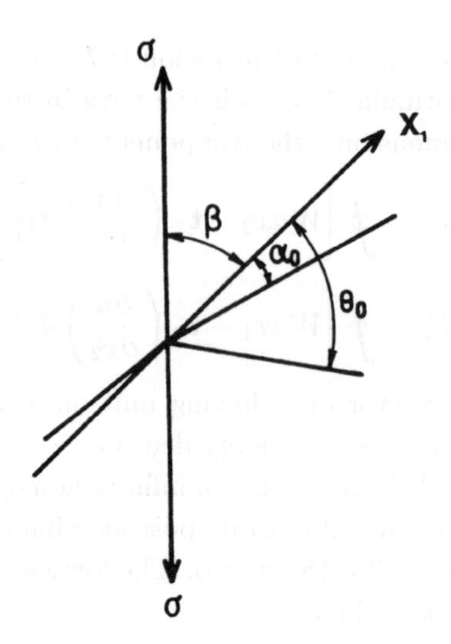


Fig. 3.17. The direction of $F_{l0}$ and the direction of the fractural plane.

with $x_1$ (Fig. 3.17),

$$\frac{\partial F_l}{\partial \alpha} = 0, \quad \alpha_0 = \arctan\left(\frac{F_2}{F_1}\right) \tag{3.6.21}$$

because $[\partial^2 F_l/\partial \alpha^2] < 0$, so that $\alpha_0$ denotes the direction of maximum $F_{lo}$,

$$F_{lo} = \sqrt{F_1^2 + F_2^2} = F_1 \sec \alpha_0 \,. \tag{3.6.22}$$

Suppose $F_{lo}(k_1, k_2) \geq F_{\text{loc}}$, the crack extends; this is the criterion of fracture under combined mode deformation. $F_{lo}$ is the maximum value of $F_1$ and $F_{\text{loc}}$ is the critical value of $F_{lo}$, which is a materials constant.

For fracture under combined mode I and mode II loading (Lung, 1976),

$$\alpha_0 = \arctan\left[-\frac{(1-2\nu)K_1K_2}{(1-\nu)(K_1^2+K_2^2)}\right] \tag{3.6.23}$$

$$F_{l_0} = \frac{1+\nu}{E}[(K_1^4+K_2^4)(1-\nu)^2 + (K_1K_2)^2(3-8\nu+6\nu^2)]^{1/2} \,. \tag{3.6.24}$$

For uniaxial applied stress,

$$K_1 = K_A \sin^2 \beta, \quad K_2 = K_A \sin\beta\cos\beta \tag{3.6.25}$$

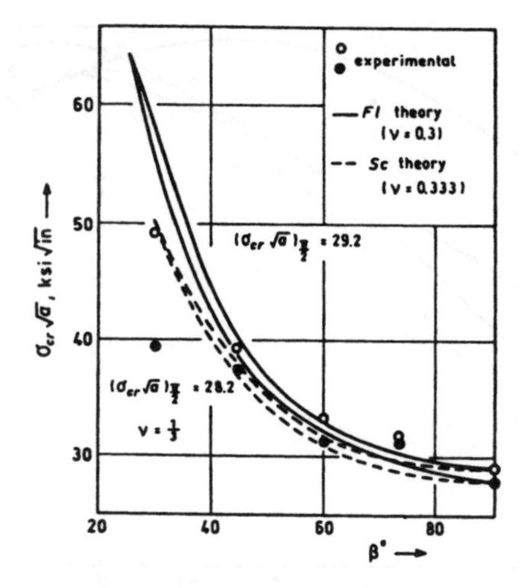


Fig. 3.18. The relationship between $(\sigma_{\rm cr}\sqrt{a})$ and $\beta$. (Lung, 1980)

where

$$K_A = \sigma^A \sqrt{\pi a} Y \, .$$

$2a$ is the crack length, $\sigma^A$ is the applied stress and $Y$ is the geometrical factor. From Eqs. (3.6.23) and (3.6.24),

$$\alpha_0 = \arctan\left[-\frac{(1-2\nu)}{2(1-\nu)}\sin 2\beta\right] . \tag{3.6.26}$$

$$F_{l_0} = \frac{1-\nu^2}{\epsilon}(K_1^2 + K_2^2)\sec\alpha_0 = \frac{1-\nu^2}{E}K_A^2 \sin^2\beta \sec\alpha_0 \, . \tag{3.6.27}$$

Equation (3.6.27) has been compared with other theories in fracture mechanics (Fig. 3.18). It was found that their differences are quite small when $\beta > 40°$.

The cracking angle $\alpha_0$ has a relation to the fracture angle $\theta_0$, in fracture mechanics,

$$\alpha_0 = \theta_0 - \beta + \frac{\pi}{2} = \theta_0 - \arctan\left(\frac{K_1}{K_2}\right) + \frac{\pi}{2} \, . \tag{3.6.28}$$

Figure 3.19 represents the theoretical curves calculated with Eqs. (3.6.26) and (3.6.28) in comparison with the finite element method, strain energy density method and maximum stress method. Five groups of experimental data are also represented. In spite of a larger difference with other theoretical curves below $\beta = 40°$, the qualitative forms of curves and the maximum cracking angles are nearly the same.

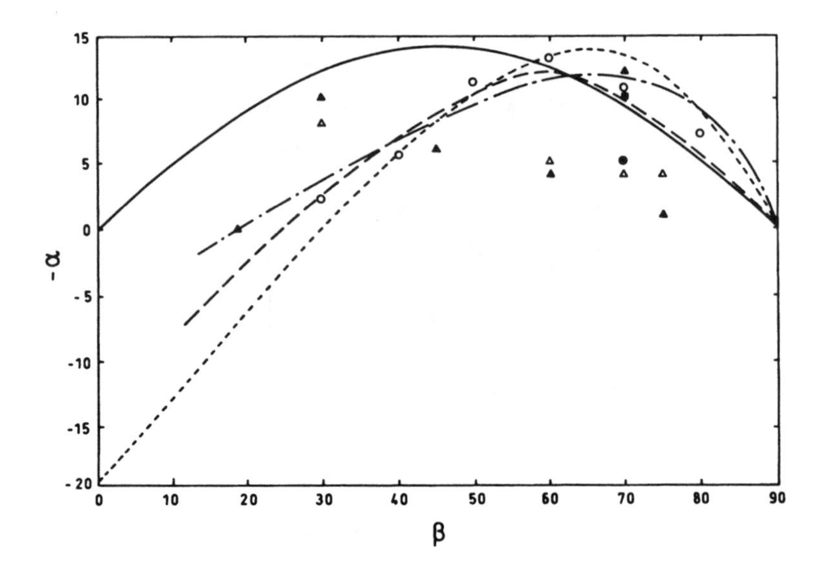


Fig. 3.19. The fracture angle $\alpha$ vs $\beta$ relationship. (Lung, 1996)

### 3.6.4. *Gauge Field Theory of Defects*

The relation between dislocation theory and non-Riemannian geometry was first recognized by Kondo (1952) and independently by Bilby *et al.* (1955). Since then, some works (Bilby, 1960; Kröner, 1960; Wit, 1981) in the subject have been done, gradually the non-Riemannian geometric continuum theory of defects has been developed and widely accepted. On the other hand, Golebiewska-Lasota (1979) first discovered that dislocation equations have gauge invariance. From then on, some researchers engaged in the subject of gauge theory of defects. Kadic and Edelen (1983) imitated the electromagnetic theory to build up a gauge theory of dislocations and disclinations; Duan and Duan (1986), having applied gauge field theory of defects continuum theory, also formulated a gauge field theory of a continuum with dislocations and disclinations (Comparing to the usual translation dislocations, a rotation dislocation defined by an extended Burgers circuit is called disclination; see Friedel, 1979; and also Appendix 3.1 of this book.). Some other contributions to the subject have been made by Gairola (1981), Kröner (1986) and Kunin (1986). New various gauge theories of defects not only developed the defects theory, but also extended our scope of understanding to gauge field theory itself and the underlying relation between gauge field theory and defects theory. These theories need to be further developed and perfected.

Deng *et al.* (1991) combine the non-Riemannian geometric continuum theory of defects with the classical gauge field theory to propose a new gauge field theory of a continuum with dislocations and disclinations.

The key point of non-Riemannian geometric continuum theory of defects is that all the structural quantities in the plastically deformed manifold should be defined by comparing the deformed material elements with those in an idealized state, i.e. by mapping the deformed material in to the idealized state. The plastic manifold constitutes a non-Riemannian space and the metric tensor, the torsion tensor and the Riemannian-Christoff curvature tensor have corresponding relations to the strain tensor, the dislocation density tensor and the disclination density tensor.

The strain tensor of the material body is:

$$E_{AB} = \frac{1}{2}(g_{AB} - \delta_{AB}) \tag{3.6.29}$$

where $g_{AB}$ is metric tensor of non-Riemannian apace, and the capital letters A, B, etc. run from 1 to 3. Deng *et al.* decomposed the strain tensor $E_{AB}$ into two parts, one is the elastic strain tensor $E^{el}_{AB}$, the other is the plastic strain tensor $E^{pl}_{AB}$,

$$E^{el}_{AB} = \frac{1}{2}(h_{AB} - \delta_{AB}), \quad E^{pl}_{AB} = \frac{1}{2}(g_{AB} - h_{AB}) \tag{3.6.30}$$

where $h_{AB} = \delta_{ij}\partial_A x^i \partial_B x^j$.

The reference state, the elastic deformed state (first added by Deng *et al.* 1991) and the final state are referred as the r-state, e-state and the f-state respectively.

In gauge field theory, the covariant derivative should be substituted for the ordinary partial differential in order to keep the field quantity differential having same form with the field quantity under the gauge transformation, i.e.

$$\partial_A \to D_A = \partial_A - \omega_A \tag{3.6.31}$$

where $\omega_A$ is called the gauge potential or gauge connection which has relation to the gauge group.

Deng *et al.* set the distortion tensor $\phi_{Bi}$, as field quantity. The covariant derivative of $\phi_{Bi}$, etc. is:

$$\begin{aligned} D_A\phi_{Bi} &= \partial_A\phi_{Bi} - \omega^j_{Ai}\phi_{Bj} \\ D_A\phi^B_i &= \partial_A\phi^B_i - \omega^j_{Ai}\phi^B_j \end{aligned} \tag{3.6.32}$$

where $i, j$, etc. run from 1 to 3. Treating the $\phi_{Bi}$ as a vector in non-Riemannian space and keeping the index $i$ unchanged, affine connection $\Gamma^c_{AB}$ can be defined through the conventional covariant derivative of $\phi_{Bi}$ with respect to the index $A$ as:

$$\nabla_A \phi_{Bi} = \partial_A \phi_{Bi} - \Gamma^c_{AB} \phi_{ci} \,. \tag{3.6.33}$$

Considering both affine connection $\Gamma^c_{AB}$ and gauge connection $\omega^j_{Ai}$, after some arguments, it is obvious that the total covariant derivatives of $\phi_{Bi}$ and $\phi^B_i$ are:

$$\begin{aligned} D_A \phi_{Bi} &= \partial_A \phi_{Bi} - \Gamma^c_{AB} \phi_{ci} - \omega^j_{Ai} \phi_{Bj} \\ D_A \phi^B_i &= \partial_A \phi^B_i + \Gamma^B_{Ac} \phi^c_i - \omega^j_{Ai} \phi^B_j \,. \end{aligned} \tag{3.6.34}$$

Usually, one assumes that the total covariant derivative of $\phi_{Ai}$ is identically zero in non-Riemannian geometry, i.e.

$$D_A \phi_{Bi} = \partial_A \phi_{Bi} - \Gamma^c_{AB} \phi_{ci} - \omega^j_{Ai} \phi_{Bj} = 0 \,.$$

It is obvious that:

$$\begin{aligned} D_A \phi_{Bi} &= \Gamma^c_{AB} \phi_{ci} \\ \nabla_A \phi_{Bi} &= \omega^j_{Ai} \phi_{Bj} \end{aligned} \tag{3.6.35}$$

i.e.

$$\begin{aligned} \Gamma^c_{AB} &= \phi^{ci} D_A \phi_{Bi} \\ \omega^j_{Ai} &= \phi^{Bj} \nabla_A \phi_{Bi} \,. \end{aligned} \tag{3.6.36}$$

The torsion tensor is defined as:

$$T^c_{AB} = \frac{1}{2}(\Gamma^c_{AB} - \Gamma_B A^c) \,. \tag{3.6.37}$$

It can be used to define the dislocation tensor:

$$\alpha^{AB} = {}^{(g)}\epsilon^{ACD} T^B_{CD} \tag{3.6.38}$$

where ${}^{(g)}\epsilon^{ACD}$ is the permutation symbol derived by $g^{1/2}$ and $g = \det(g_{AB})$. $\alpha^{AB}$ is invariant under the gauge transformation. The curvature tensor is defined as:

$$R^D_{ABC} = \partial_A \Gamma^D_{BC} - \partial_B \Gamma^D_{AC} + \Gamma^D_{AE} \Gamma^E_{BC} - \Gamma^E_{AC} \Gamma^D_{BE} \,. \tag{3.6.39}$$

It can be used to define the dislocation density tensor:

$$\theta^{AB} = {}^{(g)}\epsilon^{ACD(g)}\epsilon^{BEF} R_{CDEF} \,. \tag{3.6.40}$$

To derive the governing equations of the system, the Lagrangian density was assumed:

$$L = \frac{1}{2}\rho_0 \delta_{ij} \dot{x}^i \dot{x}_j - W(x^A;\, x^i;\, \phi_{Ai};\, \phi_{Ai,B};\, \omega^j_{Ai};\, \omega^j_{Ai,B}) \tag{3.6.41}$$

where $\rho_0$ is the mass density of the material body, $\dot{x}^i$, $\phi_{Ai,B}$ and $\omega^j_{Ai,B}$ represent the time differential of $x^i$, and $\phi_{Ai}$ and $\omega^j_{Ai}$ the differential with respect to $x^B$.

The action integral is:

$$I = \int_{f_1}^{f_2} dt \int_{E_3}^{L} L d^3 x\,. \tag{3.6.42}$$

Fixing the boundary conditions, the Euler equations was obtained on the invariant principle $\delta I = 0$,

$$\begin{aligned} \partial_A \sigma_i^A - f_i &= \rho_0 \ddot{x}^i \\ \partial_B \sigma_i^{AB} - f_i^A &= 0 \\ \partial_B \sigma_{ij}^{AB} - f_{ij}^A &= 0 \end{aligned} \tag{3.6.43}$$

Here, the $\sigma_i^A$, $\sigma_i^{AB}$ and $\sigma_{ij}^{AB}$ are called strain tensors; the $f_i$, $f_i^A$ and $f_{ij}^A$ the strain density tensors respectively,

$$\begin{aligned} \sigma_i^A &= \frac{\partial W}{\partial x^i_{,A}}\,, \quad \sigma_i^{AB} = \frac{\partial W}{\partial \phi_{Ai,B}}\,, \quad \sigma_{ij}^{AB} = \frac{\partial W}{\partial \omega^j_{Ai,B}} \\ f_i &= \frac{\partial W}{\partial \phi_i}\,, \quad f_i^A = \frac{\partial W}{\partial \phi_{Ai}}\,, \quad f_{ij}^A = \frac{\partial W}{\partial \omega^j_{Ai}}\,. \end{aligned} \tag{3.6.44}$$

Equation (3.6.43) represents the linear momentum equations, the dislocation balance equations and the disclination balance equations. They comprise 21 equations: the same number as quantities needed for describing the motion and deformation of the material body.

Dong *et al.* (1984) gave Noether's symmetrical theorem and the conservation laws on a gauge theory of defèct continuum and applied to plastic zone near a crack tip. They proposed a dynamical criterion for crack propagation in the defect continuum. The critical driving force or energy release rate of a crack in the presence of dislocations and disclinations is $F_{kc}$, a local material constant. When the total driving force acting on the crack can be expressed by an integral $F_k$ and

$$F_k > F_{kc} \tag{3.6.45}$$

the crack starts to propagate. In the case of defect free materials (perfect brittle materials), the $F_k$ integral can be reduced to the same form of Rice's $J$ integral* in fracture mechanics; the extension force acting on a crack tip. The $F$ integral is the generalization of the $J$ integral in the existence of dislocations and disclinations.

### 3.6.5. *Interactions of Dislocations with Crack*

Rice and Thomson (1973) and Asaro (1975) have shown that the interaction between a semi-infinite long crack and a straight dislocation is of a remarkably simple form. They pointed out that the image force (interaction) exerted by the crack on the dislocation is given by $F_r = -E^{\infty}/r$, where $E^{\infty}$ is the prelogarithmic energy factor of the same dislocation in the infinite uncracked medium. Asaro used a simple conformal mapping technique which may be improved in some points: (i) Generally a crack is of finite length. The crack with semi-infinite length is true only in the limiting case. (ii) The author subtracted from the potential $\Omega(z)$ the contribution of the screw dislocation located at $z_0$ in the infinite uncracked solid. This perhaps duplicates one part of the potential calculation.

Using Eshelby's expression (1979) for the image force on a dislocation near a circular hole, Lung and Wang (1984) calculated the image force on the dislocation near an elliptical hole with conformal mapping technique. Allowing the axis of the ellipse to tend to zero, the crack-limit solution is obtained. After a lengthy calculation, the image force may be expressed as:

$$4F_x = \frac{\mu b_s^2}{2\pi a} \frac{1}{\left[\left(\left(\frac{x}{a}\right)^2 - 1\right)^{1/2}\right]\left[\left(\frac{x}{a} + \left\{\left(\frac{x}{a}\right)^2 - 1\right\}^{1/2}\right)^2 - 1\right]} \tag{3.6.46}$$

where $\mu$ is the shear modulus, $2a$ is the crack length, and $x$ is the distance of the dislocation from the center of the crack. This is an expression for the image force near a finite-length crack. Taking an approximate expression, let $x = a + s$ and $s \ll a$; then, $F_s \sim Ab_s^2/4s$. That is, even in the ideal crack case, the image force is approximately half the value given by Rice and Thomson (1973) and Asaro (1975). If $s \ll \rho(\rho = b^2/a)$ or $s$ is even smaller than the radius of curvature of the ellipse at the end of the major axis, it can be proved that $F_s \sim Ab_s^2/2s$. This is just the value given by the above authors.

---

*The $J$ integral is a path independent quantity and formally equivalent to the change in potential energy when the notch is extended by an amount $da$ in non-linear elastic cases and it corresponds to the same function as does $G$ in linear elastic theory.

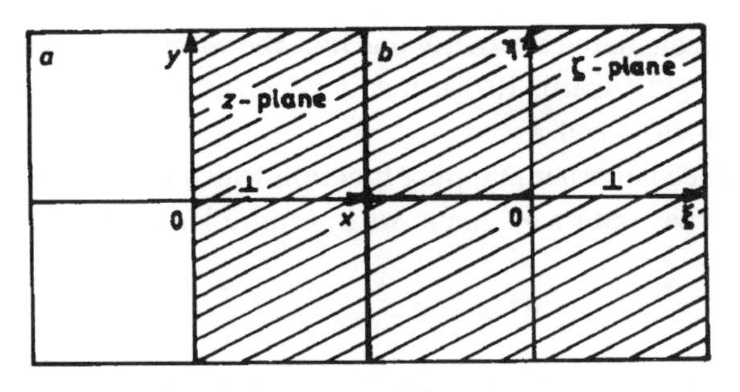


Fig. 3.20. The dislocation near (a) a free plane and (b) a crack tip. ($x$ = Re $z$, $y$ = Im $z$; $\xi = \operatorname{Re}\zeta$, $\eta = \operatorname{Im}\zeta$)

Lung calculated the image force again (1984a; 1984b). He used a simple mapping function:

$$\zeta = Z^2 \tag{3.6.47}$$

which gives a one-to-one mapping of the right half-plane of the complex variable $Z$, onto the $\zeta$-plane with an infinitely long slit or a crack along the negative real axis (Fig. 3.20). The change of the potential energy due to the interaction between a dislocation and its relevant free surface is:

$$E(\zeta) = E[\zeta(Z)] = E(Z)\,. \tag{3.6.48}$$

The image force between the dislocation and the crack in the $\zeta$-plane is:

$$F_\zeta = -\frac{dE}{d\zeta} = -\frac{dE}{dZ}\left|\frac{dZ}{d\zeta}\right| = F_Z\left|\frac{d\zeta}{dZ}\right|^{-1}. \tag{3.6.49}$$

It is well known that the image force on the dislocation near a free plane is:

$$F_Z = -\frac{Ab^2}{2\operatorname{Re} Z} \tag{3.6.50}$$

where $A = \mu/2\pi$, **b** is the Burgers vector, and $\mu$ is the shear modulus. Therefore, the image force on the dislocation near a semi-infinite crack is:

$$F_\zeta = -\frac{Ab^2}{4|Z|\operatorname{Re} Z} = -\frac{Ab^2}{4|\zeta^{1/2}|\operatorname{Re}(\zeta^{1/2})}\,. \tag{3.6.51}$$

In general, $F_\zeta$ depends on the argument of the complex variable $\zeta$. In particular, if the dislocation is on the real axis ($\xi \neq 0$ and $\eta = 0$), the image force on a dislocation exerted by the semi-infinite crack is:

$$F_\xi = -\frac{Ab^2}{4\xi} = -\frac{\mu b^2}{8\pi\xi}\,. \tag{3.6.52}$$

The image force calculated is exactly one half the value of previous works (Rice and Thomson, 1973; Asaro, 1975), but it is consistent with the approximate value of Eq. (3.6.46) when $s \ll a$.

Lung introduced another mapping function which maps the $\zeta$-plane with an infinitely long crack onto a $W$-plane with a finite length crack. The mapping function is

$$W = \frac{a\zeta + b}{c\zeta + d} = \frac{a'\zeta + b'}{c'\zeta + 1} \tag{3.6.53}$$

where $a'$, $b'$, $c'$ are equal to $a/d$, $b/d$ and $c/d$, respectively, if $d \neq 0$. The coefficient $a'$, $b'$, $c'$ can be determined from the condition that the points $-\infty$, $-1$ and 0 in the $\zeta$-plane are conformally equivalent to the points $-a$, $-a/2$ and 0 on the $W$-plane, respectively ($a$ is the crack length).

From this condition one can obtain:

$$W = \frac{a\zeta}{1-\zeta} \tag{3.6.54}$$

or:

$$\zeta = w/(a+w)\,.$$

In analogy to Eq. (3.6.49):

$$F_w = \frac{-Ab^2 a}{4(a+w)^2 \left|\left(\frac{w}{a+w}\right)^{1/2}\right| \mathrm{Re}\left(\frac{w}{a+w}\right)^{1/2}}\,. \tag{3.6.55}$$

Then, the image force on a dislocation on the real axis exerted by the finite length crack is:

$$F_s = \frac{-Ab^2 a}{4s(s+a)} \tag{3.6.56}$$

where $s = \mathrm{Re}\, W$.

From Eq. (3.6.56), one may see that:

(i) $F_s$ is a function of crack length a.

(ii) If $s \ll a$; then $F_s \neq -\mu b^2/(8\pi s)$. This expression is consistent with Eqs. (3.6.52) and (3.6.46) when $s \ll a$.

In Fig. (3.21) the dependence of the dislocation image force on the distance between the dislocation and a crack tip is shown.

The differences among the curves 2, 3 and 4 are quite small when both $\zeta$ and $s/a$ are smaller than 0.1. When both the distances $\xi$ and $s/a$ are larger than 0.1, the differences among them are large.

At a distance $s/a < 0.005$, all the image forces calculated by Eqs. (3.6.46), (3.6.52) and (3.6.56) approach the same value $y = 1/(4\zeta)$.

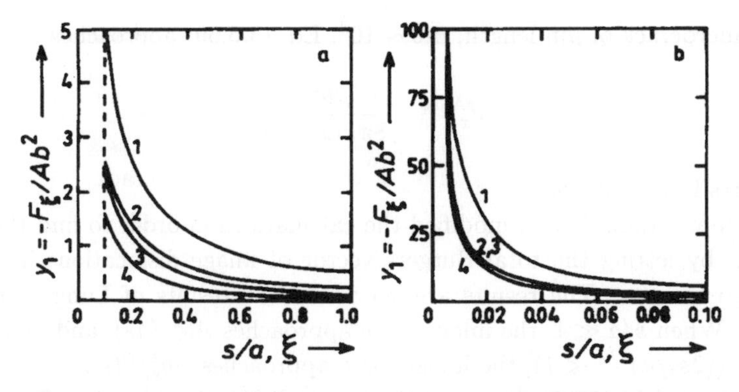


Fig. 3.21. The relationship of dislocation image force with the distance of the dislocation from a crack tip. 1. Rice and Thomson, 1973; Asaro, 1975; 2. Lung and Wang, 1983; 3. Eq. (3.6.52) 4. Eq. (3.6.56). (Lung, 1984)

Using the non-local elasticity theory developed by Kröner, Erlinger and others, the interaction between a finite-length crack and a straight screw dislocation parallel to the crack plane has been investigated by Pan (1994). The non-local image force on a screw dislocation due to a finite-length crack was calculated. The result showed that there is no singularity of the image force calculated previously and there is a maximum of the force at the crack,

$$F_x^N = -\frac{Aac}{x(x+c)}\left[1 - \exp\left(-\frac{4k^2x(a+c)}{a(x+c)}\right)\right]. \qquad (3.6.57)$$

A finite value of the dislocation image force can be obtained, when the distance of the dislocation from the crack tip tends to zero.

$$F_{x_0}^N = -4k^2A\left(1 + \frac{a}{c}\right) \qquad (3.6.58)$$

where $A = \mu b^2/(8\pi a)$, $a$ is the lattice parameter, $c$ is the crack length and $k$ is a constant in the expression of a non-local kernel function, which has been given by comparing dispersion relations for plane waves with those from lattice dynamics. From Eq. (3.6.58), a maximum of the image force $|F_{x_0}^N|/A = 3.8877$ is obtained with $c/a = 10$.

When the lattice parameter $a$ tends to $b (a \sim b)$, the Burgers vector of the dislocation, Eq. (3.6.57) becomes:

$$F_x^N = -\frac{\mu b^2 c}{8\pi x(x+c)}\left[1 - \exp\left(-\frac{4k^2x(b+c)}{b(x+c)}\right)\right]. \qquad (3.6.59)$$

For a microcrack of $\mu$m length, $c/b \sim 10^4$, Eq. (3.6.59) approaches,

$$F_x^N \cong -\frac{\mu b^2 c}{8\pi x(x+c)} . \tag{3.6.60}$$

It reverts to Eq. (3.6.56).

Zhou and Lung (1988) modified the calculations in order to find the exact solution by letting the total Burgers vector of image dislocations inside the crack go to zero. The results are very near the results of Lung and Wang (1984). When $s/a \ll 1$, the image force approaches $Ab_s^2/(4s)$; and, if s is even smaller $((2s_s/a)^{1/2} \ll 1)$, the image force approaches $Ab_s^2/(2s)$.

Zhang and Li (1989) discussed the image dislocations inside a finite crack. They pointed out that it is not necessary to add positive screw images at the center of the hole if the dislocations are emitted from the crack. On the other hand, if a dislocation moves from another source into the vicinity of the crack which is originally stress-free, a Burgers circuit enclosing only the crack would be closed. Hence in this case, the total Burgers vector of the image dislocation must be zero. Comparing to the previous case, an extra distribution of positive image dislocations should be added to compensate for all the negative ones, as has been done by Eshelby (1979).

### 3.6.6. *Dislocation Distribution Function in Plastic Zone at Crack Tip*

The problem of dislocation distribution in the plastic zone at a crack tip was first studied by Bilby *et al.* (BCS-type crack model) (1963). They described the plastic zone by an array of dislocations co-planar with the crack and derived the same relationship between the crack size (c), the plastic zone size (a–c), and the applied stress ($\sigma^\infty$) as Dugdale (1960) had done macroscopically before. Kobayashi and Ohr (1980) investigated crack propagation and the structure of the plastic zone formed ahead of a crack with the help of *in situ* electron microscope fracture experiments of the b.c.c. metals molybdenum and tungsten. They found that a part of the plastic zone immediately ahead of the crack tip is free of dislocations. Weertman (1981) proposed that in many real cases, dislocations are not emitted directly from the crack tip, and that pairs of partial or perfect dislocations of opposite sign are created on the slip plane immediately ahead of the crack tip. Dislocations of the positive ones move away from the tip region leaving dislocations of negative sign piled up against the crack tip. When the applied stress is sufficiently large the piled up negative dislocations can enter the crack tip and blunt it.

The dislocation density function in the original BCS-type crack model monotonically decreases with the distance from the cracktip, all the dislocations are of positive sign. Lung and Xiong (1983) calculated this problem from different points of views. Unlike the original BCS-type crack model, they treated this problem within a homogeneous and continuous system. Actually, they only consider the plastic zone which does not include both the crack and plastic zone in a whole system to avoid discontinuity and inhomogeneity. We know that the crack plane is a free surface and the resistance to motion of the dislocations in the crack region ($\sigma_0$) is different from that in the plastic zone ($\sigma_1$). They avoided treating the problem in a composite system and just considered the crack effect as a boundary applied stress.

At first, they did not consider crack blunting. Suppose one of the dislocation with Burgers vector of magnitude $b$ is located at a distance $x$ from a crack of length $2c$ (Fig. 3.22). In this case the resultant shear stress on it is zero when the system is in equilibrium (Bilby *et al.*, 1963) This condition leads to the following integral equation for the density function $D(x)$ of the dislocation:

$$A\int_0^D \frac{D(x')}{x-x'}dx' - \sigma_1 + \sigma^\infty + \sigma^c = 0 \tag{3.6.61}$$

where $A = \mu b/(2\pi(1-\nu))$, $\mu$ is the shear modulus, **b** is Burgers vector. The first term is the interaction between this dislocation and the others. $\sigma_1$ is the frictional stress for the dislocation motion (which is taken as the yield stress), and $\sigma^\infty$ is the applied stress field taken as constant, and $\sigma^c$ is the elastic stress field due to the crack tip calculated with linear elastic fracture mechanics. The image stress at the crack tip was ignored in that calculation. For small scale yielding, one may write:

$$\sigma^c = \frac{K}{(2\pi x)^{1/2}}, \quad K = \sigma^\infty(\pi a)^{1/2}. \tag{3.6.62}$$

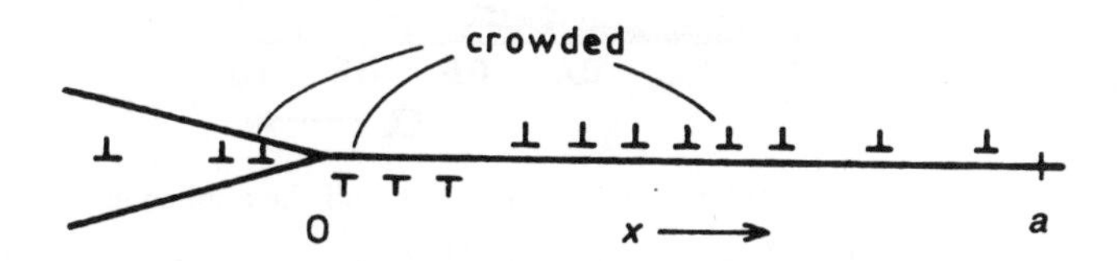


Fig. 3.22. Schematic figure of dislocations in the plastic zone at a crack tip.

Let $u = x'/a$, $w = x/a$ where $a$ is the plastic zone size, $u$ and $w$ are dimensionless parameter, then from Eq. (3.6.61),

$$A\int_0^1 \frac{D(u)}{w-u}du - \sigma_1 + \sigma^\infty + \sigma^c = 0\,. \tag{3.6.63}$$

This singular integral equation can be solved by finite Hilbert transformation:

$$D(u) = \frac{\sigma_1}{\pi^2 A}\left(\frac{1-u}{u}\right)^{1/2}(1-\alpha)\pi\left[1 - \frac{1}{4\sqrt{1-u}}\ln\left(\frac{1+\sqrt{1-u}}{1-\sqrt{1-u}}\right)\right] \tag{3.6.64}$$

where $\alpha = \sigma^\infty/\sigma_1$.

The conditions for it to exist lead to the relation (Lung and Xiong, 1983),

$$\frac{a}{c} = \left[\frac{2\sqrt{2}\alpha}{\pi(1-\alpha)}\right]^2. \tag{3.6.65}$$

As expected, $a \to 0$, when $\alpha = 0$; that is $\sigma_1 \to \infty$ (infinite resistance to dislocation motion) or $\sigma^\infty \to 0$ (no applied stress on dislocations). Again, $a \to \infty$, that is, yielding spreads across the infinite plate, when $\alpha \to 1$, this means $\sigma^\infty \to \sigma_1$ (applied stress tends to resistance $\sigma_1$ to dislocation motion) (Fig. 3.23).

Unlike the original BCS-type crack model, $D(u)$ can change sign from $u = 0$ to $u = 1$. It results from the solution of Eq. (3.6.63). This means, there are opposite sign dislocations existing in the plastic zone simultaneously as shown in Fig. 3.24(a). In this type of $\sigma^c$, $D(u) = 0$ at $u_0 = 0.08$, and it seems

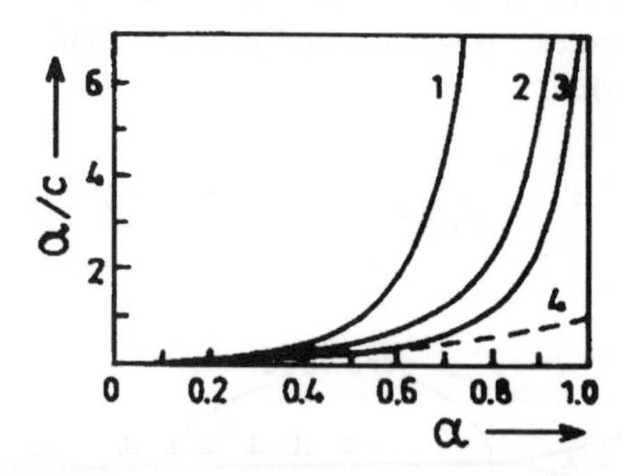


Fig. 3.23. Relationship between the plastic zone size $a/c$ and the applied stress $\alpha = \sigma^\infty/\sigma_1$. (1) $\frac{a}{c} = \left[\frac{2\sqrt{2}\alpha}{\pi(1-\alpha)}\right]^2$; (2) BCS: $\frac{a}{c} = \sec\left(\frac{\pi}{2}a\right) - 1$; (3) $a = \frac{\pi}{2\sqrt{2n}+\arccos\frac{2n-1}{2n+1}}$; (4) $\frac{a}{c} = \alpha^2$. (Lung and Xiong, 1983)

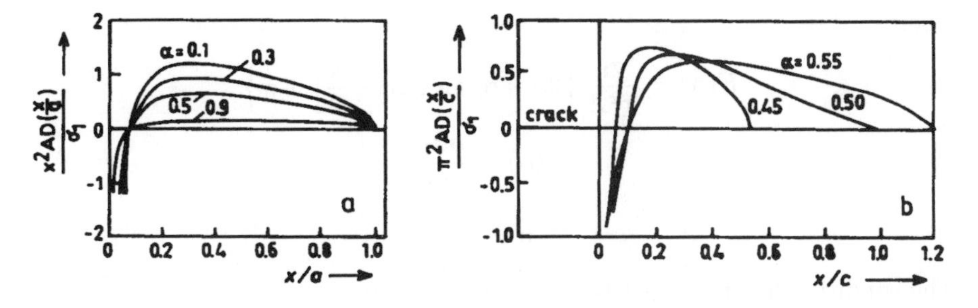


Fig. 3.24. Dislocation distribution in the plastic zone under small scale yielding. (a) $x/a$ as unit, (b) $x/c$ as unit. (Lung and Xiong, 1983)

independent of $\alpha$. That is, the dislocations always change their sign at $x_0 = 0.08a$. The fact that $D(u)$ is negative in Eq. (3.6.64) means that the sign of the Burgers vector of the dislocations relative to that of crack dislocations which create a positive stress field is negative.

Figure 3.24(b) is the plot of $D(x/c)$ vs $x/c$, where the authors used the length of crack as unit. It is shown that the higher the applied stress level, the larger the area under the distribution function curve; and then, the larger the COD* of the cracked material.

If the size of plastic zone at fracture is not very much less than the crack length, Eq. (3.6.62) is too simple for describing the stress field near the crack tip accurately. The following expression may be a refinement:

$$\sigma^e = \frac{\sigma^\infty}{\sqrt{1 - (\frac{c}{x+c})^2}} - \sigma^\infty . \tag{3.6.66}$$

The solution is:

$$D(u) = E(\alpha)\sqrt{\frac{1-u}{u}} - F(\alpha)\frac{u+n}{\sqrt{u(u+2n)}} \ln \frac{[\sqrt{2n} + \sqrt{(2n+u)(1-u)}]^2 + u^2]}{u(2n+1)} \tag{3.6.67}$$

*COD means the crack opening displacement as a parameter which might be treated as a characteristic of crack-tip region, for a given material tested under a given set of conditions. The critical COD value for a given material at fracture is sensibly constant.

where

$$E(\alpha) = \frac{\sigma_1}{\pi^2 A}\left\{2\alpha\left[\arctan\sqrt{2n} - \frac{\pi}{2}\right] + \pi\right\}$$

$$F(\alpha) = \frac{\alpha\sigma_1}{\pi^2 A}$$

$$n = \frac{c}{a}.$$

The condition for it to exist leads to the relation:

$$\alpha = \frac{\pi}{2\sqrt{2n} + \arccos\left(\frac{2n-1}{2n+1}\right)}. \tag{3.6.68}$$

The size of plastic zone as a function of various applied stress levels ($\alpha$) is shown in Fig. 3.23. The distribution function of dislocations in the plastic zone as a function of distance from the crack tip is given in Figs. 3.25(a) and (b). Similar to the small scale yielding case, the dislocation distribution function has a maximum value and a range of negative values.

Equation (3.6.61) ignored the image stress at the crack tip which may change the distribution. Lung and Deng (1989), and Liu *et al.* (1987) calculated the problems including the image effects and showed qualitatively similar distribution functions to Lung and Xiong (1983). The possibility that negative dislocations may exist at the crack tip was first discussed in Lung and Xiong's model (1983). Such behaviour was observed near a crack tip in NaCl (Narita and Takamura, 1985). The work in stainless steel confirmed the existence of negative dislocations at the immediate vicinity of the crack tip (Ohr, 1987). Antishielding dislocations are found to exist near the crack tip of I and II modes in bulk aluminium single crystal (Xu, 1991).

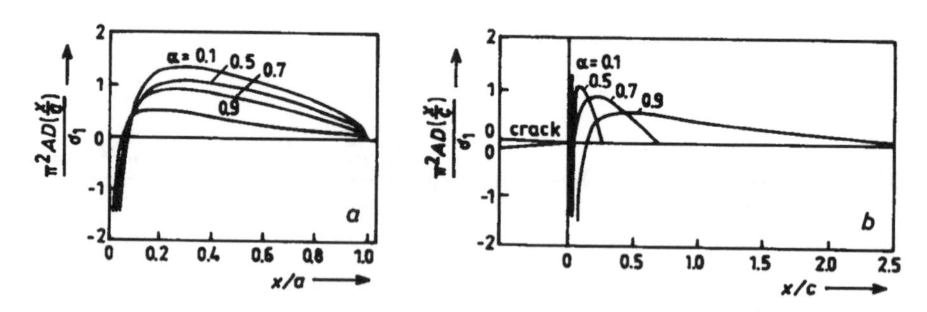


Fig. 3.25. Dislocation distribution in the plastic zone under elastic plastic stress field. (a) $x/a$ as unit, (b) $x/c$ as unit. (Lung and Xiong, 1983)

In the case of Eq. (3.6.61), only the equilibrium condition of stresses has been considered. Let us consider the thermal dynamical equilibrium condition of the whole system. Lung (1990) found that the system is unstable because of its higher total energy.

We know that positive crack dislocations and negative crystal dislocations prefer to attract and to annihilate with each other. This is due to the smaller amount of energy expended in creating new surfaces ($\gamma_s b \approx \frac{Eb^2}{20}$) compared to the energy of the existing dislocations ($\sim \mu b^2$). In principle, only when the size of negative dislocation zone ($x_0$ in Sec. 3.6.6) is larger than that of the dislocation-free-zone ($\delta_3$ in Sec. 3.6.9), can negative dislocations remain at the crack tip.

### 3.6.7. *Dislocation Emission from Cracks*

A material is said to be intrinsically ductile (or brittle) if a sharp crack emits a dislocation at stress-intensity levels below that where crack cleavage occurs (or vice versa), because except in special geometries, the emitted dislocation blunts the crack and renders it uncleavable. The intrinsic stability of the bonds at a crack tip against shear breakdown and dislocation formation is thought to be the underlying reason why certain broad classes of materials such as the fcc metals are generally ductile, and others such as ceramics are generally brittle (Zhou *et al.*, 1994).

Let us discuss the case of a specimen with a crack at the edge. A dislocation near the crack tip may be attracted to the crack tip by the image force between the dislocation and the crack. The dislocation may be acted on by another force due to the $K$ stress field at the crack tip. In the case that this force is smaller than the image force, it will have the same direction as the frictional force.

In mode III, the condition for zero force on a reference dislocation $i$ at $x$ in the pile-up near a crack tip is given by (Thomson, 1983):

$$f_d = \frac{K_{III} b}{\sqrt{2\pi x}} - \frac{\mu b^2}{4\pi x} + \sum_{j\neq 1}{}' \frac{\mu b}{2\pi(x - x_j)} \frac{x_j}{x} \,. \tag{3.6.69}$$

For simplicity, we have used the expression of image force for the semi-infinite long crack. For the dislocation at $x \approx 0$, $x$ is much smaller than the plastic zone size. Then,

$$f_d \cong \frac{k_{III} b}{\sqrt{2\pi x}} - \frac{\mu b^2}{4\pi x} \tag{3.6.70}$$

and:

$$k_{III} = K_{III} - \sum_{j=i}{}' \frac{\mu b}{(2\pi x_j)^{1/2}}$$

where $f_d$ is the force acting on the reference dislocation. $K_{III}$ is the stress intensity factor due to the crack tip, $x_j$ is the distance of the $j$ dislocations piling up from the crack tip and $k_{III}$ is called the effective stress intensity factor which modifies the original $K_{III}$ with the effects of dislocations near the crack tip.

The condition for zero force on the dislocation is given by:

$$\frac{k_{III}}{\sqrt{2\pi\delta_2}} - \frac{A}{2\delta_2} + \sigma_1 = 0 \qquad (3.6.71)$$

where $A = (\mu b)/(2\pi)$. The solution for $\delta_2$ is

$$\delta_2(k_{III}) = \frac{A}{2} \frac{1}{\left(A\sigma_1 + \frac{k_{III}^2}{2\pi}\right) \pm \left[\left(A\sigma_1 + \frac{k_{III}^2}{2\pi}\right)^2 - A^2\sigma_1^2\right]^{1/2}} . \qquad (3.6.72)$$

It is reasonable to take the plus sign in the solution, because $\delta_2$ is smaller than $\delta_1$, $(= A/(2\sigma_1))$, the solution for $k_{III} = 0$.

For simplicity, we change $k$ to a new parameter:

$$k_{III} = (2\pi A\sigma_1)^{1/2}\beta .$$

Replacing in (3.6.72),

$$\delta_2(\beta) = \frac{A}{2\sigma_1} \frac{1}{1 + \beta^2 + [(1 + \beta^2)^2 - 1]^{1/2}} . \qquad (3.6.73)$$

Here $\delta_2(\beta)$ is the position of the dislocation where $f_d = 0$.

If $\delta_2(k_{III}) \approx b$, $k_{III} = k_e$ and for $k > k_e$, a dislocation at any position in the material sees a repulsive elastic force, and it never traverses an attractive regime; then, dislocation emission will occur spontaneously. $K_{IIIe}(k_{IIIe})$ is called the critical stress-intensity factor for emission. The above description is the same as Thomson (1983) though it is simpler.

The cleavage/emission criterion is obtained by combining (3.6.70) with the Griffith criterion for cleavage. Now we use the mode I symbol for convenience.

The Griffith criterion for cleavage is:

$$k_{Ic} = 2\sqrt{\frac{\mu\gamma}{1 - \nu}} . \qquad (3.6.74)$$

The criterion for cleavage/emission in pure Mode I then becomes:

$$k_{IE} < k_{Ic} \text{ emission}$$
$$k_{IE} > k_{Ic} \text{ cleavage}. \tag{3.6.75}$$

There have been several attempts (Weertman, 1981; Schoek, 1991; Rice, 1992; Zhou *et al.*, 1994) to develop analytic, continuum-based estimates of the critical stress intensity factor for emission, $K_{Ie}$, There have also been several atomistic calculations (Hoagland, 1990; Cheung, 1991). The most successful continuum based analysis has recently been performed by Rice (1992) who found that for plane stress the critical stress-intensity factor for emission, $K_{Ie}$, is given by:

$$K_e = \sqrt{2\gamma_{us}\mu(1+\nu)Y} \tag{3.6.76}$$

where $\nu = 0.25$ is Poisson's ratio, $Y$ is a geometric factor given by the angle of dislocation emission, and $\gamma_{us}$ is the 'unstable stacking energy.' The latter is defined as the maximum energy barrier encountered when two semi-infinite blocks of material are sheared relative to one other, and is thus a measure of the theoretical shear strength of the material. For simple lattices, the maximum energy barrier corresponds to a relative displacement of $b/2$ between the blocks, where $b$ is the magnitude of the dislocation's Burgers vector. In Rice's analysis, this implies that the crack-tip dislocation is precisely half formed at the critical point of emergence.

Zhou *et al.* (1994) used a form of the force law which generalizes the universal binding energy relation (UBER) form:

$$F^{\alpha} = -kue^{-(u/l)^{\alpha}}. \tag{3.6.77}$$

Here, $u$ is the radial displacement between two atoms relative to their equilibrium position, $l$ is a length-scale parameter, and $\alpha$ varies the shape of the force law. They performed a series of atomistic calculations to establish criteria for dislocation emission from a crack in a model hexagonal lattice. From this, they proposed a new ductility criterion for materials which does not depend on the intrinsic surface energy, but contains only the unstable stacking fault parameter.

In the range of calculation, the crossover ($K_{Ie} = K_{Ic}$) is very different from that predicted by the continuum theory. It corresponds fairly closely to $\gamma_{us}/ka = 0.005$ or, using $b = a$ and the easily derived relationship $\mu = (3^{1/2}/4)k$,

$$\gamma_{us}/\mu b = 0.012. \tag{3.6.78}$$

Thus the crossover is essentially independent of $\gamma_s$. This does not imply that $K_{Ie}$ is independent of $\gamma_s$, since the crossover is determined by the ratio of $K_{Ie}$ and $K_{Ic}$, not by $K_{Ie}$ itself. The observed $\gamma_{us}$ independence of the crossover requires that the dependence of $K_{Ie}$ on $\gamma_s$ be the same as that of $K_{Ic}$,

$$K_{Ie} = K_{Ic} f(\gamma_{\mu s}) = 2\sqrt{\gamma_s \mu (1+\nu)} f(\gamma_{us}) \tag{3.6.79}$$

where $f(0.012\mu b) = 1$. The physical interpretation of this is that $K_{Ie}$ is dominated by the surface energy required to blunt the crack. Further, since the ledge surface is created during a shear dislocation formation process, emission will involve both $\gamma_{us}$ as well as $\gamma_s$, unlike pure cleavage which involves only $\gamma_s$. Because both cleavage and emission events produce surface energy, the surface energy cancels out in the ductility criterion, which involves a ratio of the two separate criteria for emission and cleavage. In order to gain confidence in this prediction, Thomson and Carlsson (1994) expose a continuum elastic description of the emission process, which can give insight into the physical basis for, and limit to, the new criterion. The models confirm the prediction of Zhou *et al.*, but suggest that at lower values of the intrinsic surface energy than explored by Zhou *et al.*, there exists a regime where the new criterion breaks down, and the ductility criterion reverts to the older predictions.*

### 3.6.8. *Dislocation Pair Creation and Separation from the Crack Tip*

Weertman (1981) treated smeared-out rather than discrete dislocations in an analysis of a mode II shear crack, and showed that a partial dislocation can be emitted at the crack tip and that one or more pairs of partial or perfect dislocations are created ahead of the crack tip. Positive perfect dislocations move into the crystal lattice away from the crack tip and negative partial or perfect dislocations move into the tip and cause the crack tip to advance by one or more interatomic distances. If the crack is a mode I crack and the dislocations are on inclined slip planes, the entry of dislocations into the tip blunts the crack. He believed that perfect dislocations are not emitted directly from the crack tip into the lattice, although the net result is the same as if they had been directly emitted. When the ratio $\sigma_{\max}/\tau_{\max}$, where $\sigma_{\max}$ is the theoretical tensile strength and $\tau_{\max}$ is the theoretical shear strength, is very large, a number of dislocations will be piled-up against the crack tip before the lead dislocation enters the crack tip.

---

*Atomistic and continuum approaches to fracture toughness are reviewed by A. E. Carlsson and R. M. Thomson in Solid State Physics *51*, 233 (1998).

Table 3.1. Values of $\sigma_{\max}/\tau_{\max}$.

| Material | $\sigma_{\max}/\tau_{\max}$ |
|---|---|
| Gold | 34 |
| Silver | 30 |
| Copper | 28 |
| Nickel | 22 |
| Iridium | 10 |
| Potassium | 10 |
| Iron | 6.75 |
| Tungsten | 5.04 |
| Diamond | 1.16 |

When the ratio $\sigma_{\max}/\tau_{\max}$ is equal to or larger than about 7 a mode I crack tip should blunt before brittle propagation occurs. The common fcc metals satisfy this condition. The bcc metals iron and tungsten have $\sigma_{\max}/\tau_{\max}$ ratio somewhat smaller than 7 and thus cracks in these metals may not blunt. Table 3.1 lists values of this ratio for various materials given by Kelly (1973).

Zhou and Lung (1989) discussed the influence of external dislocations on the crack tip process using the superdislocation pair model with numerical calculation, and analytical treatment. These results showed that negative external dislocation annihilation at a crack is much easier than dislocation emission from the crack tip. In other words, an external dislocation can be annihilated at a crack before a dislocation can be emitted from the crack tip if the source is sufficiently near the crack tip. Since negative dislocation is nearer to the crack tip than the positive one. The net effect of an external dislocation pair on the crack is anti-shielding. Hence, external dislocations can play a significantly promotive role in dislocation emission. This might be one of the reasons that the measured values of $K_{IIIe}$, which are about 50 kPam$^{1/2}$ in LiF (Burns, 1986) were much lower than the theoretical values, which are about 230 kPam$^{1/2}$, evaluated by using the theories of Rice and Thomson (1973).

In Secs. 3.6.7 and 3.6.8, we have discussed the fracture properties of metals with simple composition and structure. For real materials, the relation between composition and properties is complicated. For example, it has been shown that the addition of hydrogen to Ti alloys may have certain beneficial effects on hot working properties. Further experimental studies might be helpful in prompting theoretical explanations (Gong *et al.*, 1993; Xu *et al.*, 1992).

### 3.6.9. *Dislocation Free Zone at Crack Tip*

Kobayashi and Ohr (1980) investigated the use of *in situ* electron microscope fracture experiments of the bcc metals molybdenum and tungsten. They found that a part of the plastic zone immediately ahead of the crack tip is free of dislocations. This is called dislocation free zone (DFZ) ahead of the crack tip. There are three problems: (i) Does the DFZ really exist in bulk materials? (ii) How can it form? (iii) How can it influence the fracture property of the material?

Since the paper (Kobayashi and Ohr, 1980) was published, there was certain doubt about the existence of DFZ in bulk materials (Robertson and Birnbaum, 1985). Ohr repeated his experiments with thicker specimens (Fe single crystals of 1 $\mu$m in thickness) by HVEM technique.* The existence of DFZ was reconfirmed. However, 1 $\mu$m in still quite different from the bulk, especially in the state of loading. Ha *et al.* (1990) used specimens of $30 \times 10 \times 2$ mm$^3$ dimensions of high purity (99.999%) Al single crystal to investigate DFZ at the crack tip. They have not only verified the existence of such a DFZ, but also showed its three-dimensional characteristics in bulk crystals. Its appearance may be influenced by many factors, such as crystal orientation, length and sharpness of the crack, pinning condition of dislocations in the lattice, applied load and the method of crystal preparation, etc.. The size of the DFZ they observed is much larger than that observed under TEM.† Ha *et al.* attribute this mainly to the large pre-crack tip radius.

The problem on the formation of DFZ at a crack tip is open. Image force on positive dislocations near a crack may be one of the mechanisms for the formation of DFZ. Discussions along this line have led to many important concepts such as dislocation emission, brittle and ductile transition, etc.. It is well known that any free surface produces image forces on the nearby dislocations. A dislocation near the free surface may move out the material due to the attraction of image forces (Friedel, 1964). As a result of that process, a DFZ may be left in the material near the surface. Near a free plane surface, the size of the DFZ due to image forces may be estimated by the following arguments. The image force due to a free plane surface is given by:

$$\sigma^I = A/(2x) \tag{3.6.80}$$

*See Appendix 2.1.

†See again Appendix 2.1.

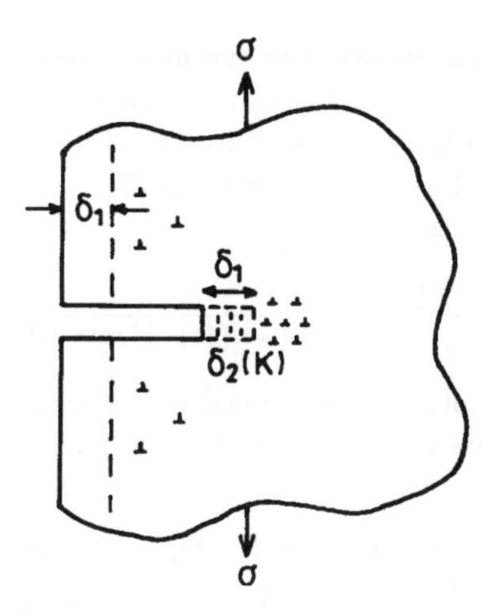


Fig. 3.26. Schematic figure for the dislocation free zone at a crack tip and a free plane.

where $x$ is the distance of the dislocation from the free surface, $A = \mu b/(2\pi)$. $\mu$ is the shear modulus and $b$ is the Burgers vector. If $\sigma^I$ is larger than the frictional stress $\sigma_1$ the dislocation is attracted by the surface and may move out from the free surface. The furthest distance that the dislocation can move out of the free surface is given by $\sigma^I = \sigma_1$. The size of the DFZ near a free plane results in (Fig. 3.26):

$$\delta_1 = x_1 = A/(2\sigma_1) \tag{3.6.81}$$

For the crack tip, the stress concentration must be considered. The dislocation symbol in Fig. 3.26 may be taken to indicate any type (edge or screw); the subsequent analysis is in terms of screw dislocations, but an equivalent edge dislocation model may of course be developed. In addition to the image force, the dislocation may be acted on by another force due to the $K$ stress field at a crack tip. In the case that this force is smaller than the image force it will have the same direction as the frictional force. According to this analysis, the crack should inhibit the formation of a DFZ. However, observation of the original photograph from Kobayashi and Ohr (1980), shows the DFZ at the edge is not larger than the one at the crack tip. Therefore, the image force theory seems to disagree with Kobayashi and Ohr's experiment. We extend this argument further below (Lung, 1990). In mode III, the condition for zero force on a reference positive dislocation $i$ at $x$ in the pile-up near a crack tip is given by Eq. (3.6.69).

As the DFZ is much smaller than the plastic zone, we can assume $x \approx 0$ (or $x \ll x_j$), which is a valid approximation. Then, $f_d$ is given by Eq. (3.6.70). The condition for zero force on the dislocation is Eq. (3.6.71). The solution for $\delta_2(k_{III})$ is given in Eqs. (3.6.72) and (3.6.73).

Replacing $A/2\sigma_1$ with $\delta_1$ in Eq. (3.6.73):

$$\frac{\delta_2(\beta)}{\delta_1} = \frac{1}{1+\beta^2+[(1+\beta^2)^2-1]^{1/2}} \tag{3.6.82}$$

where $\delta_2(\beta)$ is the size of above mentioned DFZ due to the image force mechanism. Figure 3.27 shows the $\delta_2(\beta)-\beta$ curve. From Fig. 3.27 and Eq. (3.6.82), we may conclude:

(1) $\partial\delta_2/\partial\beta < 0$ or $\partial\delta_2/\partial k_{III} < 0$. The crack inhibits the formation of DFZ. This conclusion, consistent with above qualitative argument, is contrary to Kobayashi and Ohr's observation (1980). In their photograph, the DFZ at the crack tip is much larger than that at the edge of a free plane, where the DFZ is too small to be seen.

(2) $K = 0$, $\delta_2(0) = A/(2\delta_1) = \delta$. It reduces to the size of the corresponding zone near a free plane.

(3) If $\delta_2(k) \propto b$, $k = k_e$ and for $k > k_e$ there is no DFZ near the crack tip. This is the condition for dislocation emission at the crack tip. However, to our knowledge, no experimental result showed that the DFZ size decreases with increasing $k$.

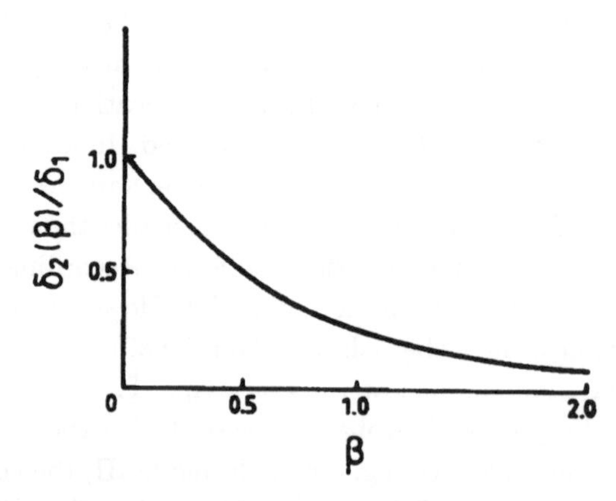


Fig. 3.27. The relationship of the DFZ size with applied stress $\beta(K)$ according to image force theory. (Lung, 1990)

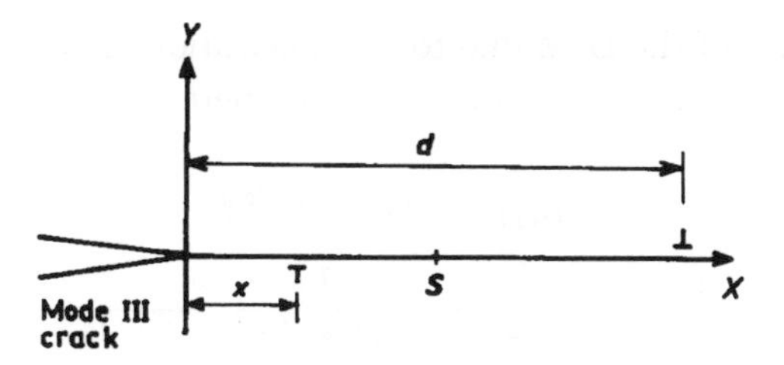


Fig. 3.28. A pair of screw dislocations emitted from an external source labelled S.

We propose another mechanism to explain the formation of DFZ. In the vicinity of a crack, pairs of dislocations with opposite Burgers vectors may be created by various multiplication mechanisms at dislocation sources. Positive ones with shielding Burgers vectors are repelled by the crack tip, while the negative ones, with antishielding Burgers vectors, are attracted. The negative ones may be observed and annihilated by the open cleavage surface producing steps on this surface. This qualitative picture has been pointed out by Tetelman and McEvily (1967) and Thomson (1983).

Following the analysis of Li (1981) we obtain the force exerted on the negative external dislocation at $x$ from the crack tip (Fig. 3.28).

$$f_d = \frac{K_{III}(-b)}{\sqrt{2\pi x}} + \frac{\mu(-b)}{2\pi}\left[\frac{-(-b)}{2x} + \frac{(-b)}{d-x}\sqrt{\frac{d}{x}}\right] \tag{3.6.83}$$

where $d$ and $x$ are the distances of the positive and negative dislocations from the crack tip respectively.

The simplest assumption is that $d \gg x$, or $d \to \infty$

$$f_d \cong \frac{-k_{III}b}{\sqrt{2\pi x}} - \frac{Ab}{2x}, \quad k_{III} = K_{III} + \frac{\mu b}{\sqrt{2\pi d}}. \tag{3.6.84}$$

The boundary of a DFZ is given by the condition for zero force on the dislocation at $\delta_3$,

$$\frac{k_{III}}{\sqrt{2\pi\delta_3}} + \frac{A}{2\delta_3} - \sigma_1 = 0. \tag{3.6.85}$$

The physical solution of $\delta_3$ (for $\sigma_1 > 0$) is given by

$$\delta_3(k_{III}) = \frac{A^2}{2}\frac{1}{\left(A\sigma_1 + \frac{k_{III}^2}{2\pi}\right) - \left[\left(A\sigma_1 + \frac{k_{III}^2}{2\pi}\right)^2 - A^2\sigma_1^2\right]^{1/2}}. \tag{3.6.86}$$

$\delta_3(k_{III})$ is the size of the DFZ due to the dislocation annihilation mechanism at the crack tip. Let us again change the parameter $k_{III}$ to $\beta$

$$k_{III} = (2\pi A\sigma_1)^{1/2}\beta$$

$$\frac{\delta_3(\beta)}{\delta_1} = \frac{1}{1+\beta^2 - [(1+\beta^2)^2 - 1]^{1/2}} . \tag{3.6.87}$$

The relationship between Eqs. (3.6.82) and (3.6.87) is

$$\delta_3(\beta)\delta_2(\beta) = \delta_1^2 \tag{3.6.88}$$

or $[\delta_3(\beta)/\delta_1][\delta_2(\beta)/\delta_1] = 1$. Therefore,

$$\frac{d\delta_3}{d\beta} = -\frac{\delta_3}{\delta_2}\frac{d\delta_2}{d\beta} > 0 \tag{3.6.89}$$

or

$$\frac{d\delta_3}{dk_{III}} > 0\,, \quad \left(\frac{d\delta_2}{dk_{III}} < 0\right) .$$

Figure 3.29 shows the curve of $\delta_3(\beta) - \beta$. We know that $K \propto c^{1/2}$. Assuming that $\beta_i = \beta_0(c_1/c_0)^{1/2}$, we can obtain a series of curves for different crack lengths. There is a rough similarity between the curves calculated by us and the experimental ones measured by Ha *et al.* (1989) [57] (Fig. 3.30). Due to the

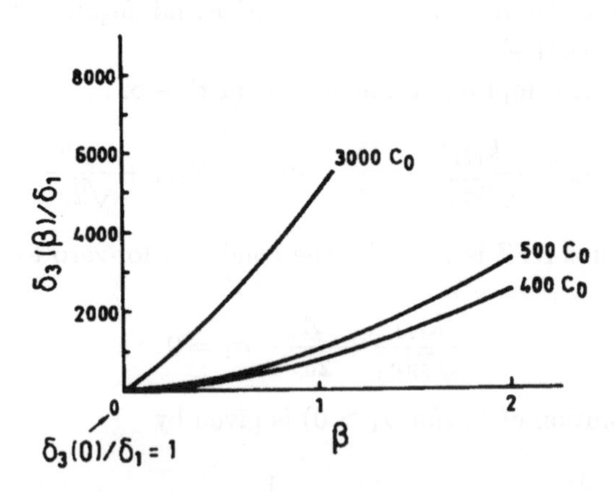


Fig. 3.29. The relationship of the DFZ size with applied stress $\beta(K)$ according to dislocation annihilation mechanism. (Lung, 1990)

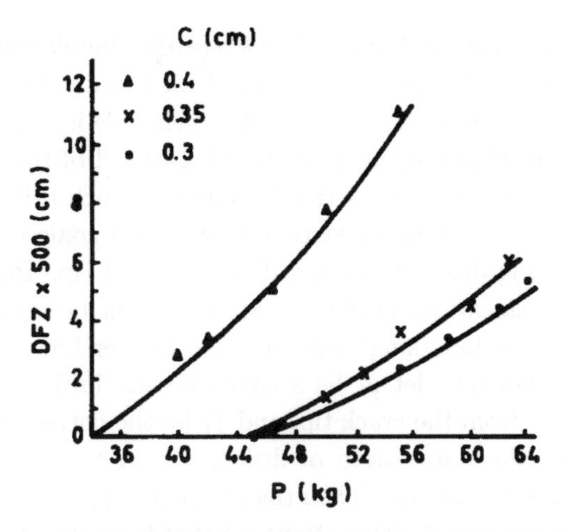


Fig. 3.30. Dependence of the size of DFZ on crack length and applied load in Fe–3% Si single crystal (Ha, 1989). DFZ = 0–0.2 mm. $\partial(\mathrm{DFZ})/\partial p > 0$, $\partial(\mathrm{DFZ})/\partial a > 0$, $\partial(\mathrm{DFZ})/\partial K > 0$. (Lung, 1990)

simplicity of our model, it makes no sense to compare the two curves numerically. From Fig. 3.29 and Eq. (3.6.87), we may see:

(i) $\partial\delta_3/\partial k_{III} > 0$. The crack would promote the formation of a DFZ. It seems that this mechanism is consistent with Kobayashi and Ohr's original experimental results qualitatively, which showed a much larger DFZ at the crack tip than at the free plane.

(ii) K = 0, $\delta_3 = A/(2\sigma_1) = \delta$. It reduces to the size of the DFZ near a free plane.

(iii) $\delta_3(k) > \delta_1 > \delta_2(k)$. Usually $\delta_3(k)$ is larger than $\delta_1$ and $\delta_2(k)$. From the above calculations, we conclude that the mechanism of annihilation of negative dislocations by the open cleavage surface is the most likely one for the formation of a DFZ. Without completely neglecting the effect of image forces on positive dislocations as a possible mechanism, we believe that most experimentally found DFZs are due to annihilation of external dislocations by the open cleavage surface.

Since the experimental work showed much bigger DFZ (Ha, 1990), the effect on fracture properties of metals can not be ignored. Brede and Haasen (1988) suggested that a local stress intensity factor must be introduced

$$k_{III} = \frac{3k}{2\pi}\left(\frac{c}{d}\right)^{1/2}\left(\ln\frac{4d}{c} + \frac{4}{3}\right) \qquad (3.6.90)$$

where $c$ is the size of the DFZ, $d = 2X = k^2/(\pi\sigma_y^2)$, double the size of the plastic zone. Ha *et al.* (1994) proposed a model for the ductile-brittle transition (DBT) in bulk single crystals. The essential principle is that the brittle fracture initiates by an elastic opening of the DFZ ahead of the crack tip, so the prerequisite for brittle fracture is the formation of such a DFZ. Ductile fracture is characterized by void nucleation growth and coalescence through the activation of existing dislocations ahead of the crack tip. Both the crack tip and the existing dislocations near the crack tip might be regarded as effective dislocation sources in bulk single crystals. No matter how complicated these mechanisms may be, they let $v_1$ be a characteristic initial velocity by which dislocations escape from the crack tip, and $f_1$ be the corresponding local driving force exciting the unit length of dislocation in the slip plane along the slip dislocation into the lattice. Then the product $v_1 f_1$ will be the energy consumption rate of unit length dislocations emitted from the crack tip. Similarly, they let $v_2$ be the average velocity of a dislocation excited in the centre of the plastic zone, and $f_2$ be the corresponding local driving force acting upon the unit length of the dislocation in the slip plane along the slip direction. Then the product $v_2 f_2$ will be the energy consumption rate of unit length dislocation moving from the centre of the plastic zone. Thus if

$$v_1 f_1 > v_2 f_2 \tag{3.6.91}$$

ductile fracture is more favourable than brittle fracture from the energy consumption rate point of view; similarly, if

$$v_1 f_1 < v_2 f_2 \tag{3.6.92}$$

brittle fracture is more favourable. Consequently, the DBT temperature can be found from the following equality

$$v_1 f_1 = v_2 f_2 \,. \tag{3.6.93}$$

One of their conclusions obtained from this model is that the effect of dislocation density on DBT temperature is negligibly small. This has been verified by their experimental results in bulk Fe–3% Si single crystals on the DBT temperature (Ha *et al.*, 1995).

## 3.7. *Geometrical Aspects of Grain Boundaries*

As set out by, for example by Christian, (1975; see also Vitek, 1995) any interface may be described as a network of dislocations. Suppose the two adjoining crystals are obtained from a reference lattice by transformations which are represented by $\mathbf{R_1}$ and $\mathbf{R_2}$. Denoting an arbitrary vector in the interface by $\mathbf{P}$, the total Burgers vector of dislocations crossing this vector can be written (Christian, 1975; Christian and Crocker, 1980):

$$\mathbf{B} = (\mathbf{R_1^{-1}} - \mathbf{R_2^{-1}}) \cdot \mathbf{P}\,. \tag{3.7.1}$$

If the interface is then a grain boundary separating grains misoriented by angle $\theta$, then the two transformations are equal and opposite rotations by an angle $\theta/2$ about a common direction, say $\mathbf{e}$. Equation 3.7.1 then leads to the Frank formula (Frank, 1950)

$$\mathbf{B} = 2\sin\left(\frac{\theta}{2}\right)(\mathbf{P} \otimes \mathbf{e})\,. \tag{3.7.2}$$

This result of Frank may appear to determine the dislocation content in this boundary. As Vitek (1995) stresses, neither the choice of the reference lattice nor that of the transformation matrix is unique, and therefore the dislocation content representing a given interface is not unique. However, certain choices for the dislocation content may have a particular physical significance. Vitek (1995) cites in this context, the fact that the short-range elastic field associated with the interface may be identified with the elastic field of a specific dislocation network that satisfies the Burgers vector Eq. (3.7.2). In such a case, that part of the interfacial energy stored in the elastic field can be calculated by means of such a model. Vitek (1995) refers specifically to the Read-Shockley model of low angle grain boundaries (see Appendix 3.4) which are viewed as networks of lattice dislocations (Read and Shockley, 1950).

Vitek (1995) further emphasizes that grain boundaries may not possess any periodicities. However, he notes that boundaries with periodic structures have frequently been observed and, at least in cubic metals, any grain boundary may be approximated usefully by a boundary with periodic structure. This then leads to the simplification that, in atomistic studies of grain boundaries (see Chap. 8) one can impose periodic boundary conditions. Such periodic boundaries can be characterized and geometrically constructed using the content of the coincidence site lattice first formulated by Brandon *et al.* (1964; see also the summary in Appendix 3.4).

### 3.7.1. $\sum = 3$ *Tilt Boundary with a* $\langle 112 \rangle$ *Rotation Axis in Cu*

Atomistic studies by Schmidt *et al.* (1995) have revealed that a particular $\sum = 3$ tilt boundary with a $\langle 112 \rangle$ rotation axis stabilized Cu in the bcc configuration. Cu has earlier been observed in the bcc form either as in bcc iron by Celinski *et al.* (1991) or as small precipitates in a bcc Fe matrix (see eg, Jenkins *et al.*, 1991). The study of Schmidt *et al.* (1995) demonstrates that the bcc structure can also occur under purely internal constraints in grain boundaries. The boundary under discussion is an asymmetrical tilt configuration with inclination of the boundary plane 84° with respect to the usual $(111)/(11\bar{1})$ coherent twin. Laub *et al.* (1994), by thermal grooving experiments obtained indications that the boundary energy of $\sum 3\langle 112\rangle$ tilt boundaries has a minimum at this inclination of the boundary plane.

The atomistic calculation, using Finnis-Sinclair type potentials for Cu (Ackland *et al.*, 1987; Ackland and Vitek, 1990) demonstrated that a layer of bcc structure forms in this boundary and $\{110\}_{bcc}$ planes connect the $\{111\}_{fcc}$ planes of each grain (see also Vitek, 1995). Vitek stresses that the structure of the phase boundaries between bcc and fcc is a near coincidence type, involving a compromise in lattice strains. It is best suited to this specific boundary inclination and accounts for its lower energy relative to neighboring inclinations. However, an essential prerequisite is that the energy difference between bcc and fcc Cu is especially small (0.023 ev/atom: see Vitek, 1995). This was, in fact, demonstrated by the *ab initio* calculations of Paxton *et al.* (1993) and of Kraft *et al.* (1993). The structure discussed above has been confirmed experimentally by HREM* (Schmidt *et al.*, 1995).

### 3.7.2. *Grain Boundary Phases in Copper-Bismuth Alloys*

As emphasized by Vitek (1995), the copper-bismuth system is well suited to study segregation and embrittlement phenomena.

Some of the experimentally observed facts are briefly summarized in Appendix 2.1: a remarkable phenomenon is the segregation induced faceting (Ference and Balluffi, 1988). It appears (Vitek, 1995) that this phenomenon is associated with the formation of a new two-dimensional phase. This motivated a combined atomistic theoretical and HREM investigation, which not only revealed the existence of this phase but also determined its detailed structure (Luzzi, 1991; Yan *et al.*, 1993).

Finnis-Sinclair type many-body potentials were set up. The Cu-Bi interaction was fitted to (i) the lattice parameter (ii) bulk modulus of the theoretical $Cu_3Bi$ compound in the $Ll_2$ structure and (iii) the enthalpy of mixing for

*See Appendix A2.1.

the Cu-Bi liquid solution at 1200K. The parameters for the $Ll_2Cu_3Bi$ structure were found from an *ab initio* LMTO calculation (Yan *et al.*, 1993; Vitek, 1995).

Vitek (1995) describes the structure of the $(111)/(11\bar{1})$ facets containing Bi; first extracted from HREM observations (see Appendix 2.1) and then confirmed as the stable structure by molecular statics calculations, as follows. Using the 'ABC' notion for the stacking of (111) atomic planes in fcc structures, the $\sum = 3(111)/(11\bar{1})$ twin in pure Cu can be depicted as

$$\begin{array}{c} | \\ A - B - C - A - B - C \quad - \quad B - A - C - B - A - C \\ | \end{array}$$

where the vertical lines show the position of the boundary. The structure with Bi could be represented also as above, except that in the marked layer C one third of the Cu atoms are replaced by hexagonally arranged Bi atoms. However, as Vitek (1995) stresses, the Bi atoms are centred outside this atomic plane so that a physically more appropriate representation of the model structure is

$$\begin{array}{c} | \\ A - B - C - A - B - C' \quad - \quad C(\mathrm{Bi}) - B - A - C - B - A \\ | \end{array}$$

Here $C'$ stands for a plane of Cu atoms in which one third of the atoms are replaced by hexagonally arranged vacancies, and $C$(Bi) denotes the plane of Bi atoms positioned above these vacancies. Vitek (1995) points out that the structure can thus be interpreted, to a useful aproximation, as splitting of one of the $\{111\}$ planes, $C$, into two $C'$ and $C$(Bi) planes, containing Cu and Bi respectively. While the Bi atoms (i.e. the whole layer $C$(Bi)) are contracted towards the plane of Cu containing vacancies ($C'$), there is a substantial overall expansion between the Cu layer $C'$ and the Cu layer above the Bi layer.

A detailed comparison between calculated and observed structures showed an almost perfect match between theoretical and experimental images. Quantitative features such as total expansion, i.e. the relative displacement of the two grains in the direction, perpendicular to the boundary, are shown by Yan *et al.* (1993) to be within the experimental limits of accuracy.

As Vitek (1995) discusses, this study demonstrates that a two-dimensional ordered phase may form at the $(111)/(11\bar{1})$ twin boundaries in Cu-Bi provided a sufficient amount of Bi is available at this boundary. He stresses that these results provide concrete evidence that empirical many-body potentials are very useful even in a relatively complex Cu-Bi system, provided sufficient relevant input is employed to construct them. He emphasizes in the present example the importance of *ab initio* electronic structure calculations, which sample atomic configurations not attainable in the laboratory. Such input allows the size of a Bi atom when surrounded by Cu atoms to be correctly incorporated.

### 3.7.3. *Grain Boundary in NiAl with N-body Empirical Potentials*

Following the HREM image of the $\sum = 5$, (310) [001] grain boundary in NiAl (see Fig. 1 of Fonda *et al.*, 1995) reproduced in Appendix 3.4, atomistic calculations were also performed by Fonda *et al.* (1995). They used N-body empirical potentials constructed following the Finnis-Sinclair (1984) approach (compare Ackland *et al.*, 1987). The total energy $E$ of the system of $N$ atoms is expressed as

$$E = \frac{1}{2}\sum_{i,j=1}^{N} V_{s_i s_j}(R_{ij}) - \sum_{i=1}^{N}\left(\sum_{i}\phi_{s_i s_j}(R_{ij})\right)^{\frac{1}{2}}. \qquad (3.7.3)$$

The first term on the right-hand side represents the energy of direct interaction between two atoms, while the second term accounts for the many-body attractive part of the cohesive energy. The quantities $V$ and $\phi$ are, of course, dependent on the species $S_i$ and $S_j$, $R_{ij}$ being the separation of atom $i$ and $j$. Both $V$ and $\phi$ are empirical fitted pair potentials, as used in the study of Fonda *et al.* (1995).

Specifically, the Ni-Ni potential was adjusted to fit (i) to experimental lattice parameter (ii) cohesive energy (iii) elastic constants and (iv) vacancy formation energy of pure Ni (Ackland *et al.*, 1987). The procedure for constructing the Al-Al potential was analogous to that for Ni-Ni (Vitek *et al.*, 1991). For separations less than the first nearest-neighbour distance in the Al fcc structure, the Al-Al repulsive interaction was increased in order to make the Al antisite defect on Ni site energetically unfavorable relative to the formation of double vacancies on Ni sites. Fonda *et al.* (1995) base this requirement on the evidence that Al enrichment in NiAl is associated with the formation of constitutional vacancies on Ni sites (Bradley and Taylor, 1937; Wasilewski, 1968).

The Ni-Al interactions were constructed as follows (Fonda *et al.*, 1995). $\phi_{\rm NiAl}$ was written as the geometric mean of the $\phi$ potentials for pure Ni and pure Al; this is consistent with its interpretation in terms of hopping integrals (Ackland *et al.*, 1988). The direct interaction $V_{\mathrm{NiAl}}$, finally, was constructed to reproduce (i) the known lattice constant (ii) cohesive energy and (iii) elastic constants of the NiAl B2 compound. At atomic separations smaller than the first nearest-neighbour distance, $V_{\mathrm{NiAl}}$ was adjusted by Fonda *et al.* to obey equation of state requirements. With these potentials, the NiAl B2 structure is energetically favoured over the $\mathrm{Ll}_0$ structure and is also stable up to 2000K. In addition, these potentials give the energy of the [110] antiphase boundary to within the experimentally known range (Mirade, 1993).

### 3.7.4. *Example (Schematic) of Low-Energy Grain-Boundary Structure as Determined by Atomistic Simulation*

The relaxed structure which had the lowest energy and also was in accord with the experimental images (see Appendix 3.4) was produced by ordering the antisite defects on sites $G$ and $H$ with equal numbers of defects on each site as

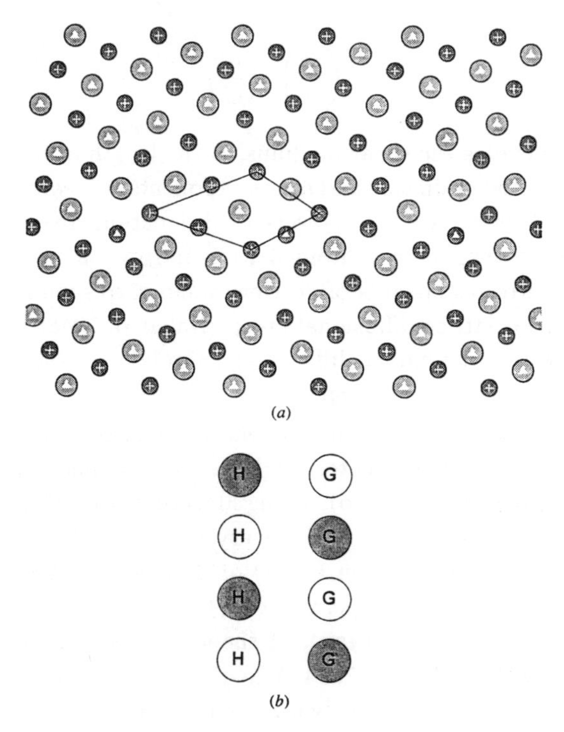


Fig. 3.31. (a) and (b)

shown in Figs. 3.31(a) and (b) (Fonda *et al.*, 1995). This particular ordering removes all Al-Al interactions along [001] for these sites. The grain-boundary energy was calculated by Fonda *et al.* (1995) to be 618 $\mathrm{mJm^{-2}}$, which is only $\sim$ 50% of the grain-boundary energy calculated for the stoichiometric structure. The final structure obtained by these workers contains a rigid-body displacement along the grain boundary of 0.39 Å and a grain-boundary expansion of 0.27 Å, in good agreement with experiment. Images simulated from this grain-boundary structure correlate well with the experimental images (Fonda *et al.*, 1995).

## 3.8. *Creep: Example of Single Crystal of Lead*

Wang (1994) has re-examined data on creep in single crystals of Pb, rutile ($TiO_2$) and ice: the metal crystal Pb will be focus of the summary below. Wang first notes that experimental studies of creep processes in crystalline materials have shown that the creep rate $\dot{\epsilon}$ is related to the stress $\sigma$ and the grain size $d$ by

$$\dot{\epsilon} \propto \left(\frac{B}{d}\right)^p \left(\frac{\sigma}{G}\right)^n . \tag{3.8.1}$$

In Eq. 3.8.1, $G$ denotes the shear modulus, $B$ the length of the Burgers vector while $p$ and $n$ are the grain size and stress exponents respectively. At intermediate stresses, Wang asserts that power-law dislocation creep is the dominant process, leading to exponent $n$ from 3 to 5 and exponent $p$ equal to zero. However at low stresses creep is often Newtonian corresponding to $n = 1$. In this case, $p = 2$ for lattice diffusional creep (Nabarro, 1948; see also Herring, 1950), $p = 3$ for grain-boundary diffusional creep (Coble, 1963), and $p = 0$ for Harper-Dorn (1957) ($H - D$) creep.

Wang records many subsequent observations of $H - D$ creep in both single crystals and in polycrystalline materials (he records for metals references on Al, Pb, Sn, $\alpha$-Ti, $\alpha$-Fe, $\alpha$-Zr and $\beta$-Co). Using literature data, Wang (1994) argues for the operation of $H - D$ creep in single crystals of Pb. He uses, in particular the data of Gifkins and Snowden (1967) on Pb, deformed at a temperature $T = 295\mathrm{K}$ ($\sim \frac{1}{2}$ of melting temperature $T_m$). Their data, along with results on polycrystalline Pb ($d = 200$ $\mu$m), at the same temperature, are displayed in Fig. 3.32, taken from Wang (1994, Wang's Fig. 1). This figure lends some support to Wang's assertion that, below a transition stress $\sigma_t \sim 1.4$ MPa, the single crystals exhibited Newtonian creep behaviour.

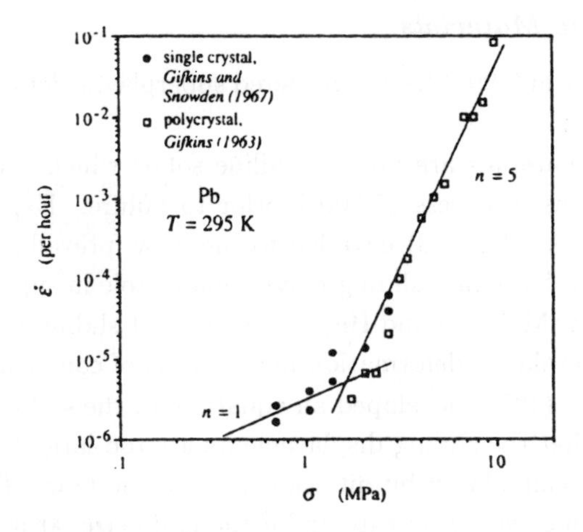


Fig. 3.32. Variation in strain rate with stress in lead. A Newtonian regime in single crystals at stresses lower than 1.4 MPa is indicated (Wang, 1994).

Wang also presented data on rutile and on ice and argued that all three materials deform by Newtonian creep under certain conditions. He noted that such Newtonian behaviour is probably induced by the operation of $H-D$ creep which involves dislocation motion. Diffusional creep is also invoked to account for the deformation of single-crystal specimens.

Wang (1993) had earlier shown that the stress marking the transition from power-law creep at higher stresses to $H-D$ creep at lower stress values ($\sigma_t$) is determined by the Peierls stress, $\tau_p$ say, of the material: the greater $\tau_p$, the larger is $\sigma_t$. A relation of the same kind between $\sigma_t$ and $\tau_p$ also holds for single crystals of Pb, rutile and ice (Wang, 1994).

Kosevich (1979) writes the equation for $\tau_p$ as

$$\tau_p = \frac{2G}{1-\gamma}\exp(-2\pi\eta) : \eta = \frac{3-2\gamma}{4(1-\gamma)}\frac{L}{B} \tag{3.8.2}$$

where $L$ is the distance between atomic planes. In estimating $\tau_p$ for crystals, $L$ is taken as the distance between the closest-packed atomic planes in the lattice, and $B$ as the atomic distance in the close-packed direction on the closest-packed planes (Wang, 1993). Wang (1994) asserts that the values of $\tau_p$ thereby determined represent the lower limit of the true Peierls stress of a crystal. Wang (1994) plots $\sigma_t/E$ vs $\tau_p/G$, $E$ being Young's modulus and for the three materials Pb, $TiO_2$ and $H_2O$ (ice) it is clear that $\sigma_t/E$ increases as $\tau_p/G$ increases.

## 3.9. *Superplastic Materials*

Mabuchi and Higashi (1994) have considered superplastic deformation in metal matrix composites.

Superplastic materials are polycrystalline solids which can undergo large uniform elongations in excess of 200% prior to failure. Experiments of Lee (1969) and of Lin *et al.* (1988) have led to the view, prevailing at the time of writing, that grain-boundary sliding plays a major role in superplastic flow for metallic materials. Mabuchi and Higashi assert that sliding processes are also important in superplastic deformation mechanisms of composites too.

Raj and Ashby (1971) developed an equation for the sliding rate, based on the assumption that the sliding displacements are too large to be elastic, and the sliding is accommodated by diffusional or plastic flow. The sliding rate, written as $\dot{\varepsilon}d$ with $\dot{\varepsilon}$ the strain rate and $d$ the grain size, at a plane boundary containing an array of discrete impermeable particles, is given by the above workers as

$$\dot{\epsilon}d = \frac{1.6\tau_a\Omega}{k_BT}\left\{\frac{\lambda^2}{a^3}D_L\left[1+\frac{5\delta D_{GB}}{aD_L}\right]\right\}. \tag{3.9.1}$$

Here $\tau_a$ is the local stress, $\Omega$ the atomic volume, $k_BT$ the thermal energy and $\lambda$ the particle spacing, $D_L$ denotes the lattice diffusion constant, $a$ is the reinforcement size, $\delta$ the grain-boundary thickness and $D_{GB}$ is the grain-boundary diffusion coefficient. Equation 3.9.1 is readily rearranged, of course, to yield the local stress around the matrix-reinforcement interfaces.

## 3.10. *Mechanical Properties of Fatigued fcc Crystals and their Dislocation Arrangements*

Holzwarth and Etzmann (1994) have emphasized that fatigued fcc metals exhibit characteristic dislocation arrangements which depend on the applied amplitude of resolved plastic shear strain. Experimental work on these structures has been brought together in a number of reviews (e.g. Grosskreutz and Mughrabi, 1995; Mughrabi *et al.*, 1979; Brown, 1981; Basinski and Basinski, 1992). What the experiments reveal is that, at a certain stress level, the so-called matrix structure transforms into the peculiar dislocation arrangement of persistent slip bands (see, for example, Fig. 2 of Holzwarth and Etzmann, 1994). The study of Holzwarth and Etzmann had as its aim to test whether it would be possible to retransform the characteristic ladder-like dislocation arrangement of persistent slip bands into a matrix-like structure. Their conclusion, using transmission electron microscopy, was that such a retransformation did not occur.

Following Kê's earlier work in 1962, Wang and Kê (1981) studied the early stage of fatigue damage by means of energy loss studies with low frequency internal friction method. Later on, a method for measuring the charge of ultrasonic attenuation in the course of push-pull loading was developed. This method is very sensitive to the instantaneous variation of dislocation mobility. Gremaud *et al.* (1981), Vincent *et al.* (1986) and Zhu and Fei (1990) studied ultrasonic attenuation effects in polycrystalline Al; they used models of interaction between attenuation under low strain amplitude ($< 3 \times 10^{-5}$). Zhu and Fei (1990) found that the average stress field of dislocation created by the applied alternative load induces the redistribution of point defects and thus a stable saturation stage would be reached with the increasing of cyclic number.

Heinz and Neumann (1990) studied twin boundary (TB) cracking in austenitic stainless steel polycrystals and proposed that the elastic anisotropy of crystals can induce an additional shear stress near the TBs, and that the plastic strain due to the stress could be localized at the TBs near the fatigue limit. Llanes and Laird (1992) also reported that the TBs can act as stress concentrators due to elastic strain incompactibility.

In closing this chapter, we recommend the reader to consult the various Appendix 3.1–3.4 as supplementary reading.

# Chapter 4

# Some Characteristic Features of Fractals

## Introduction

Following the ideas of Mandelbrot (1982), a wide range of complex structures of interest to scientists have been quantitatively characterized using the idea of a fractal dimension: a dimension that corresponds in a unique fashion to the geometrical shape under study, and often is not an integer. The key to this progress is the recognition that many random structures obey a symmetry such that objects look the same on many different scales of observation. The purpose of this section is to provide a brief introduction to fractal phenomena. For the details, the reader may refer to special books and articles (for example, Mandelbrot, 1983; Feder, 1988; Pietronero and Tosatti, 1986; Aharony and Feder, 1989; Herrmann and Roux, 1990). Although there are many different types of fractal phenomena, we shall concentrate on mechanical properties and especially on fracture* (Mandelbrot *et al.*, 1984; Lung, 1986; Herrmann, 1990; Milman *et al.*, 1994).

## 4.1. *Self-Similarity and Fractals*

Before discussion on the definition of fractals, we introduce the law of self-similarity which was ignored in Euclidean geometry and which was called irregular shapes for many years. Regular and irregular things are in a relative sense. Fractals may be considered as *regular things which obey the law of self-similarity*. What is self-similarity?

*For the reader already acquainted with fractals, the application to fractured surfaces begins in earnest in Section 4.13.

Let us take the von Koch curve as an example to illustrate it. A simple line segment is divided into three and the middle segment is replaced by two equal segments forming part of an equilateral triangle. At the second stage, in the construction, each of these 4 segments is replaced by 4 new segments with length $\frac{1}{3}$ of their parents according to the original pattern. Repeated over and over, a von Koch curve is yielded (Fig. 4.1).

In this figure, each small portion of the curve, when magnified, can reproduce exactly a larger portion. The curve is said to be invariant under changes of scale. We may say that the curve possesses as an exact self-similariity

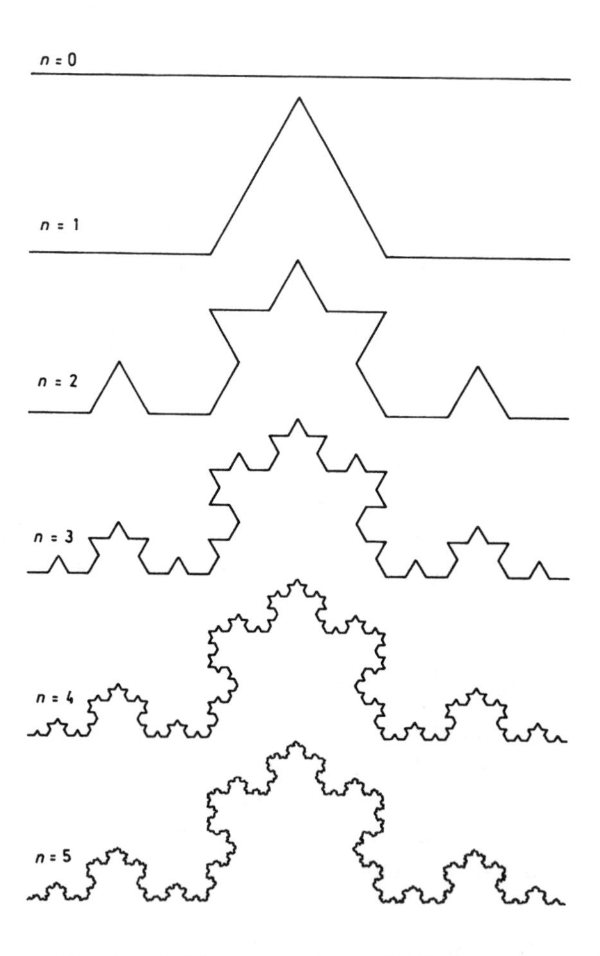


Fig. 4.1. Construction of the triadic Koch curve.

property. At each stage in its construction the length of the curve increases by a factor of $\frac{4}{3}$. Thus, the limiting curve is of infinite length as it is of infinite generations. In nature, many curves modelled with Koch curves are only of finite generations (see later sections).

The property of self-similarity, or scaling, is one of the central concepts of fractal geometry. Mandelbrot (1986) proposed the definition of a fractal:

*A fractal is a shape made of parts similar to the whole in some way.*

Previously, Mandelbrot (1982) offered another tentative definition of the fractal:

*A fractal is by definition a set for which the Hausdorff-Besicovitch dimension $D$ (see Eq.* 4.2.3) *strictly exceeds the topological dimension $D_T$ (see Sec.* 4.3).

The previous definition is too mathematical, though correct and precise. However, an ideal definition of fractals is still lacking. Fractals are a decidedly modern invention. However, it has been recognized as useful to natural scientists only since around 1980.

Euclidean geometry provided an accurate description of shapes of objects which were called regular (lines, curves, circles and so on). Euclidean dimension has a rigorous definition; such as:

(i) A point has no volume,
(ii) A line is a breadthless length,
(iii) A surface has length and breadth only.

Fractals, like the Koch curve, possess no characteristic sizes, whereas regular shapes have them, or at most a few, characteristic sizes or length scales (the radius of a sphere, the side of a cube etc.). Fractal shapes are said to be self-similar or scale invariant. Some regular shapes also possess this property but not all of them. We can divide a cube into self-similar parts but cannot do so for circles, ellipses and other complex Euclidean shapes. Fractals provide an excellent description of many natural shapes and have already given computer imaginary natural landscapes whereas Euclidean geometry provides concise accurate descriptions of man-made objects such as mechanical parts of a machine. Finally, fractals, in general, are the result of a construction procedure of algorithm that is often recursive and ideally suited to computers, whereas Euclidean shapes are usually described by simple algebraic formula (e.g. $A = a^2$ defines an area of a square of edge length $a$). In short, fractal geometry is more appropriate for natural shapes than Euclidean geometry.

Fractals are still approximate descriptions of objects in nature though many scientists deem appropriate for the irregular shapes of the real world. Mandelbrot found the law of self-similarity to describe shapes for which Euclidean geometry is inappropriate. Then he expanded the concept of *regular shapes* to more broad geometrical objects. We can say that fractals may describe shapes by iteration of a very simple rule of self-similarity.

## 4.2. *Self-Similarity and Dimension*

An object, say two-dimensional, such as a square area in the plane, can be divided into $N$ self-similar parts, each of which is scaled down by a factor $r = \frac{1}{\sqrt{N}}$, or $N = r^{-2}$. For a line segment, (one dimensional), $N = r^{-1}$, and a solid cube, (three dimensional), $N = r^{-3}$. This concept can be generalized to fractal dimension with self-similarity. *A D-dimensional self-similar object can be divided into N smaller copies of itself each of which is scaled down by a factor r*;

$$N = \left(\frac{1}{r}\right)^D . \tag{4.2.1}$$

Equation (4.2.1) is a power-law relationship. If we change $r$ to $br$, a self-similarity relationship can be obtained

$$N(br) = b^{-D} N(r) . \tag{4.2.2}$$

Then, the fractal or similarity (or scaling) dimension is given by

$$D = \frac{\ln N}{\ln(\frac{1}{r})} . \tag{4.2.3}$$

The fractal dimension does not need to be an integer. For the above example, the Koch curve, $N = 4$, $r = \frac{1}{3}$. Its fractal dimension is $D = \frac{\ln 4}{\ln 3} = 1.2618\ldots$. This non-integer dimension, greater than one but less than two, reflects the fact that the curve fills more of space than a line ($D = 1$), but less than a Euclidean area of a plane ($D = 2$).

The form of Eq. (4.2.1) can be changed to read

$$\frac{1}{N} = r^D = v . \tag{4.2.4}$$

In other words, if the generalized volume of a geometrical object is enlarged (or reduced) by a factor $v$ with the enlargement (of reduction) of a factor $r$ in linear dimension; then, the fractal dimension is given by

$$D = \frac{\ln v}{\ln r}. \tag{4.2.5}$$

For a line segment (one dimension), $\frac{L}{L_0} = r^1$; a square area in the plane, $\frac{A}{A_0} = r^2$, and a solid cube, $\frac{V}{V_0} = r^3$. For the Koch curve:

(i) $v = \frac{V}{V_0} = 4, r = 3$ (enlargement); or
(ii) $v = \frac{V}{V_0} = \frac{1}{4}, r = \frac{1}{3}$ (reduction); therefore, the fractal dimension of Koch curve is $\frac{\ln 4}{\ln 3}$, as above.

## 4.3. *Hausdorff-Besicovitch Dimension*

Now, we turn back to the tentative definition of fractals given by Mandelbrot (1982). This definition requires previous definitions of Hausdorff-Besicovitch dimension and topological dimension.

The Hausdorff dimension $D$ of a subset $S$ of Euclidean space arises from asking 'how big is $S$' for very general sets. The answer comes from counting the number of open balls* needed to cover the set $S$. For each $r > 0$, let $N(r)$ denote the smallest number of open balls of radius $r$ needed to cover $S$. One can show that the limit (see Hastings and Sugihara, 1993)

$$D = \lim_{r \to 0} \frac{-\ln N(r)}{\ln r} \tag{4.3.1}$$

exists. The value of $D$ is called the Hausdorff dimension of $S$. Since $\ln r \to -\infty$ as $r \to 0$; the negative sign is needed in order that $D$ should be positive.

Equation (4.3.1) is equivalent to an approximate power-law relationship

$$N(r) \approx \text{const.}\ r^{-D}. \tag{4.3.2}$$

Assume that $S$ can be decomposed into $n$ rescaled copies of itself, each contracted by a linear factor $k$. From Eq. (4.2.3)

$$D = \frac{\ln n}{\ln k} \tag{4.3.3}$$

where $k = \frac{1}{r}$. Now, we suppose that $S$ can be covered by $N(r_0)$ open balls of radius $r_0$. Each reduced copy can be covered by $n$ "rescaled" open balls

---

*Open balls or open boxes (boxes without their boundaries) are used to define the Hausdorff dimension and topological dimension by the properties of suitably minimal open covering.

of radius $\frac{r_0}{k}$. Then, $S$ can be covered by $nN(r_0)$ open balls of radius $\frac{r_0}{k}$, (Eq. 4.3.2). At least approximately,

$$N\left(\frac{r_0}{k}\right) = nN(r_0)\,. \tag{4.3.4}$$

The iteration of the above process $m$ times gives

$$N\left(\frac{r_0}{k^m}\right) = n^m N(r_0),\ (m = 1, 2, 3, \ldots)\,. \tag{4.3.5}$$

Equation (4.3.5) means that

$$\begin{aligned}
\lim_{m\to\infty}\left[\frac{-\ln N(\frac{r_0}{k^m})}{\ln(\frac{r_0}{k^m})}\right] &= \lim_{m\to\infty}\left[\frac{-\ln n^m N(r_0)}{\ln(\frac{r_0}{k^m})}\right] \\
&= \lim_{m\to\infty}\frac{[(m\ln n + \ln N(r_0)]}{(m\ln k - \ln r_0)} \\
&= \ln n/\ln k \\
&= D\,.
\end{aligned} \tag{4.3.6}$$

Therefore, the scaling dimension is equivalent to the Hausdorff dimension. The condition that the limit of Eq. (4.3.1) exists is equivalent to finding a real value $D$ such that, for the set $S$, the $d$-measure is infinite if $d < D$ and vanishes if $d > D$. $D$ is called the Hausdorff dimension of $S$, and is also called the Hausdorff-Besicovitch dimension when non-integer values of $D$ are included.

Topology is a study of continuity (see Mumkres, 1975; Mansfield, 1963). Two geometrical figures are topologically equivalent, or homomorphic, if each can be transformed into the other by a continuous deformation. That means, the intrinsic qualitative property does not change when the object under consideration is subjected to stretching, contracting and bending but without tearing. For example, if a piece of plasticine is moulded into various different shapes without making breaks or joins, then all the associated geometrical figures are topologically equivalent. Thus the surfaces of a sphere, an ellipsoid, a cube and a tetrahedron are all topologically equivalent. Lines of finite length can be contracted continuously into a point and all the elements of the set are not connected. Thus their topological dimension is zero (Cantor set). The perimeter of an island cannot be contracted continuously into a point and the dimension of its arbitrary small neighborhood is zero. Thus, their topological dimension is unity (Koch island). The Sierpinski gasket and carpet (see

Appendix 4.1) can be contracted continuously into lines. Their topological dimension is also unity. Similarly, that of the Menger sponge (see Appendix 4.1) is two.

It is interesting to discuss which shape is fractal according to the definitions of 1982 and of 1986 by Mandelbrot. It seems that some shapes are fractals according to the definition of 1986, but not according to the definition of 1982 (for instance, see Fig. 4.2(a)). At order one, the unit square is divided into nine equal-sized smaller squares with $r = 1/3$. At order two, the remaining squares are divided into nine smaller equal-sized squares with length, $r^2 = 1/9$.

If $N_1 = 1$, $N_2 = 1$, then, $D = \frac{\ln 1}{\ln 3} = 0$. $D_T = 0$, $D = D_T$. (see also: Turcotte, 1992, Fig. 2.2(a) and Fig. 3.2, $f = 1/8$). It is not a fractal according to the definition of 1982, even the shapes are self-similar. However, if $N_1 = 9$, $N_2 = 81$, $D = \frac{\ln 9}{\ln 3} = 2$, $D > D_T$ (Fig. 4.2(b)) (see also: Turcotte, 1992, Fig. 2.2(e)), it is a fractal even though the shapes are regular in Euclidean geometry.

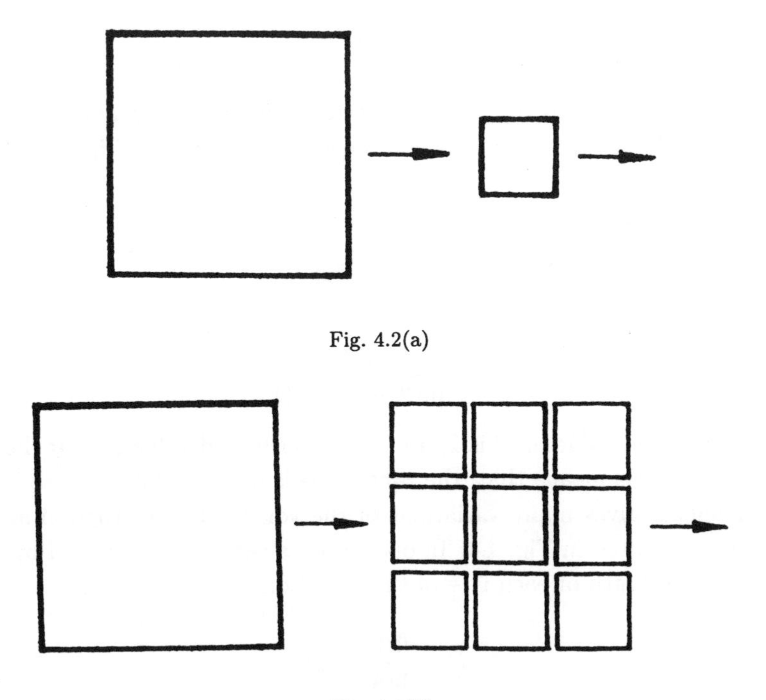

Fig. 4.2(a)

Fig. 4.2(b)

It also seems that some objects are fractals according to the definition of 1982 but not according to the definition of 1986.

Suppose an area of any shape is divided into nine equal-area smaller areas without limitations to their shapes, and more than one smaller areas are retained. Then, if $N_1 > 1$, (say $N = 2$), $D = \frac{\ln 2}{\ln 3} = 0.6309$. $D > D_T = 0$. It is a fractal according to the definition of 1982, but it does not possess self-similarity property in shapes.

Mathematics, be it qualitative or quantitative, must be based upon precise definitions (see Mansfield, 1963). A definition should be as broad as possible, so that it would include all special cases as the various examples were useful in mathematics; but the definition would also be required to be narrow enough to distinguish it from other fields. This is always the problem when one is trying to formulate a new mathematical concept to decide how general its definition should be. The definition finally settled on may seem a bit abstract, but as one works through the various ways of constructing fractal structures (Stanley, 1991; Feder, 1988; Voss, 1988; Vicsek, 1989; and Lung, 1992, ...), one will get a better feeling for what the concept means.

## 4.4. *The Koch Curves*

The triadic Koch curves represent interesting examples of fractal figures. We have already discussed a typical one in the above paragraphs (Fig. 4.1: see also Fig. 4.17).

Now, if $N = 2$, and

$$r = \frac{L_1}{L_0} = \frac{1}{2}\csc\left(\frac{\theta}{2}\right),$$

then

$$D = \frac{\ln 2}{[\ln 2 + \ln\sin(\frac{\theta}{2})]}. \tag{4.4.1}$$

The upper bound value of $D$ is 2; and the lower bound value is 1. If the angle $\alpha = \frac{1}{2}(\pi - \theta)$, this means that, the range of $\alpha$ is $0 < \alpha < 45°$.

Mandelbrot gives many variations of the von Koch construction and two others are presented in Fig. 4.3. In one figure, a segment is replaced by 8 new segments, the length of each is $\frac{1}{4}$ of the initial one, and

$$D = \frac{\ln 8}{\ln 4} = 1.5.$$

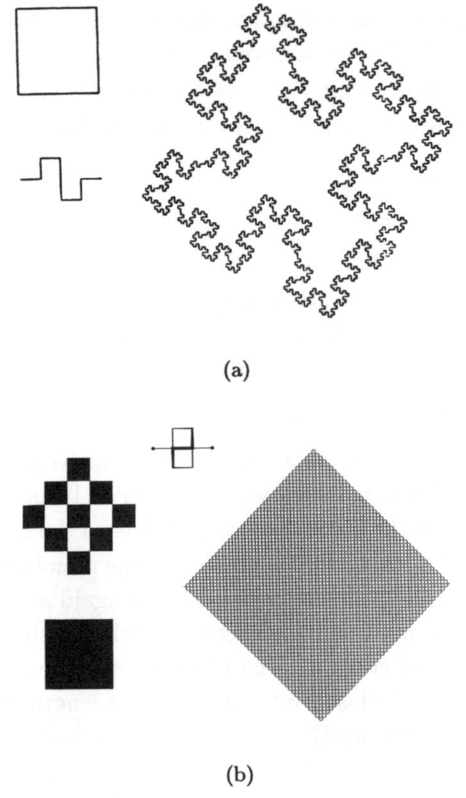


Fig. 4.3 Construction of the quadratic Koch curves. (a) $N = 8, r = \frac{1}{4}, D = \frac{3}{2}$; (b) $N = 9, r = \frac{1}{3}, D = 2$.

In another figure, each segment is replaced by 9 new segments, the length of each is $\frac{1}{3}$ of the initial one, and

$$D = \frac{\ln 9}{\ln 3} = 2\,.$$

The total length of a Koch curve is given by

$$L_F(\varepsilon) = N(\varepsilon) \cdot \varepsilon = \varepsilon \cdot \varepsilon^{-D} = \varepsilon^{1-D}\,. \tag{4.4.2}$$

The total length of the residual set is given by

$$L_R(\varepsilon) = 1 - \varepsilon^{1-D} < 0\,. \tag{4.4.3}$$

The negative value means an addition of this amount of segments in the stages of construction instead of removing some segments.

Equation (4.4.3) shows that $L_R(\varepsilon)$ is not a power-law relationship. However, the generation distribution (or the total length created in one generation) of it is given by

$$L_R(\varepsilon_{n+1}) - L_R(\varepsilon_n) = \varepsilon_{n+1}^{(1-D)} - \varepsilon_n^{(1-D)} = \varepsilon^{n(1-D)}[1 - \varepsilon^{(1-D)}] = \text{const.}\ \varepsilon_n^{(1-D)}\,. \tag{4.4.4}$$

It is a power-law (fractal) relationship with the same fractal dimension $D$ as the original Koch curve. The length of the initial segment is $(1 - \varepsilon^{1-D})$.

## 4.5. *The Cantor Set*

The Cantor set is constructed as a limit of an iterative process in which the first stage, stage number 0, is the closed unit interval [1, 0], and the generation divides the interval into three equal parts and deletes the open middle third of the closed interval. This procedure is applied to the remaining two parts again. Referring to Fig. 4.4, the processes are repeated again and again. Given any stage $n$, the next stage $n+1$ is constructed by deleting the open middle third of each closed interval of stage $n$. The Cantor set is the resulting limit. The stage $n$ is the union of $2^n$ closed intervals, each of length $(1/3)^n$. Thus stage $n$ has length or linear measure $(2/3)^n$, which approaches zero as $n$ approaches infinity.

According to Eq. (4.2.3)

$$N = 2, r = \frac{1}{3}\,.$$

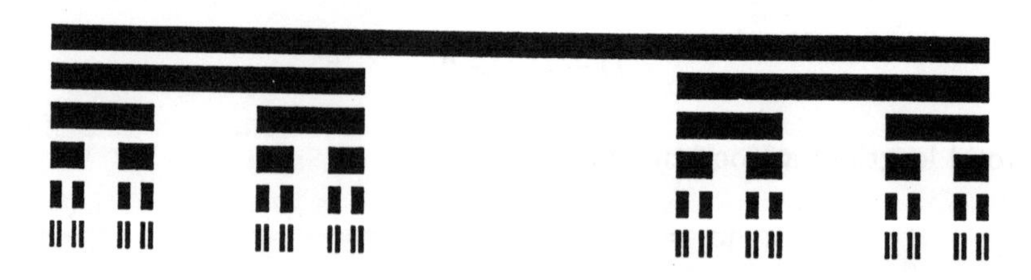

Fig. 4.4. Construction of the triadic Cantor set. The initiator is the unit interval [0, 1]. The generator removes the open middle third. The figure shows the construction of the five first generations. $D = \ln 2/\ln 3 = 0.6309$.

The fractal dimension is given by

$$D = \frac{\ln 2}{\ln 3} \sim 0.6309\,.$$

The stage $n$ has length $(\frac{2}{3})^n$; let it be $(\frac{1}{3})^{n\beta}$, then,

$$\beta = \frac{\ln 2 - \ln 3}{-\ln 3} = 1 - \frac{\ln 2}{\ln 3} = 1 - D\,. \tag{4.5.1}$$

The total length of the Cantor set at stage $n$, then, is given by

$$C_F(\varepsilon_n) = \varepsilon_n^{(1-D)}\,. \tag{4.5.2}$$

It is the same as Eq. (4.4.2). The length of the residual set is given by

$$C_R(\varepsilon_n) = 1 - \varepsilon_n^{(1-D)} = 1 - \left(\frac{2}{3}\right)^n > 0\,. \tag{4.5.3}$$

The positive value means deletion of this amount of segments in the stages of construction. Equation (4.5.3) shows that $C_R(\varepsilon)$ is not a power law (fractal) relationship. However, the generation distribution (or the total length created in one generation) of it is given by

$$C_R(\varepsilon_{n+1}) - C_R(\varepsilon_n) = \left(\frac{2}{3}\right)^n \left(1 - \frac{2}{3}\right) = \left(\frac{1}{3}\right)\left(\frac{1}{3}\right)^{n(1-D)}\,. \tag{4.5.4}$$

Equation (4.5.4) is power-law relationship with the same fractal dimension $D$ as the original Cantor set.

A generalized model of a triadic Cantor set is assumed such that the deleted part is $C_k$ ($k$ stage) where $0 < C_k < 1$. The length of the elements in the $n$th generation is (see Vicsek, 1989),

$$\varepsilon_n = \frac{1}{2^n}\prod_{k=1}^{n}(1 - C_k)\,. \tag{4.5.5}$$

The total length of the $n$th generation is

$$C_n = \varepsilon_n^\beta \tag{4.5.6}$$

and the residual set is given by

$$C_{R,n} = 1 - \varepsilon_n^{\beta}. \tag{4.5.7}$$

$$\beta = \lim_{n\to\infty} \frac{\ln(1 - C_{R,n})}{\ln \varepsilon_n}$$

$$= \lim_{n\to\infty} \frac{\ln \prod_{k=1}^{n}(1 - C_k)}{n \ln(\frac{1}{2}) + \ln \prod_{k=1}^{n}(1 - C_k)}.$$

$$(C_{R,\infty} = 1,\ C_{R,n} = 1 - 2^n \cdot \varepsilon_n) \tag{4.5.8}$$

Assuming $C_1 = C_2 = C_3 = \cdots = C_k = \cdots = C$; and $\ln \prod_{k=1}^{n}(1 - C_k) = \sum_{k=1}^{n} \ln(1 - C_k)$

$$\beta \approx \frac{\ln(1 - C)}{\ln(1 - C) - \ln 2} \equiv 1 - D.$$

where $D$ has the meaning of fractal dimension.

For a regular Cantor set, $C = \frac{1}{3}$

$$\beta = 1 - \frac{\ln 2}{\ln 3} = 1 - D.$$

Indeed, $D$ is the fractal dimension of the Cantor set.

An extension of a triadic Cantor set is $N$ and $r$ being various numbers. For example

$$N = 2,\ r = \frac{4}{15},\quad \left(\text{or } C = \frac{7}{15}\right)$$

$$D = \frac{\ln 2}{\ln(\frac{15}{4})} \cong 0.524.$$

Another discussion on fractals is the following: From Eq. (4.2.1), if we choose $D$ as a definite value, $N$ and $r$ can vary according to Eq. (4.2.1),

$$r = \left(\frac{1}{N}\right)^{\frac{1}{D}} \tag{4.5.9}$$

$$r^{-1} = N_0^{\frac{1}{E}} = N^{\frac{1}{D}}. \tag{4.5.10}$$

Let

$$N_0 = m^E,\ N = m^D \tag{4.5.11}$$

where $N_0$ is the number of smaller size copies which the parent is divided into and $E$ is the Euclidean dimension. $N$ is the number of smaller size copies remaining. $N$ and $m$ should be positive integers. Many values of $N$ and $m$ can form fractals with the same fractal dimension. Morever, any changes of the distribution of $N$ which is important for the construction and properties, do not change the value of $D$. This means that for a definite value $D$, the configuration of fractals is not unique. Fractal dimension, as a parameter can characterize the *continuous change* of a configuration, but we cannot overestimate its role to request a one to one correspondence between $D$ and any construction.

## 4.6. *The Residual Set and "Fat Fractals"*

For a fractal, the total generalized volume is given by

$$V_F(\varepsilon) = \varepsilon^\beta \tag{4.6.1}$$

where $\varepsilon = r\varepsilon_0$ and $\varepsilon_0$ is the length of the initiator, $\beta = E - D$. $E$ and $D$ are the Euclidean and fractal dimensions respectively.

For the residual set, the total volume is given by

$$\begin{aligned} V_R(\varepsilon) &= V_F(1) - \varepsilon^\beta \\ V_R(0) &= V_F(1) = 1\,. \end{aligned} \tag{4.6.2}$$

It is not a power-law relationship. However, the generation distribution (or the volume created in one particular generation) is given by

$$V_R(\varepsilon_{n+1}) - V_R(\varepsilon_n) = \varepsilon_n^\beta(1 - \varepsilon^{E-D}) = \text{const. } \varepsilon_{\text{n}}^\beta\,. \tag{4.6.3}$$

It is a power relationship, and

$$\begin{aligned} \beta &= \lim_{\varepsilon\to 0} \frac{\ln[V_F(1) - V_R(\varepsilon)]}{\ln \varepsilon} \\ &= \lim_{\varepsilon\to 0} \frac{\ln[V_R(0) - V_R(\varepsilon)]}{\ln \varepsilon} \\ &= E - D \end{aligned} \tag{4.6.4}$$

$$\beta = E - D \tag{4.6.5}$$

$$V_F(\varepsilon) + V_R(\varepsilon) = V_F(1) = 1\,. \tag{4.6.6}$$

Furthermore, we let

$$V_R(\varepsilon_{n+1}) - V_R(\varepsilon_n) = V_R(n+1, n)$$

$$V_F(\varepsilon_n) + V_R(\varepsilon_n) = V_F(\varepsilon_n) + \sum_{i=1}^{n} V_R(n, n-1) = 1\,.$$

Generally, the fat fractal proposed by Mandelbrot and Farmer (see Farmer, 1986) can be expressed in the form

$$V(\varepsilon) \approx V(0) + A\varepsilon^{\beta} \tag{4.6.7}$$

Equation (4.6.2) can be expressed as

$$V_R(\varepsilon) = V_R(0) - \varepsilon^{\beta}\,. \tag{4.6.8}$$

Thus the residual set of a fractal is a kind of "fat fractal" in the case of $A = -1$. Fat fractals are in general not self-similar objects; they are not power-law relationships. However, the volume created is given by

$$V(\varepsilon_{n+1}) - V(\varepsilon_n) = A\varepsilon_n^{\beta}(1 - \varepsilon^{\beta}) = \text{const. } \varepsilon_n^{\beta} \tag{4.6.9}$$

where

$$\beta = E - D\,. \tag{4.6.10}$$

It is a fractal with the fractal dimension $E - \beta$.

## 4.7. *Statistical Self-Similarity*

The above exact fractals may be considered as approximate models of nature. For example, upon magnification segments of the coastline look like segments at different scales. The property of self-similarity is in a statistical sense. The property that objects can look statistically similar while at the same time different in detail at different length scales, is the central feature of fractals in nature. Thus, the hierarchic structure should be experimentally verified. Secondly, the results of measurements should indicate power-law relationships within experimental error over a sufficiently wide range of scale.

For exact fractals, $N$ should be a positive integer, but no limitation exists for a fractal in the statistical self-similarity sense. For exact fractals only discrete values for $N$ and $\frac{1}{r}$ are allowed. This is perhaps the reason why sometimes we can only obtain discrete values of data on a straight line in double logarithmic plots. For statistical self-similarity fractals nearly continuous data with a certain degree of scatter would be obtained on a straight line in double logarithmic plots.

However, only one of the factors mentioned above is insufficient to chara-terize a fractal structure. There are some complicated cases, which will be discussed later.

## 4.8. *Brownian Motion and Time Series*

The central concept of Brownian motion that has played an important role in both physics and other sciences is the self-affinity, or non-uniform scaling property of the traces and probability distribution.

Consider the random walk in one-dimensional line, the particle jumps a step of $+\xi$, or $-\xi$ every $\tau$ seconds. In the diffusion process, $\xi$ may be regarded as some microscopic length and $\tau$ as microscopic time interval — the collision time.

The step length, $\xi$ has a Gaussian or normal probability distribution

$$P(\xi,\tau) = \left(\frac{1}{\sqrt{4\pi D\tau}}\right) \exp\left(\frac{-\xi^2}{4\pi D\tau}\right) \tag{4.8.1}$$

where $P(\xi,\tau)$ is the probability of finding the particle in the range $\xi$ and $\xi+d\xi$ and $\tau$.

In Eq. (4.8.1), if we replace $\xi^* = b^{\frac{1}{2}}\xi$ and $\tau^* = b\tau$, then, it becomes

$$P^*(\xi^*,\tau^*) = b^{\frac{-1}{2}} P(\xi,\tau) \tag{4.8.2}$$

and

$$\int_{-\infty}^{\infty} P^*(\xi^*,\tau^*)\, d\xi^* = b^{\frac{1}{2}} \int_{-\infty}^{\infty} b^{\frac{-1}{2}} P(\xi,\tau)\, d\xi = 1\,. \tag{4.8.3}$$

Thus, the distribution does not change under change of $\xi$ to $b^{\frac{1}{2}}\xi$ and $\tau$ to $b\tau$.

The variance of this process is given by

$$\langle \xi^2 \rangle = \int_{-\infty}^{\infty} \xi^2 P(\xi,\tau)\, d\xi = 2D\tau \tag{4.8.4}$$

where $D$ is called the diffusion coefficient. We change the scales of $\xi$ and $\tau$ again, ($\xi^* = b^{\frac{1}{2}}\xi; \tau^* = b\tau$). We obtain

$$\begin{aligned}
\langle \xi^{*2} \rangle &= \int_{-\infty}^{\infty} \xi^{*2} P^*(\xi^*, \tau^*)\, d\xi^* \\
&= b \int_{-\infty}^{\infty} \xi^2 P(\xi.\tau) \cdot b^{\frac{-1}{2}} d(b^{\frac{1}{2}}\xi) \\
&= 2D(b\tau) \\
&= 2D\tau^* .
\end{aligned} \tag{4.8.5}$$

It is also scale invariant but in different scale change ratios for $\xi$ and $\tau$. The expressions in Eqs. (4.8.3) and (4.8.5) indicate that the system has anisotropic scaling. This is not a homogeneous fractal. It is called a self-affine fractal.

An extension of the central concept of Brownian motion was proposed by Mandelbrot and Van Ness (1968). It is called fractional Brownian motion (fBm). Many computer fractal simulations are based on an extension of fBm to higher dimensions such as landscape. It is also a good starting point for understanding anomalous diffusion and random walks on fractals.

A fBm, $V_H(t)$, is a single valued function of one variable, $t$ (usually time). The typical change in $V(\Delta V = V(t_2) - V(t_1))$, to the time interval $\Delta t = t_2 - t_1$ is given by

$$\Delta V \propto \Delta t^H . \tag{4.8.6}$$

$H$ is a parameter which characterizes the scaling behaviour of the different traces of fBm: $0 < H < 1$. $H = \frac{1}{2}$ corresponds to a trace of Brownian motion.

Wang (Wang and Lung, 1990; Wang, 1992) investigated the dynamical mechanisms from the starting point of the generalized Langevin equation and Fokker-Planck equation, and established the bridge between the fBm, anomalous diffusion and the long-time correlation effects. It has been shown that a kind of dynamical mechanism for anomalous diffusion is the long-time correlation effects. Wang (1992) studied biased diffusion and found the probability density function for finding the Brownian particle at displacement $X$ and $t$. The probability density function has Gaussian distribution for displacement $X$. Furthermore, Wang (1994) investigated the diffusive motion of a Brownian particle which is acted upon by both a friction force with memory effect and a noise with long-range correlation effects. Due to the long-range correlation effects, the effective diffusion coefficient is dependent on both the displacement and time, and the probability density for finding the Brownian particle at displacement $X$ and $t$ is a non-Gaussian distribution.

The fBm problem is related to Hurst's empirical law and rescaled range analysis, ($R/S$ analysis). For details, one may refer to Feder's book (1988). The model of fBm has been applied to describe the fractal structure of fractured surfaces. We will return to this problem in later sections.

## 4.9. *Self-Similarity and Self-Affinity*

It is important to distinguish self-similarity and self-affinity. Consider a set $S$ of points at positions

$$\mathbf{x} = (x_1, x_2, \ldots, x_E)$$

in Euclidean space of dimension $E$ under a similarity transform with real scaling ratio $0 < r < 1$, the set becomes $rS$ with points at

$$r\mathbf{x} = (rx_1, rx_2, \ldots, rx_E)\,.$$

A bounded set $S$ is self-similar when $S$ is the union of $N$ non-overlapping distinct subset each of which is congruent to $rS$, where congruent means identical under translations and rotations.

The set $S$ is also self-similar if each of the $N$ subsets is scaled down from the whole by a different similarity ratio $r_n$. In this case, $D$ is given by (Feder, 1988)

$$\sum_{n=1}^{N} r_n^D = 1\,. \tag{4.9.1}$$

In some cases, $r_n$'s are correlated. For example, the ratios of $\frac{r_n}{r_1}$ are a series of constant ratio.

$$\sum_{n=1}^{N} \left(\frac{r_n}{r_1}\right)^D = r_1^{-D}, \left(\frac{r_{n+1}}{r_n} = a < 1\right) \tag{4.9.2}$$

$$\sum_{n=1}^{N} \left(\frac{r_n}{r_1}\right)^D = \frac{1-a^{ND}}{1-a^D}\,. \tag{4.9.3}$$

For infinite number of $N$, $N \to \infty$ it gives

$$\frac{1}{1-a^D} = r_1^{-D}$$

$$r_1^D + a^D = 1\,. \tag{4.9.4}$$

The equation has been simplified even with a large value of $N$. In general, numerical solution of Eq. (4.9.4) for $D$ is needed. Long *et al.* (1995) calculated the fractal dimension of martensite structure with this method subsequently.

If the $E$ coordinates of $\mathbf{x}$ may be scaled by a different ratio $(r_1, r_2, \ldots, r_E)$, the set $S$ is transformed to $r(S)$ with points at $r(\mathbf{x}) = (r_1 x_1, \ldots, r_E x_E)$. It is called self-affinity. A bounded set $S$ is self-affine when $S$ is the union of $N$ distinct (non-overlapping) subsets each of which is congruent to $r(S)$. $S$ is also statistically self-affine when $S$ is the union of $N$ distinct subsets each of which is congruent in distribution to $r(S)$. For example, Brownian motion in Sec. 4.8 is self-affine. It is scaling in different scale change ratios for $\xi$ and $\tau$.

## 4.10. *The Relation of D to H*

The fractal dimension of even the simplest self-affine fractals is not uniquely defined. The difficulties have been illustrated with the case of fBm by Voss (1985) and Pietronero (1987).

Suppose the time span is divided into $N$ equal intervals, each with $\frac{\Delta t}{t_0} = N^{-1}$, where $t_0$ is the unit of time. Each of these intervals will contain one portion of vertical range

$$\frac{\Delta V_H}{V_0} = \left(\frac{\Delta t}{t_0}\right)^H = N^{-H}. \tag{4.10.1}$$

The occupied portion of each interval will be covered by

$$\frac{\left(\frac{\Delta V_H}{V_0}\right)}{\left(\frac{\Delta t}{t_0}\right)} = \frac{1}{N^{H-1}} \tag{4.10.2}$$

square boxes of linear scale $l = N^{-1}$. Now, we have a square box of linear size $L$ on the horizonal axis. The number of square boxes is

$$N'\left(\frac{L}{t_0}\right) = \left(\frac{L}{t_0}\right)^{-1}. \tag{4.10.3}$$

The fractal volume of it will be covered by

$$N'\left(\frac{L}{t_0}\right)_F = \left(\frac{L}{t_0}\right)^{-1} \cdot \frac{1}{N'^{H-1}} = \left(\frac{L}{t_0}\right)^{H-2}. \tag{4.10.4}$$

where $F$ means fractal.

Change the scale from $\frac{L}{t_0}$ to $\frac{\lambda L}{t_0}$,

$$N'\left(\frac{\lambda L}{t_0}\right) = \lambda^{2-H} N'\left(\frac{L}{t_0}\right) = \lambda^D N'\left(\frac{L}{t_0}\right) . \tag{4.10.5}$$

Then,

$$D = 2 - H \tag{4.10.6}$$

for a trace of $V_H(t)$. Consequently, the trace of normal Brownian motion is $D = 1.5$.

The total length $L^T(H, \frac{L}{t_0})$ is given by

$$N'\left(\frac{L}{t_0}\right)_F \cdot \left(\frac{L}{t_0}\right) = \left(\frac{L}{t_0}\right)^{H-1} . \tag{4.10.4$'$}$$

Equation (4.10.4)′ reduces to

$$L^T\left(H, \frac{L}{t_0}\right) = L^T\left(D, \frac{L}{t_0}\right) = \left(\frac{L}{t_0}\right)^{1-D} \tag{4.10.4$''$}$$

as $H = 2 - D$. It is of a similar form to Eq. (4.4.2).

Voss introduced another method to estimate $D$ for a trace of fBm from the "coastline" method. One may divide the curve into $N$ segments by walking a ruler of size $l$ *along* the curve. The length *along* each segment is

$$\frac{l}{t_0} = \sqrt{\left(\frac{\Delta t}{t_0}\right)^2 + \left(\frac{\Delta V}{V_0}\right)^2} . \tag{4.10.7}$$

Using Eq. (4.10.1) and $l_0 = \sqrt{2} t_0$, $V_0 = t_0$,

$$\frac{\sqrt{2} l}{l_0} = \frac{\Delta t}{t_0} \sqrt{1 + \left(\frac{\Delta t}{t_0}\right)^{2H-2}} . \tag{4.10.8}$$

When $\frac{\Delta t}{t_0} \ll 1$, the second term in Eq. (4.10.8) dominates and

$$\left(\frac{l}{l_0}\right) \approx \left(\frac{\Delta t}{t_0}\right)^H ;$$

$$N = \left(\frac{\Delta t}{t_0}\right)^{-1} .$$

Thus,

$$N\left(\frac{l}{l_0}\right) = \frac{t_0}{\Delta t} \approx \left(\frac{l}{l_0}\right)^{-\frac{1}{H}}$$

$$D = \frac{1}{H}\,. \tag{4.10.9}$$

On the other hand, on larger scales with $\frac{\Delta t}{t_0} \gg 1$, the first term in Eq. (4.10.8) dominates and

$$D = 1\,.$$

Thus, the same $V_H(\frac{t}{t_0})$ trace can have an apparent self-similar dimension $D$ either 1, $\frac{1}{H}$. or $2 - H$ depending on the measurement technique and arbitrary choice of length scale. The same conclusion has been reached by Pietronero in taking normal Brownian motion as an example.

Moreover, the total length $L$ of the "coastline" is given by

$$L\left(H, \frac{l}{l_0}\right) = \frac{Nl}{l_0} = \left(\frac{\Delta t}{t_0}\right)^{-1}\left(\frac{l}{l_0}\right)$$

$$\approx \left(\frac{l}{l_0}\right)^{1-\frac{1}{H}}\,. \tag{4.10.10}$$

The fracture toughness may be analyzed based on Eq. (4.10.10) if the fractured surface is considered as self affine. Equation (4.10.10) reduces to

$$L\left(H, \frac{l}{l_0}\right) \approx \left(\frac{l}{l_0}\right)^{1-D} \tag{4.10.10$'$}$$

when $\frac{\Delta t}{t_0} \ll 1$ and $H = \frac{1}{D}$. Equation (4.10.10)$'$ is of similar form to Eq. (4.4.2).

In general, if we have

$$\Delta V = \Delta t^H$$

$$tg\alpha = \frac{\Delta V}{\Delta t} = \Delta t^{H-1}\,.$$

$\alpha$ depends on $\Delta t$; then, the fractal dimension which depends on $\alpha$ is variable due to the change of $\Delta t$ or that of the scale. If $\Delta t \approx \infty, \alpha \approx 0$ (because $H < 1$); then, the curve looks flat ($D = 1$). On the other hand, if $\Delta t$ is small,

$\alpha$ may be large; then, the curve looks rough ($D > 1$) (see Eq. (4.4.1)). We will discuss this problem again in Sec. 4.13.2.

## 4.11. *Multifractal Measures*

Multifractal measures are related to the study of the distribution of physical or other quantities on a geometric support. The support may be an ordinary plane, the surface of a sphere or a volume, or it could itself be a fractal. The idea that a fractal measure may be represented in terms of intertwined fractal subsets having different scaling exponents allows substantial progress for the applications of fractal geometry to physical systems.

Let us consider the triadic Cantor bar. A bar of unit length and mass is divided into two halves of equal mass $\mu_i$ and hammered to reduced length $\delta_i$, so that the density $\rho$ increases. Repeating this process, we have $N = 2^n$ small bars in the $n$th generation, each with a length, $\delta^n$ and a mass, $\mu_i = 2^{-n}$. Note that the process conserves the mass so that

$$\sum_{i=1}^{N} \mu_i = 1 \,. \tag{4.11.1}$$

The mass of a segment of length $\delta_i$ when this is very small, is given by

$$\mu_i = \delta_i^{\alpha} \,. \tag{4.11.2}$$

Here the scaling exponent $\alpha$ is given by

$$\alpha = \frac{-\ln 2}{\ln \delta} \,. \tag{4.11.3}$$

The density of each of the small pieces is

$$\rho_i = \frac{\rho_0 \mu_i}{\delta_i} = \rho_0 \delta_i^{\alpha - 1} \tag{4.11.4}$$

which diverges as $\delta_i \to 0$. The scaling exponent $\alpha$ is called the Lipschitz-Hölder exponent. This exponent controls the singularity of the density. From Eq. (4.11.4), $\rho = \rho_0$, a constant, if $\alpha = 1$; $\rho$ has a derivative if $\alpha > 1$; and is singular if $0 \leq \alpha < 1$.

The singularities of the measure of physical quantities, $M(x)$ are characterized by $\alpha$. The measure $M(x)$ is characterized by the $S_\alpha$, (see Feder, 1988):

$$M(x) = \sum_{i}^{x} \mu(i) = \sum_{i=0}^{x.2^n} \mu_i = \sum_{i=0}^{x.2^n} \delta_i^{\alpha} \,. \tag{4.11.5}$$

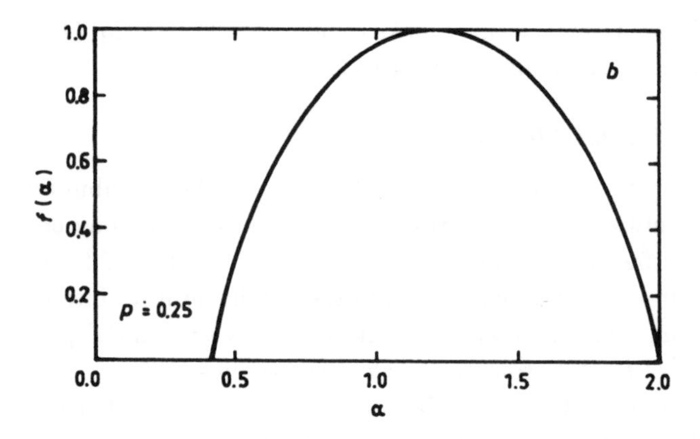


Fig. 4.5. Fractal subsets of the measure generated by a binomial multiplicative process with $p = 0.25$. The fractal dimension of subsets $S_\alpha$ having a Lipschitz-Hölder exponent $\alpha$, as a function of $\alpha$.

It is convenient to use $\alpha$ as the parameter and then the measure is given by

$$M_d(S_\alpha) \approx \delta^{-f(\alpha)}\delta^d\,. \tag{4.11.6}$$

The measure has a mass exponent $d = f(\alpha)$ for which the measure neither vanishes nor diverges as $\delta \to 0$. Then $f(\alpha)$ is the fractal dimension. The $f(\alpha)$ vs $\alpha$ curve is (Fig. 4.5) the measure of the population generated by the multiplicative process (Feder, 1988).

For more complicated processes, such as the coastline folding back and forth so that it crosses a given 'box' a number of times $n_i$, that box still contributes only 1 to the number of boxes needed to cover the set in box-counting method which counts the number of $N(\delta)$ of cubes that contain at least one point of the set $S$. We need a sequence of mass exponents $\tau(q)$ for the set depending on the moment order $q$ chosen. The measure is then characterized by a whole sequence of exponents $\tau(q)$ that controls how the moments of the probabilities $\mu_i$ scale with $\delta$. The weighted number of boxes $N(q, \delta)$ has the form

$$N(q,\delta) = \sum_i \mu_i^q \approx \delta^{\tau(q)} \tag{4.11.7}$$

and the mass exponent is given by

$$\tau(q) = -\lim_{\delta\to 0}\frac{\ln N(q,\delta)}{\ln\delta}\,. \tag{4.11.8}$$

(i) q = 0, $N(0,\delta) = N(\delta)$. It is simply the number of boxes needed to cover the set, and $\tau(0) = D$ equals the fractal dimension of the set.
(ii) q = 1. $N(1,\delta) = 1$, $\tau(1) = 0$ from Eq. (4.11.8).

Taking the multiplicative binomial process as a typical example we find that

$$N(d,\delta) = \sum_{k=0}^{n} \binom{n}{k} p^{qk}(1-p)^{q(n-k)} = [p^q + (1-p)^q]^n \,. \tag{4.11.9}$$

With the relation $\delta = 2^{-n}$, $\tau(q)$ is given by

$$\tau(q) = \frac{\ln[p^q + (1-p)^q]}{\ln 2} \,. \tag{4.11.10}$$

In this equation, $\tau(0) = 1$, which is the dimension of the support, i.e. the unit interval.

In general, the support may be a fractal; then what are the fractal dimensions in this system? What is the relation between $\tau(q)$ and $f(\alpha)$?

Now we choose $\alpha$, the Lipschitz-Hölder exponent, as the parameter. The fractal set $S$ is the union of fractal subsets $S_\alpha$ which have fractal dimensions $f(\alpha) < D$, the fractal dimension of the complete set $S$. The number $N(\alpha,\delta)$ of segments of length $\delta$ needed to cover the subset $S_\alpha$ in the range $\alpha$ to $\alpha + d\alpha$ is

$$N(\alpha,\delta)d\alpha = \rho(\alpha)d\alpha\delta^{-f(\alpha)} \tag{4.11.11}$$

where $\rho(\alpha)d\alpha$ is the number of sets from $S_\alpha$ to $S_{\alpha+d\alpha}$. The measure for the set $S$ may be written as

$$M_d(q,\delta) = \int \rho(\alpha)d\alpha\delta^{-f(\alpha)}\delta^{\alpha q}\delta^{-\tau(q)} \,. \tag{4.11.12}$$

This integral is determined by the term which has its minimum exponent, the condition of which is given by

$$\frac{d}{d\alpha}\{q\alpha - f(\alpha) - \tau(q) + d\} = 0 \,. \tag{4.11.13}$$

The solution of this equation is $\alpha = \alpha(q)$. Then,

$$\tau(q) = q\alpha(q) - f(\alpha(q)) \,. \tag{4.11.14}$$

If we know $\tau(q)$, we may determine $\alpha(q)$ and $f(\alpha(q))$ by

$$\frac{d\tau(q)}{dq} = \alpha(q) \tag{4.11.15}$$

and Eq. (4.11.14). In fact the pair of Eqs. (4.11.14) and (4.11.15) can be obtained by Legendre transformation. Usually we calculate the mass exponent $\tau(q)$, say for the binomial process as an example, Eq. (4.11.10). Then, we may obtain $\alpha$ and $f(\alpha(q))$ by using the pair of equations (Fig. 4.5).

## 4.12. *Percolation Models of Breakdown*

Diffusion process as random walks have been discussed in Sec. 4.8. Another type of randomness is frozen into the medium itself: the percolation process discussed by Broadbent and Hammersley (1957). A diffusing particle may reach any position in the medium. Percolation processes are different in that there exists a percolation threshold, below which the spreading process is confined to a finite region. The typical example is the spread of blight from tree to tree in an orchard where the trees are planted on the intersections of a square lattice (see Stauffer, 1985, Feder, 1988). If the probability for infecting a neighboring tree falls below a critical value, $p_c$ where the spacing between the trees is increased) then the blight will not spread over the orchard. For site percolation, on a square lattice, $p_c = 0.59275$.

The concept of percolation has been applied to crack problems in materials. A critical threshold for the concentration of cracks is expected. Duxbury and Li (1990) briefly reviewed progress on the percolation model that plays a central role in extreme properties, such as brittle failure, impact strength, yield strength and creep. Based on percolation theory introduced by Stauffer (1985), they defined a volume fraction of removed bonds, $f$. On application of an external stress, one or more of the bonds of the network may fail, and in this way the random network may become progressively more damaged and, for sufficiently large external load, will eventually fail. For tensile fracture of brittle porous materials, the result is (Duxbury *et al.*, 1988; 1987a; 1987b; Sieradski and Li, 1986),

$$\sigma_b(f) \approx \frac{\sigma_0 \xi^{\frac{-x}{\nu}}}{1 + k_m \left( \frac{d \ln(\frac{L}{\xi})}{\xi_1} \right)^{\alpha}} \tag{4.12.1}$$

where $d$ is the dimension, $\sigma_b(f)$ is the strength of a brittle material of size $L$ containing volume fraction $f$ of random pores, and $\sigma_0 = \sigma_b(0)$, $k_m$ is an undetermined constant and $x$ lies in the range $(d-1)\nu < x < d\nu$, $\xi$ is the percolation correlation length, while $\xi_1 = \ln(f)$. $\alpha$ lies in the range $\frac{1}{2(d-1)} < \alpha < 1$.

These workers have studied critical scaling in the mechanical case, by using perforated sheets, and have found qualitative agreement with predicted critical exponents (Sieradski and Li, 1986).

Percolation models may show a transition from brittle to ductile behaviour (Duxbury and Leath, 1987; Li and Duxbury, 1988).

Although the qualitative feature of percolation models of breakdown was given, the dependence of size distribution on fracture, it seems, cannot be ignored (Sieradski, 1989).

## 4.13. *Fractal Description of Fractures*

Mandelbrot *et al.* (1984) were the first to show that fractured surfaces are fractals in nature and that the fractal dimensions of the surfaces correlate well with the toughness of the material. Later, one of the present authors (Lung, 1986) analyzed the critical crack extension force with the fractal model and pointed out that the true areas of the fractured surfaces in materials are actually larger than that indicated by the data obtained by macroscopic measurements. The effective critical extension force in the linear elastic fracture case would thereby be larger than that calculated from a flat fractured surface. Since then, many authors have found that the fractal dimension depends on the fracture properties of materials (Lung, 1986; Pande *et al.*, 1987; Lung and Mu, 1988; Xie and Chen, 1988; Mu and Lung, 1988; Wang *et al.*, 1988; Peng and Tian, 1990; etc.) but the values of fractal dimension seem to lie in a narrow range for measurements with a resolution down to the micron scale. Herrmann (1990), in his theoretical analysis, modelled fractures on a square lattice and found that the patterns of cracks calculated can be fractal even without including noise due only to interplay of anisotropy and memory. The shapes of the cracks calculated can be compared to the ones found experimentally for stress corrosion in a qualitative sense. It seems that the scaling hypothesis on fractured surfaces has been firmly established by many studies and for many materials.

Fractal description of fractures is a question of technological importance and also an interesting theoretical problem. Further, Bauchaud *et al.* (1990), Maløy *et al.* (1992) and Thør Engøy *et al.* (1994) reported that fracture surfaces possess a statistically self-affine scaling property. They proposed that the roughness exponent could be universal (independent of the material for a range of materials). Implementing the molecular dynamics simulation technique, Abraham *et al.* (1994) found that the crack tip initially propagates straight, then the onset of the crack instability begins as a roughening of the created surfaces which eventually results in the zigzag tip motion at 30° from the mean crack direction. From the profiles, the roughness exponent was found to be 0.81. They supported the assumption on the universality of the roughness exponent.

Three questions arose:

— Do fracture surfaces possess a statistically self-similar or self-affine scaling property? Which fractal model is better to describe them?
— What is the relationship between the fracture properties and fractal structures ($D$ or $H$)? Is the roughness exponent universal?
— How can the fractal structure form?

Different points of view have appeared in some related studies (Milman *et al.*, 1993; 1994; and Hansen *et al.*, 1993). We will comment on these in the following sections.

Before we discuss the fractal description of fracture in materials, we should know some specialities of fractals in materials. Fractals in nature are approximate models. The difference between fractals in nature and rigorous ones are (Liu, 1990; Vicsek, 1989):

— The range of scaling in which self-similarity holds is bounded from above by the size of the object and from below by the size of the smallest building block.
— They usually appear random,but are self-similar in a statistical sense.

Fractals in materials are more complicated and sometimes not so typical as the sea shore, snow flowers and trees even in a statistical sense. The range of length in which self-similarity holds is small. Some authors pointed out that a constant value of fractal dimension in a certain range of scale is a necessary prerequisite for self-similarity of a structure; e.g. the number of generations should be larger than three and the range of scale should be observable for more than one or two orders of magnitude (Hornbogen, 1989).

The approximation to self-similarity in materials is poor even in a statistical sense. There may be many physical sources of self-similarities in some ranges of scale and sometimes they overlap each other which leads to more complexity.

The problem is how to find out the decisive one corresponding to the property studied.

Fractal analysis has by now been applied to macro- and microstructural elements in materials which were usually described by their integer Euclidean dimension; for example, macroscopic crack lines, or planes, vacancies, dislocations, grain boundaries, dispersive particles etc. However, a fractal structure should be geometrically scaling.

Self-similarity implies that a similar morphology appears in a wide range of magnification in the analysis. Fractured surfaces require careful metallographic analysis to determine not only a $D$ value but also to establish their self-similarity property. How wide should be the range is an open problem. For instance, the total length, $L_n(\varepsilon)$ of a Koch curve is given by

$$\frac{L_n(\eta)}{L_0} = L(\varepsilon) = \varepsilon^{1-D} = \left(\frac{\eta}{L_0}\right)^{1-D} \tag{4.13.1}$$

where $\eta$ is the yardstick length (or the smallest step of the crack), $\varepsilon$ is the *normalized yardstick length* with respect to $L_0$, the length of the initiator. $\varepsilon = \frac{\eta}{L_0}$; $D$ is the fractal dimension of the Koch curve. We know that $\varepsilon_n = r^n$, where $r$ is the reduction ratio in scale by one iteration of $n$. Now, if we restrict the range of scale of $\varepsilon$ only one or two decades then that means only two or five generations are enough to expose the self-similarity in these ranges of scale if $r = \frac{1}{3}$. However, two, even five generations seem not adequate to be considered as a fractal at least for measuring the fractal dimension with the slit-island method (Lung, 1988; Lung and Zhang, 1989). Wang *et al.* (1990, 1993) simulated the perimeter-area relationship and found that the fractal dimension measured with SIM (see Section 4.13.1) approaches the real value when the number of generations is over 18.

Usually, $n$ and $r$ take discrete values; then from Eq. (4.13.1) the length of the yardstick cannot change continuously. The measured $\ln L_n(\eta)$ vs $\ln \eta$ plot would be wavy rather than a straight line if the yardstick length is changed continuously.

Many techniques of fractal analysis of surfaces have been developed recently: the Richardson plot, slit island method (SIM) (Mandelbrot *et al.*, 1984), perimeter-diameter scaling method (Mu *et al.*, 1993), return probability histogram method (Mal$\phi$y *et al.*, 1992) variation method, (Dubuc *et al.*, 1989), direct surface area measurement (Denley, 1990; Friel and Pande, 1993), scanning tunneling microscopy (STM) method (Blumenfeld and Ball, 1993) and other physical and chemical methods such as: structure factor measurement (Kjen, 1991), Faradaic current dependence (Imre,1992); secondary electron line scanning, (Huang *et al.*, 1989), and positron annihilation, (Lung, 1993). An introduction can be found in the review article by Milman *et al.* (1994) and related papers. In the following paragraphs, we will discuss mainly the former two methods which at the time of writing are the most widely used.

### 4.13.1. *Slit-island Method (SIM) and Difficulties*

The slit-island method was proposed by Mandelbrot *et al.*, (1984) and has been widely used for fractal dimension measurements (Pande *et al.*, 1987; Mu and Lung, 1988; Wang *et al.* 1988; Xie and Chen, 1988 and etc.). In spite of various explanations on the dependencies of toughness on fractal dimension, Lung and Mu (1988) based on experiments, pointed out that the fractal dimension determined by the slit-island method is dependent on the yardstick chosen at least over a certain range of scale. The measured value would not be the real fractal dimension of fractured surfaces when the yardstick length is not *sufficiently small.*

The theoretical basis of the slit-island method is that the ratio

$$\alpha_D(\varepsilon) = \frac{[L(\varepsilon)]^{\frac{1}{D}}}{[A(\varepsilon)]^{\frac{1}{2}}} \tag{4.13.2}$$

is size independent, but it does depend on yardstick chosen, $\varepsilon$ (Feder, 1988). Islands similar in shape satisfy the following perimeter-area relation due to Mandelbrot (see Feder, 1988)

$$L(\varepsilon) = C\varepsilon^{(1-D)}\sqrt{A(\varepsilon)}^{D} \tag{4.13.2$'$}$$

where $C$ is the constant of proportionality. This holds for any given yardstick $\varepsilon$ small enough to have the measured value of $A(\varepsilon)$ approaching the limiting finite value. Then, the local fractal dimension can be determined by the slope of the straight line fitted data of various islands similar in shape to $\ln L(\varepsilon)$ figure. However, if $\varepsilon$ is not small enough (not only for accurate measurement of the smallest island), $C$ would not be a constant, the measured value of $D_m$ by Eq. (4.13.2′) may depend on $\varepsilon$ as Lung and Mu (1988), Lung and Zhang (1989), Wang *et al.* (1990, 1993) and Shi *et al.* (1996) have analyzed.

Usually, the selection of islands similar in shape is guided by eye and in practice by checking the linearity in $\ln P$ vs $\ln A$ figure. In principle, one may keep the relative error of the ratios of the axes in different directions of various islands within a chosen small value to keep selected island similar in shape more rigorously.

Actually, in practical measurements, it is more convenient to keep $\eta$ constant as has been done in many measurements. We did not keep $\varepsilon$, which is equal to $\frac{\eta}{L_0}$, constant, due to different sizes of the Koch islands, $(L_0)$.

In this case, actually, for the Koch perimeter, we have (Lung, 1986; 1992)

$$L_n(\eta) = L_0^D \eta^{1-D} . \tag{4.13.3}$$

The ratio of the Koch island is

$$\alpha_n(\eta) = \frac{L_n(\eta)^{\frac{1}{D_m}}}{A_n(\eta)^{\frac{1}{2}}} = L_0 \eta^{\frac{(1-D)}{D}} A_n(\eta)^{\frac{-1}{2}} . \tag{4.13.4}$$

In general, from Eq. (4.13.4), we may see that $\alpha_n(\eta)$ is dependent on the size of the Koch island $(L_0)$. The relationship is complicated between logarithmic values of $A_n's$ and $L_n's$ of different size similar Koch islands with a constant yardstick length $\eta$.

Furthermore, $\varepsilon_i = \frac{\eta}{L_{0i}}$. For the larger island, $\varepsilon_i$ is smaller. The smaller normalized yardstick sees more generations. In addition, $\alpha_n(\varepsilon)$ is yardstick dependent, $\alpha_i(\varepsilon_i) \neq \alpha_j(\varepsilon_j)$. The measured values of $D_m$ must be yardstick dependent. As was expected, the experimental data verified the above conclusion (Lung and Mu, 1988) (Fig. 4.6).

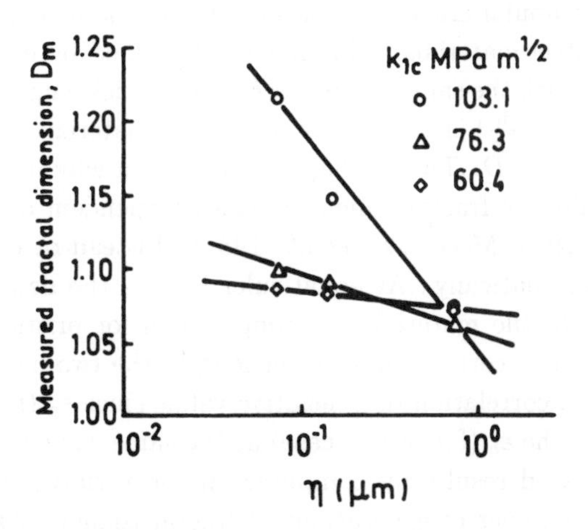


Fig. 4.6 The dependence of measured $D_m$ on yardstick $\eta$ (Lung and Mu 1988).

We notice that

$$\alpha_0 = \frac{L_0}{A_0^{\frac{1}{2}}}, \text{ for } D = 1, \text{ and } \eta = L_0 \,.$$

From Eq. (4.13.4)

$$\alpha_n(\eta) = \alpha_0 \left[ \frac{A_0}{A_n(\eta)} \right]^{\frac{1}{2}} \eta^{\frac{(1-D)}{D}} \,. \tag{4.13.5}$$

$\frac{A_0}{A_n(\eta)}$ would approach a constant value as $n > n_c$ ($n_c \approx 20$ for a quadric Koch island, (Lung and Mu, 1988; Lung, 1992); $n_c \approx 100$ for a triadic Koch island (Lung and Zhang, 1989) and then $\alpha_n(\eta)$ is size independent approximately. In the limiting case, one can obtain

$$D_m(\eta) \approx D \,. \tag{4.13.6}$$

Fractals in materials with only several self-similar generations ($n < n_c$) are not appropriate for measuring $D$ in general with the slit-island method. The approximation would be poor.

In Fig. 4.6, the $D$ vs $\eta$ relation is an empirical one; it is not necessarily a straight line, if one considers the limiting value due to Eq. (4.13.6). The reason why the line for high toughness material has higher slope is not clear. However, it is not unreasonable to assume $L_0$ to be the distance of one step of crack propagation from initiation to temporary arrest and the wavy path as due to crack instability, intergranular cracking, transgranular cracking, etc. Then, $L_0$ would be longer for brittle materials. This has been verified by experiments (Lung and Mu, 1996). With the same value of $\eta$ (the yardstick length) the normalized yardstick length $\varepsilon(= \frac{\eta}{L_0})$ is smaller for the brittle materials. Then, it is nearer to the limiting value, $D$. This perhaps is the reason why in brittle materials, the measured values of fractal dimension is less dependent on the ruler length (Lung and Mu, 1988; Mecholsky *et al.*, 1989). Instead of Fig. 4.6, one may draw Fig. 4.7 schematically. At equal ruler length, the dependence of $D$ on ruler length for ductile materials is stronger than for brittle materials. One may expect that at a certain larger ruler length, the two curves can intersect to change positive correlation to a negative value, since $\varepsilon_p$ (for ductile case) is much larger than the $\varepsilon_b$ (for brittle case) at the same value of $\eta$. Figure 4.7(a) shows the calculated results by computer on the variation of $D_m$ and $\alpha$ as functions of the number of generations of fractal islands (Wang *et al.*, 1990, 1991, 1993). These results are consistent with previous model calculations

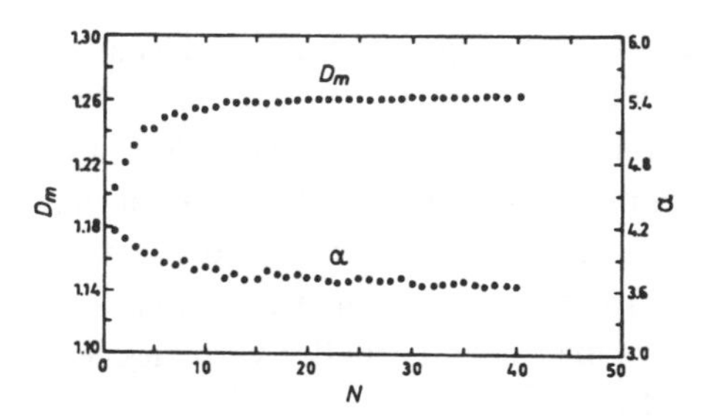


Fig. 4.7. (a) The change in fractal dimensionality $D_m$ and measurement constant $\alpha$ in measurement of many islands with $N$ different generations.

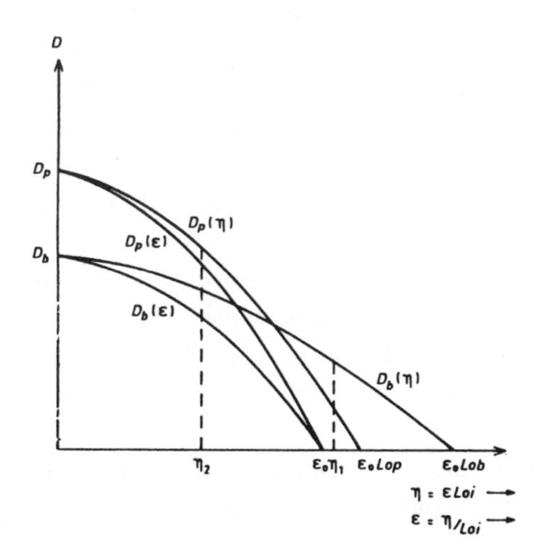


Fig. 4.7 (b) Schematic figure for the transition from a negative correlation to a positive one as the yardstick used becomes smaller.

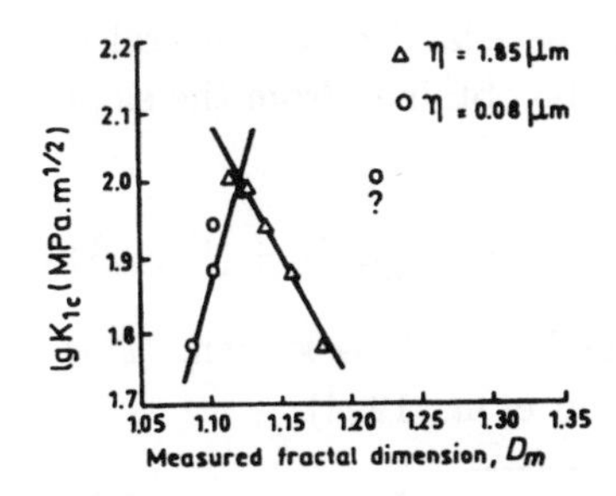


Fig. 4.7 (c) Experimental evidence for the transition from a negative correlation to a positive one as the yardstick used becomes smaller.

(Fig. 4 of Lung and Mu, 1988; Fig. 2 of Lung and Zhang, 1989). Figure 4.7(b) is a schematic figure which shows the intersection of two $D_m(\eta)$ curves. The two $D_m(\varepsilon)$ curves show the general property in Fig. 4.7(a). The reason why the two $D_m(\eta)$ curves intersect is that brittle materials have larger $L_0$ and that workers usually keep $\eta$ constant instead of $\varepsilon$ constant in SIM measurements. Therefore, at larger yardstick length, $\eta_1$, the correlation of $D_m$ with fracture toughness is a negative one; but at smaller yardstick length, $\eta_2$ it changes to be positive (see Fig. 4.7(c)).*

The condition that Eq. (4.13.6) holds is $n > n_c$; however, from an experimental point of view, the problem is how can one know that the normalized yardstick length is small enough to have satisfied the approximation condition within reasonable experimental error. This is difficult to judge.

Morever, in many cases, the scaling range in materials, $n_0$ is limited. If $n_0 < n_c$, it is not useful to reduce the yardstick length to the condition that the measured $\varepsilon(n) < \varepsilon_c(n_c) < \varepsilon_0(n_0)$; (or $n > n_c > n_0$) because $\varepsilon$ is outside the lower limit of the scaling range of the fractal (Lung and Zhang, 1989; Lung, 1992).

SIM is used in many cases due to its convenience (measurement of different sizes is easier than to change the ruler length over an order of magnitude) and the advantage of less error (Milman *et al.*, 1994). We recommend that results obtained by this method should be checked with others and we consider it to be a subsidiary method.

It seems that the best way to measure the fractal dimension of fractured surfaces may be through the relation (Lung, 1986),

$$L_n(\eta) = L_0^D \eta^{1-D} . \tag{4.13.7}$$

One may measure the total length of crack propagation with different lengths of yardstick on one island or on several islands with the same size and shape. Then, $D$ can be obtained from the slope of the linear relationship between $\ln L_n(\eta)$ and $\ln \eta$,

$$\ln L_n(\eta) = D \ln L_0 + (1 - D) \ln \eta . \tag{4.13.8}$$

$L_0$ can also be determined after $D$ is known. This relation seems not to be recommended (see Milman *et al.*, 1994).

---

*Shi *et al* (1996) analysed the effects of the area change of the island on the $D_m$ value. They showed that the area change influences the $D_m$ value greatly if the yardstick is not small enough.

Another way to measure $D$ may be with the same yardstick length on different sizes of islands ($L_{0i}$) of similar shapes. Then, Eq. (4.13.8) changes to Eq. (4.13.9),

$$\ln L_n(L_{0i}) = D \ln L_{0i} + \text{const.}. \tag{4.13.9}$$

Equation (4.13.9) is reasonable due to its scaling property with the sizes of islands. This experiment has been done (Mu *et al.*, 1993) and yields better results than the slit-island method. The fractal dimension measured is almost independent of the length of the yardstick in a certain range of the scale. It is called the perimeter-diameter scaling (PDS) method (see also Milman, 1994).

### 4.13.2. *The Roughness Exponent*

Bauchaud *et al.* (1990) and Mal$\phi$y *et al.* (1992) have reported elegant experimental studies of the roughness exponent $\zeta$ for fractured surfaces of different brittle materials. This exponent describes the scaling of roughness $w$, defined as the width of the profile, with the length $L$ of a one-dimensional cut through the surface, $w \propto L^{\zeta}$. They found that the value of $\zeta$ is universal for all brittle materials, and conjectured that this universal value may also apply for ductile fracture. In spite of controversial points of view on universality of roughness exponent (Milman *et al.*, 1993; Hansen *et al.*, 1993) we shall first discuss this problem from the standpoint of methods of measurements in this section.

Since most real surfaces scale differently in the plane of fracture and in the vertical direction, they are self-affine rather than self-similar. What will happen if one measures the self-affine surfaces with $D$, or describes the self-similar surfaces with $H$?

If the surface is self-affine, we may determine $H$ from double logarithmic plots of $\Delta V(t)$ vs $\Delta t$, where $V$ is the vertical height and $t$ is the horizonal axis. Then, $H$ might be a constant which is independent of the yardstick. However, if we measure the fractal dimension of the surface artificially, the $D$ value might be yardstick dependent. From Eqs. (4.10.7) and (4.10.8):

$$N\left(\frac{l}{l_0}\right) = \left(\frac{\Delta t}{t_0}\right)^{-1} = \frac{1}{\sqrt{2}}\left(\frac{l}{l_0}\right)^{-1}\left[1 + \left(\frac{\Delta t}{t_0}\right)^{2H-2}\right]^{\frac{1}{2}} \tag{4.13.10}$$

$$N\left(\frac{l}{l_0}\right) = \left(\frac{l}{l_0}\right)^{-D}. \tag{4.13.11}$$

Therefore,

$$D = 1 - \frac{\ln\left[\frac{1}{2}\left(1 + \left(\frac{\Delta t}{t_0}\right)^{2H-2}\right)\right]}{2\ln\left(\frac{l}{l_0}\right)}, \tag{4.13.12}$$

which becomes, with the substitution

$$\xi = \frac{1}{2}\left[1 + \left(\frac{\Delta t}{t_0}\right)^{2H-2}\right],$$

$$D\left(H, \frac{\Delta t}{t_0}\right) = 1 - \left[1 + \frac{2\ln(\frac{\Delta t}{t_0})}{\ln\xi}\right]^{-1}. \tag{4.13.13}$$

(i) When $\frac{\Delta t}{t_0} \ll 1$, $N = (\frac{\Delta t}{t_0})^{-1} \approx (\frac{l}{l_0})^{\frac{-1}{H}}$; and then $D = \frac{1}{H}$.
(ii) When $\frac{\Delta t}{t_0} = 1$, $D$ approaches the limiting value $\frac{2}{(1+H)}$.

Figure 4.8 shows the relationship of $D(H, \frac{\Delta t}{t_0})$ with $\frac{l}{l_0}$. In the double logarithm plots, the dependence of $D$ with $\frac{\Delta t}{t_0}$ becomes weaker and weaker as $H$ rises from 0.5 to 1. One cannot judge whether $D$ is dependent on the yardstick when $H$ value rises up to a certain value near unity (say $0.8 < H < 1$) within the accuracy of measurements. In this case, we cannot distinguish which one, self-similarity or self-affinity, is better to describe the fractal surface.

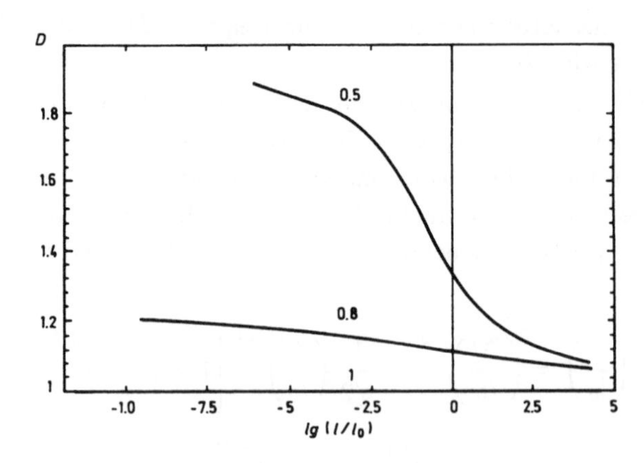


Fig. 4.8. $D$ vs $lg(l/l_0)$ curve ($H = 0.5, 0.8, 1$).

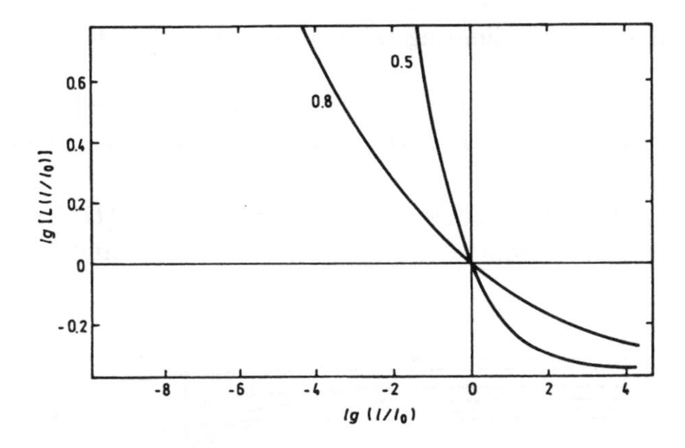


Fig. 4.9. $lgL$ vs $lg(l/l_0)$ curve ($H = 0.5, 0.8, 1$).

Figure 4.9 shows the relationship of $L(\frac{l}{l_0})$ with $\frac{l}{l_0}$. In this figure, $L(\frac{l}{l_0})$ is calculated as $(\frac{l}{l_0})^{1-D}$ where $D(l)$ is calculated by Eq. (4.13.13). Because $D$ is a function of $H$ and $l$, $L(H,l)$ and $l$ do not show linearity in double logarithmic plots. From Fig. 4.9, the deviation of linear relationship is larger when $H = 0.5$ and smaller when $H = 0.8$. $L(\frac{l}{l_0})$ is independent of $\frac{l}{l_0}$ when $H = 1$. Similarly, as $H$ approaches 1 (say $0.8 < H < 1$), we cannot judge whether it is a curve or a straight line within the range of experimental error.

Moreover, the values of $L(l)$ are points on the $L$ vs $l$ double logarithmic plots with various values of $H$. The slope of the curve does not have the meaning of fractal dimension. The real fractal dimension $(l - D)$ is the slope of the straight line connected to the point $(L/Lo, \frac{l}{l_0})$ and the zero point (0, 0).

On the other hand, if the fracture surface is of self-similar structure, one may determine $D$ from the double logarithmic plots of the $L(\frac{l}{l_0})$ and $\frac{l}{l_0}$ relationship. The value of $D$ might be a constant and is independent of $\frac{l}{l_0}$. If one measures the roughness exponent of the surface, the value of $H$ might be yardstick dependent.

Now, we have

$$H(l,D) = 1 + \frac{\ln\left[2\left(\frac{l}{l_0}\right)^{2-2D} - 1\right]}{2\ln\left(\frac{\Delta t}{t_0}\right)}. \tag{4.13.14}$$

(i) When $\frac{l}{l_0} \ll 1$

$$H(l,D) = \frac{1}{D} + \frac{\ln 2}{2\ln(\frac{\Delta t}{t_0})} \approx \frac{1}{D}. \tag{4.13.15}$$

(ii) When $\frac{l}{l_0} = 1$, the limiting value of $H(l, D) = \frac{2}{D} - 1$.

Figure 4.10 shows the double logarithmic plots of $H(D, \frac{l}{l_0})$ and $\frac{l}{l_0}$. Similar to the above, the dependence of $H$ on $\frac{l}{l_0}$ becomes weaker and weaker as $D$ approaches unity.

Figure 4.11 shows the double logarithmic plots of $\Delta V(D, \frac{l}{l_0})$ and $\frac{\Delta t}{t_0}$. As above, the deviation from a straight line is smaller and smaller as $D$ approaches unity.

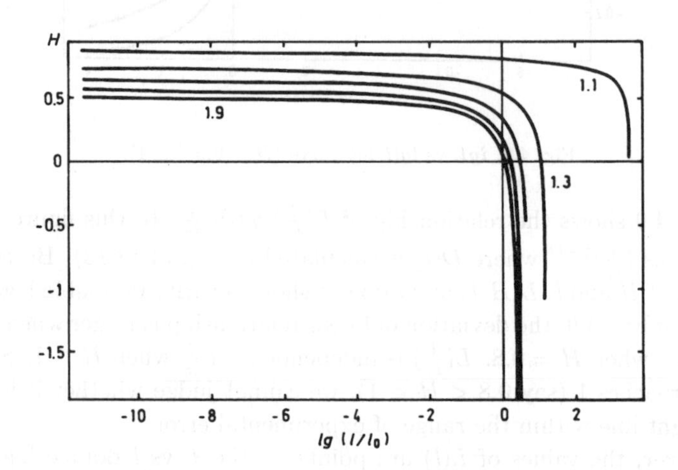


Fig. 4.10. $H$ vs $lg(l/l_0)$ curve ($D = 1.1, 1.3, 1.9$).

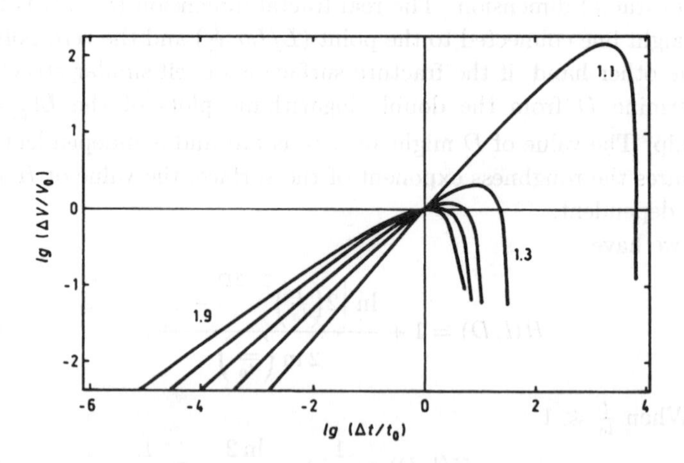


Fig. 4.11. $lg(\Delta V/V_0)$ vs $lg(\Delta t/t_0)$ curve ($D = 1.1, 1.3, 1.9$).

From the above analysis, one may draw the following conclusions:

If one describes a surface of a self-affine structure with fractal dimension, the apparent fractal dimension might be yardstick dependent. However, the dependence is very weak as the $H$ value is near unity. On the other hand, if one describes a surface of self-similar structure, with roughness exponent, the apparent $H$ value might be yardstick dependent. However, the dependence cannot be correctly appraised when the $D$ value is near unity.

In principle, comparing the linearity of $H$ vs $l$ and $D$ vs $l$ relation in double logarithmic plots, one may make an appraisal as to which structure, either self-affine or self-similar, is the correct one. However, if the surface appears to flatten, this experimental method is not sensitive; one should then adopt another experimental method to make the appraisal.

In addition, comparing Figs. 4.8 and 4.10, we can see that the dependence of $H(D, \frac{l}{l_0})$ on $\frac{l}{l_0}$ is weaker than that of $D(H, \frac{l}{l_0})$ on $\frac{l}{l_0}$. Then, the measurement of the roughness exponent is a less sensitive way to judge the deviation from self-affinity than the fractal dimension is to the deviation from self-similarity. The range of $H$ values from 0.5 to 1 is half the range of $D$ values from 1 to 2. Using the measured values of the $H$ parameter to characterize the roughness of materials, the differences among them are easy to be ignored or the universal properties are easy to be emphasized. (Deng *et al.*, 1999)

The above discussion on $D$ and $H$ is important. Considering the complicated mechanisms of fracture in materials, it is worthwhile to check if the fractured surface is self-similar or self-affine experimentally. Usually we discuss the case of mode I fracture (see Thomson, 1986), but how are mode II, mode III and Complex Mode? If we consider the fractured surface on an atomic level, say the dislocation mechanism of micro-crack nucleation; the percolation model for brittle fracture, or the fragmentation, friction, wear, corrosion and other processes, what will happen?

In the following sections, we consider the crack as a Koch line (self-similar) as a tentative model and for reasons of simplicity. The total length of it is based on Eq. (4.4.2). If one considers the crack surface as self-affine, one should use Eq. (4.10.10).

## 4.14. *Multirange Fractals in Materials*

Fractals in materials are more complicated. There are many kinds of structure able to form fractals; e.g. intergranular crack lines, transgranular crack lines, dislocation lines, vacancy clusters, etc. They are geometrically scaling in different ranges of scale (Hornbogen, 1989; Lung, 1992; 1993; 1994).

In materials, the fractal property of fractured surfaces is the total contribution of many elementary processes. Every elementary microstructure contributes its fractal or non-fractal property in its own characteristic range of scale. In principle, we should pick out the decisive one among them. If one measures the fractal dimension outside the scale range of the decisive fractal, one may find that the fractal dimension measured is insensitive to the property (Lung, 1992).

If more than two fractals exist in a material in different ranges of scale, we call them *multirange fractals*. Multirange fractals are not necessarily *multiscaling fractals* even when they have overlapping ranges of scale. For instance, if transgranular cracking and intergranular cracking are associated with fracture toughness values differing by a large amount, they preferentially occur at different levels of stress and do not mix with each other in every generation. Each one has its *own self-similar system* instead of forming *one self-similar system* with the same fraction of the population in *every generation*. Usually, they superpose in the material (Fig. 4.12). However, if $G_{Ic}^T$ (transgranular) is near to $G_{Ic}^I$ (intergranular), the probabilities of these two kinds of fracture are nearly the same, or the crack branches in every step of propagation, these two processes may mix and form one *multiscaling self-similar fractal system* (Lung, 1993; 1994).

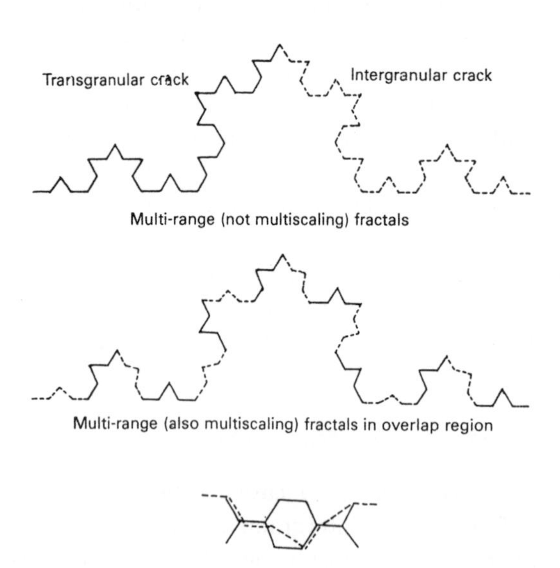


Fig. 4.12. Multirange fractals and multiscaling fractals.

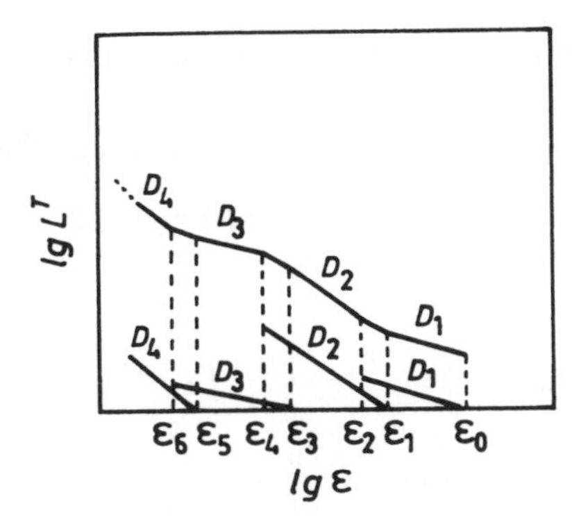


Fig. 4.13. $N$ fractals in various ranges of scale.

$N$ fractals with overlapping regions between two adjacent fractal structures with different fractal dimensions in the ranges of scale are complicated. Crack lines are considered as Koch lines. Suppose we have $N$ fractals in a wide range of measurement with overlapping regions between two adjacent fractals. The total length of the crack line can be expressed as (Fig. 4.13).

$$L^T(\varepsilon_{2N-1}) = \sum_{j=1}^{N-1} P_j \prod_{i=1}^{j}(1 - P_{i-1}) \left(\frac{\varepsilon_{-1}}{\varepsilon_0}\right)^{1-D_1} \left(\frac{\varepsilon_{2i-1}}{\varepsilon_{2i-3}}\right)^{1-D_i} \left(\frac{\varepsilon_{2j}}{\varepsilon_{2j-1}}\right)^{1-D_j}$$

$$+ \prod_{i=1}^{N}(1 - P_{i-1}) \left(\frac{\varepsilon_{-1}}{\varepsilon_0}\right)^{1-D_1} \left(\frac{\varepsilon_{2i-1}}{\varepsilon_{2i-3}}\right)^{1-D_i} \qquad (4.14.1)$$

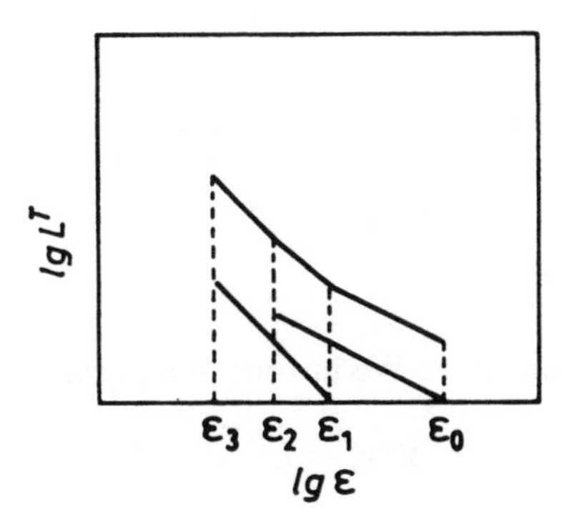


Fig. 4.14. Two fractals with overlapping region of scale.

where $D_i$ is the fractal dimension, $\varepsilon$ is the yardstick ($\varepsilon_0 = 1$) and $P_i$ is the fraction of population of segments with $D_i$ in the overlapping regions. It is easy to see that Eq. (4.14.1) holds when $N = 1$ and 2 (Fig. 14.14). For $N = 1$,

$$L^T(\varepsilon_1) = \varepsilon_1^{1-D_1} \quad (\text{if } \varepsilon_0 = 1)\,. \tag{4.14.2}$$

For $N = 2$,

$$L^T(\varepsilon_3) = P_1\varepsilon_2^{1-D_1} + (1 - P_1)\alpha^{D_1-D_2}\varepsilon_3^{1-D_2} \tag{4.14.3}$$

$$\left(\text{if } \alpha = \frac{\varepsilon_0}{\varepsilon_1} = \varepsilon_1^{-1}\right).$$

The critical crack extension force is enhanced by the $N$ fractals according to

$$\frac{G_{Ic}(D_1 \ldots D_N, \varepsilon_{2N-1})}{G_{Ic}(1,1)} = L^T(\varepsilon_{2N-1}) \tag{4.14.4}$$

where $G_{Ic}(1,1)$ is the critical crack extension force when no fractal structure is formed in the materials. In principle, for $N$ fractals in different ranges of scale, we should use Eq. (4.14.1) instead of the only one fractal dimension in a narrow range (say in microns). However, under certain conditions, if one scale range dominates, the fractal dimension in this range of scale would represent the fracture property mainly. Otherwise, it is not sensitive to the property. We will discuss this problem in a later section.

For measurement of $D_N$, we choose $\varepsilon$ as a variable in the range $\varepsilon_{2N-1} < \varepsilon < \varepsilon_{2N-2}$. Then, (compare Eq. (4.14.1))

$$L^T(\varepsilon) = \sum_{j=1}^{N-1} P_j \prod_{i=1}^{j}(1 - P_{i-1}) \left(\frac{\varepsilon_{-1}}{\varepsilon_0}\right)^{1-D_1} \left(\frac{\varepsilon_{2i-1}}{\varepsilon_{2i-3}}\right)^{1-D_i} \left(\frac{\varepsilon_{2j}}{\varepsilon_{2j-1}}\right)^{1-D_j}$$

$$+ \prod_{i=1}^{N-1}(1 - P_{i-1}) \left(\frac{\varepsilon_{-1}}{\varepsilon_0}\right)^{1-D_1} \left(\frac{\varepsilon_{2i-1}}{\varepsilon_{2i-3}}\right)^{1-D_i} \left(\frac{\varepsilon}{\varepsilon_{2N-3}}\right)^{1-D_N}$$

$$= A + B\left(\frac{\varepsilon}{\varepsilon_{2N-3}}\right)^{1-D_N} \tag{4.14.5}$$

where A and B are constants. We introduce a parameter $Y(\varepsilon)$; such that

$$Y(\varepsilon) \equiv L^T(\varepsilon) - A$$

$$\ln Y(\varepsilon) = \text{const.} + (1 - D_N)\ln\left(\frac{\varepsilon}{\varepsilon_{2N-3}}\right). \tag{4.14.6}$$

From the slope of the straight line in a double-logarithmic plot, $D_N$ can be determined. Sometimes experimentalists plot $\ln L_(\varepsilon)$ vs $\ln \varepsilon$ straight line and determine $D_N$ from the slope of the line in this double-logarithmic plot. However, this procedure is incorrect, because

$$\ln L^T(\varepsilon) \neq \text{const.} + (1 - D_N) \ln \varepsilon\,. \tag{4.14.7}$$

In the overlap region,

$$\begin{aligned}
L^T(\varepsilon) &= \sum_{j=1}^{N-2} P_j \prod_{i=1}^{j} (1 - P_{i-1}) \left(\frac{\varepsilon_{-1}}{\varepsilon_0}\right)^{1-D_1} \left(\frac{\varepsilon_{2i-1}}{\varepsilon_{2i-3}}\right)^{1-D_i} \left(\frac{\varepsilon_{2j}}{\varepsilon_{2j-1}}\right)^{1-D_j} \\
&\quad + P_{N-1} \prod_{i=1}^{N-2} (1 - P_{i-1}) \left(\frac{\varepsilon_{-1}}{\varepsilon_0}\right)^{1-D_1} \left(\frac{\varepsilon_{2i-1}}{\varepsilon_{2i-3}}\right)^{1-D_i} \left(\frac{\varepsilon}{\varepsilon_{2N-5}}\right)^{1-D_{N-1}} \\
&\quad + \prod_{i=1}^{N-1} (1 - P_{i-1}) \left(\frac{\varepsilon_{-1}}{\varepsilon_0}\right)^{1-D_1} \left(\frac{\varepsilon_{2i-1}}{\varepsilon_{2i-3}}\right)^{1-D_i} \left(\frac{\varepsilon}{\varepsilon_{2N-3}}\right)^{1-D_N} \\
&= A' + B' \left(\frac{\varepsilon}{\varepsilon_{2N-5}}\right)^{1-D_{N-1}} + C' \left(\frac{\varepsilon}{\varepsilon_{2N-3}}\right)^{1-D_N}\,.
\end{aligned} \tag{4.14.8}$$

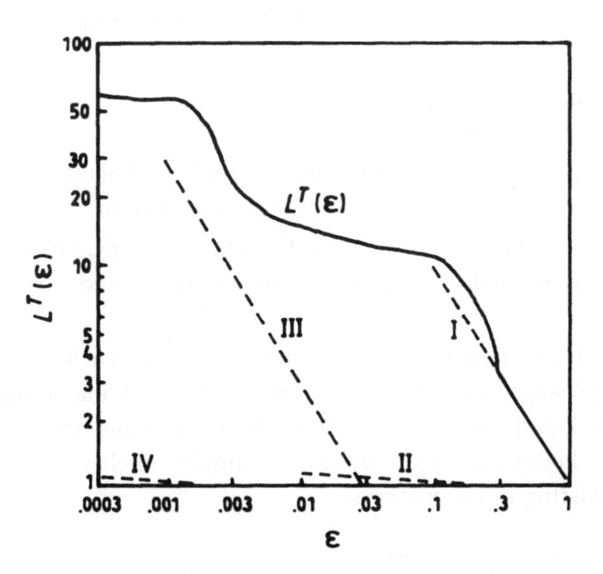


Fig. 4.15. An example of multirange fractals with $N = 4$.

This relation is not simple. However, it should be possible to solve it numerically on a computer.

Four fractals exist in a material with overlapping regions between two adjacent fractal structures as shown in Fig. 4.15.

Assuming that, $\varepsilon_0 = 1$, $\varepsilon_1 = 3\times10^{-1}$, $\varepsilon_2 = 10^{-1}$, $\varepsilon_3 = 3\times10^{-2}$, $\varepsilon_4 = 10^{-2}$, $\varepsilon_5 = 3\times10^{-3}$, $\varepsilon_6 = 10^{-3}$, $\varepsilon_7 = 3\times10^{-4}$, and $D_1 = 2$, $D_2 = 1.1$, $D_3 = 2$, $D_4 = 1.1$, and $P_1 = P_2 = P_3 = \frac{1}{2}$, the double logarithmic plots are shown in Fig. 4.15. The superposed curve is non-linear. From this simple non-linear curve, we cannot conclude that no fractal exists in this material. We should do some experimental inspections* to make sure that it is not the case of more than two fractals existing in different ranges of scale with overlapping regions.

In the case of multirange fractals composed of two fractals (Fig. 4.14), Eqs. (4.14.5), (4.14.6) and (4.14.8) can be simplified as

$$L^T(\varepsilon) = \varepsilon^{1-D_1}, (\varepsilon_1 < \varepsilon < \varepsilon_0) \tag{4.14.9a}$$

$$L^T(\varepsilon) = P_1\varepsilon^{1-D_1} + (1-P_1)\alpha^{D_1-D_2}\varepsilon^{1-D_2}, (\varepsilon_2 < \varepsilon < \varepsilon_1) \tag{4.14.9b}$$

$$L^T(\varepsilon) = P_1\varepsilon_2^{1-D_1} + (1-P_1)\alpha^{D_1-D_2}\varepsilon^{1-D_2}, (\varepsilon_3 < \varepsilon < \varepsilon_2) \tag{4.14.9c}$$

where we assume $\alpha = \frac{\varepsilon_0}{\varepsilon_1} = \frac{\varepsilon_2}{\varepsilon_3}$. The width of the overlap range, $\Delta\varepsilon$, can be expressed as

$$\Delta\varepsilon = \varepsilon_1 - \varepsilon_2 = \frac{\varepsilon_0}{\alpha} - \alpha\varepsilon_3\,. \tag{4.14.10}$$

When $\alpha = 1$, $\Delta\varepsilon = \varepsilon_0 - \varepsilon_3$, which is the case of two fractals overlapping each other in the whole range. When $\alpha = (\frac{\varepsilon_0}{\varepsilon_3})^{\frac{1}{2}}$, which is the case of two fractals without overlap, when $1 < \alpha < (\frac{\varepsilon_0}{\varepsilon_3})^{\frac{1}{2}}$, $0 < \Delta\varepsilon < \varepsilon_0 - \varepsilon_3$, i.e. the two fractals overlap each other partly. Figure 4.16 shows the calculated curves for $\alpha = 1, 3$, and 5.

For Fig. 4.16, we may see that if two fractals overlap each other in the entire range of length scale, an approximate constant value of 'one' fractal dimension in the double logarithmic plot has been obtained. Thus, from a straight line in the double logarithmic plot, we cannot simply conclude that there is only one fractal existing in the material.

---

*For example, scanning electron microscopy could possibly identify whether more than one fractal exists.

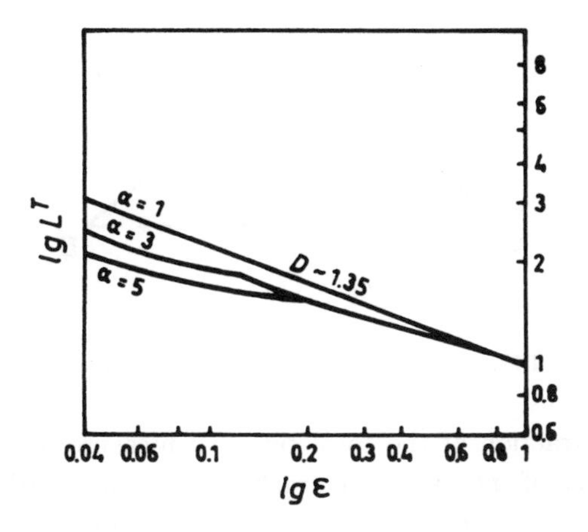


Fig. 4.16. Calculated $\log L^T(\varepsilon) - \log \varepsilon$ curves for $\alpha = 1, 3$ and 5 ($D_1 = 1.26$, $D_2 = 1.5$, $P_1 = 68\%$).

One of the present authors and his collaborators have done various experiments on fractals. One of our experiments (Long *et al.*, 1992) demonstrated that in spite of the fractured surfaces being all of mixed inter-transgranular character, an approximately constant value of the fractal dimension in double-logarithmic plots was obtained in the range $2 < \eta_i < 50$ $\mu$m, where "$\eta_i$" is the yardstick length. On the other hand, two fractal dimensions were observed in one range of yardstick lengths (Mu *et al.*, 1993). We think that the former is the case of $\alpha = 1$ in Fig. 4.16. The calculated curve is approximately linear, since $D_1$ (intergranular crack) and $D_2$ (transgranular crack) do not differ too much. The fractal average value, $\bar{D}(P, D_1, D_2)$, is between $D_1$ and $D_2$ but is not the simple average of $D_1$ and $D_2$. The latter seems to be the case of $\alpha = 5$ or 3 in which the two fractals never overlap or partly overlap.

As above, multirange fractals even having overlapping regions are not necessarily multiscaling fractals. However, if $G^T_{Ic}$ is near to $G^I_{Ic}$ i.e. the probablity of intergranular cracking is nearly the same as that of transgranular cracking, or the crack branches in every step of propagation, these two processes may mix and form one multiscaling self-similar fractal system.

In this case, the physical quantity which we want to measure is the fracture toughness of the complex system. In linear elastic fracture mechanics analysis, this is equal to twice the specific surface energy multiplied by the crack surface area or by the length of the crack line in the 2D fracture

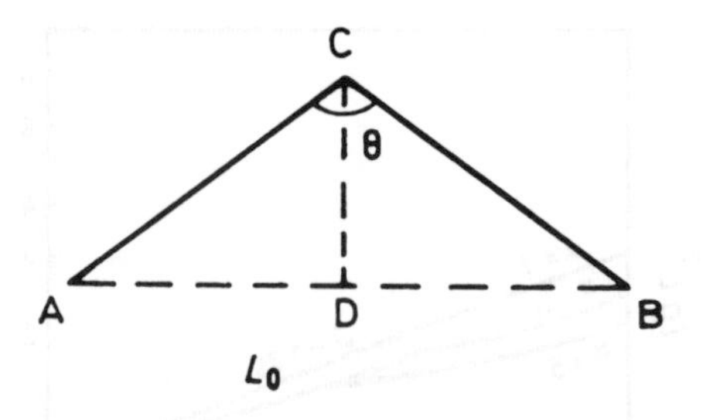


Fig. 4.17. Koch curve model for the crack line, ACB.

mechanics. The fraction of the total physical measure includes two factors, i.e. the probability of intergranular cracking and the length of the segment. $\frac{L_i}{L_0} = \frac{1}{2}[\csc(\frac{\theta_i}{2})]$ ($i = 1$, for intergranular cracking and $i = 2$ for transgranular cracking) (Fig. 4.17). We assume that

$$P_0 = \frac{G_{Ic}^T}{(G_{Ic}^I + G_{Ic}^T)} \tag{4.14.11}$$

$$P_1 = P_0 \csc\left(\frac{\theta_1}{2}\right); P_2 = (1 - P_0)\csc\left(\frac{\theta_2}{2}\right)$$

$$P = \frac{P_1}{P_1 + P_2}\,. \tag{4.14.12}$$

$P$ is the fraction of population for intergranular fracture contributing to the fracture toughness of the material.

According to Feder (1988),

$$\tau(q) = -\lim_{\delta \to 0} \frac{\ln N(q, \delta)}{\ln \delta} \tag{4.14.13}$$

$$N(q, \delta) = \sum_{k=0}^{n} \binom{n}{k} P^{qk}(1 - P)^{q(n-k)}$$

$$= [P^q + (1 - P)^q]^n \tag{4.14.14}$$

$$\alpha(q) = \frac{d\tau(q)}{dq} = \frac{-1}{\ln 2}\left[\frac{P^q \ln P + (1 - P)^q \ln(1 - P)}{P^q + (1 - P)^q}\right] \tag{4.14.15}$$

$$f(\alpha(q)) = q\alpha(q) + \tau(q)\,. \tag{4.14.16}$$

Empirically, from Long *et al.*'s experiment (1992), assuming $P = 2/3$

$$\tau(q) = \left(\frac{1}{\ln 2}\right)[\ln(2^q + 1) - q\ln 3] \tag{4.14.17}$$

$$\alpha(q) = \frac{\ln 3}{\ln 2} - \frac{2^q}{(2^q + 1)} \tag{4.14.18}$$

$$f(\alpha(q)) = \frac{\ln(2q + 1)}{\ln 2} - \frac{q2^q}{2^q + 1} \tag{4.14.19}$$

$$D(q) = \frac{1}{(1 - q)\ln 2}[\ln(2^q + 1) - q\ln 3]\,. \tag{4.14.20}$$

The function $D(q)$ is the spectrum of fractal dimensions for the fractal measure of $\frac{G^F_{Ic}}{G^0_{Ic}}$ on a geometrical set $D(0) = 1$ (a straight line) (Fig. 4.18).

We summarize the above discussion as follows:

(i) Unlike the original theoretical model of fractals with an infinite number of generations, fractals in nature are bounded from above by the size of the initiator and from below by the size of the smallest building block. There are a number of multirange fractals existing in materials. In order to study the relationship between fractal structure and mechanical properties we should first make sure which fractal structure dominates, (e.g. intergranular crack lines, transgranular crack lines or

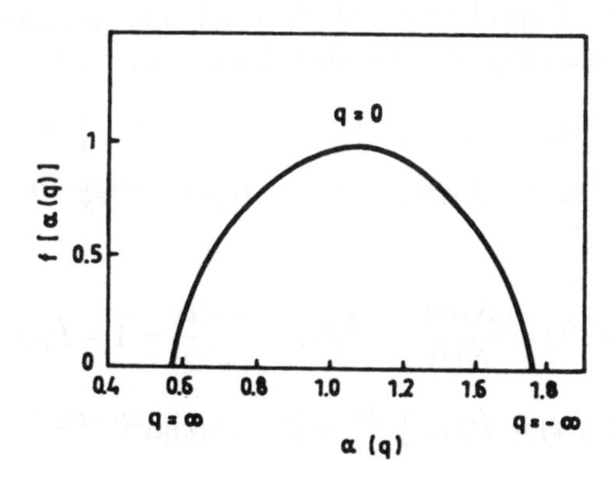


Fig. 4.18. $f(\alpha(q))-\alpha(q)$ curve for the multifractals of intergranular and transgranular cracks. (Lung, 1993; 1995)

dislocation lines, etc.) in relation to the particular mechanical property.

(ii) Multirange fractals with overlapping regions in the range of yardstick lengths are not necessarily multiscaling fractals. However, if the probabilities of different physical mechanisms are nearly the same, processes may mix and form one multiscaling self-similar fractal system.

(iii) From a straight line in a double-logarithmic plot, we cannot conclude whether there is just one fractal structure in the material or more; and also from a nonlinear curve on a double logarithmic plot, we cannot conclude that no fractal exists in the materials. Experimental inspections and measurements in the whole range of scale are to be recommended.

(iv) The concept of multirange fractals could possibly be extended to other processes, e.g. fractal cluster growth, surface dynamics, etc.

## 4.15. *Time Evolution of Multirange Fractals*

In Long *et al.*'s experiment (1992), transgranular cracking and intergranular cracking are associated. They showed that in spite of the fractured surface being all of mixed inter-transgranular character, an approximately constant value of the fractal dimension was obtained in the range $2 < \eta_i < 50\ \mu\text{m}$, where "$\eta_i$" is the yardstick length. In different stages of crack propagation, the fractal dimension measured changes. Fractographic investigation showed that the relative composition of intergranular and transgranular cracks changes in different stages. Considering that the total crack initiation does not hold constant in all stages, Eq. (4.14.3) must change in form to

$$\begin{aligned} N(t)L^T(\varepsilon_3) &= N_1(t)\varepsilon_2^{1-D_1} + N_2(t)\alpha^{D_1-D_2}\varepsilon_3^{1-D_2} \\ \dot{N}(t)L^T(\varepsilon_3) &= \dot{N}_1(t)\varepsilon_2^{1-D_1} + \dot{N}_2(t)\alpha^{D_1-D_2}\varepsilon_3^{1-D_2} \end{aligned} \tag{4.15.1}$$

$$\dot{P}_1(t) = \frac{\dot{N}_1(t)}{\dot{N}(t)}, \quad \dot{P}_2(t) = \frac{\dot{N}_2(t)}{\dot{N}(t)} = 1 - \dot{P}_1(t)$$

$$\dot{L}^T(\varepsilon_3, t) = \dot{P}_1(t)\varepsilon_2^{1-D_1} + [1 - \dot{P}_1(t)]\alpha^{D_1-D_2}\varepsilon_3^{1-D_2}\,. \tag{4.15.2}$$

As in the experiment of Long *et al.*, (1992), the double logarithmic plot of $L^T$ vs $\eta_i$ relation appears as a straight line. Empirically, a $D_{\text{eff}}(t)$ can

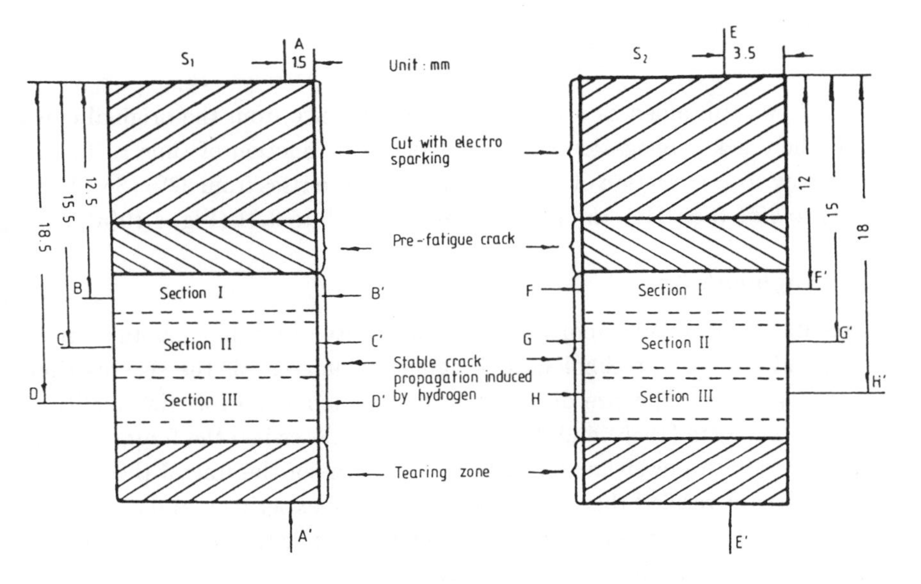


Fig. 4.19. (a) The locations of the profiles. Section I is the region just ahead of the pre-fatigue crack tip, followed by Sections II and III successively.

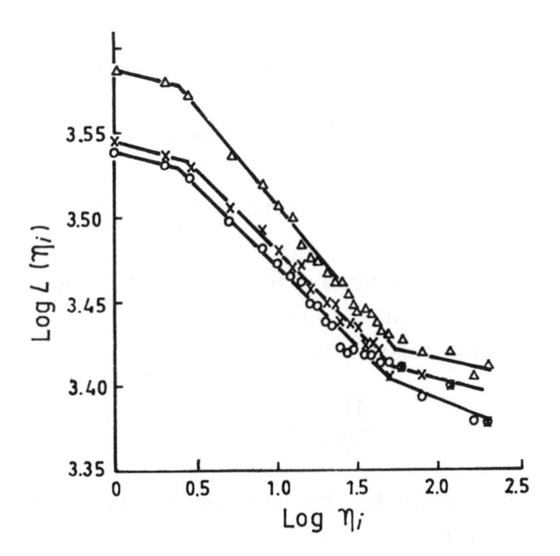


Fig. 4.19. (b) The measured relation between log L ($\eta_i$) and log $\eta_i$ for profiles EE′-1 (○), EE′-2 (×) and EE′-3 (Δ).

be defined through

$$L^T(\eta, t) \approx \text{const.}\ \eta^{1-D_{\text{eff}}(t)}\,.$$

$D_{\text{eff}}(t)$ was obtained as a function of the distance from the original crack tip, or the crack propagation distance which is a function of time $t$. In this experiment, as the fraction of transgranular fracture rises from 32% (Sec. I and Sec. II) to 55% (Sec. III), the $D_{\text{eff}}(t)$ rises from 2.19 to 2.21 (Fig. 4.19(a), (b)).

## 4.16. *Fragmentation*

Fragmentation plays an important role in a variety of machine, metallurgical, mining and geological technologies. Fragmentation controls the size distribution of fragments. The size distribution of fragments controls the quality of products. Turcotte (1986) and Zhao *et al.* (1990) described the fragmentation distributions with power-law relationships in terms of the fractal theory. Xie and Gao (1991) verified this description with more experimental data on rock failure in the laboratory. The power-law relationship is taken as evidence that the fragmentation mechanism is scale invariant.

Many experimental results indicate that there is a simple power-law relationship in size-frequency distributions of fragments.

$$N(m > m_c) = Cm_c^{-b} \tag{4.16.1}$$

where $N(m > m_c)$ is the number of fragments with a mass $m$ greater than $m_c$ and $C$ is a constant.

Another representation is the Weibull dependence:

$$\frac{M(r)}{M_T} = 1 - \exp\left[-\left(\frac{r}{\sigma}\right)^{\alpha}\right] \tag{4.16.2}$$

where $M(r)$ is the cumulative mass of fragments with a radius less than $r$, $M_T$ is the total mass, and $\sigma$ is related to the mean size. If $\frac{r}{\sigma} \ll 1$, equation (4.16.2) reduces to the power-law relationship:

$$\frac{M(r)}{M_T} \approx \left(\frac{r}{\sigma}\right)^{\alpha}. \tag{4.16.3}$$

From the definition of a fractal, the number $N$ with a characteristic linear dimension greater than $r$ is given by

$$N \propto r^{-D}. \tag{4.16.4}$$

Equation (4.16.3) gives

$$dM \propto r^{\alpha-1} dr \tag{4.16.5}$$

Taking the derivative of Eq. (4.16.4), yields

$$dN \propto r^{-D-1} dr \,. \tag{4.16.6}$$

The increment of number related to the incremental mass is given by

$$dN \propto r^{-3} dM \,. \tag{4.16.7}$$

Combining Eqs. (4.16.5), (4.16.6) and (4.16.7), one finds

$$D = 3 - \alpha \,. \tag{4.16.8}$$

The power-law mass distribution is equivalent to a fractal distribution with the relation of Eq. (4.16.8).

Many experimental studies on the frequency-size distributions of fragmentations showed fractal distributions. But, how can we describe them with fractal?

A simple model of how fragmentation can result in a fractal distribution has been proposed by Turcotte (1992). It is based on the concept of renormalization. A zero-order cube with dimension $h$ is divided into eight zero cubic elements each with dimension $h/2$ (Fig. 4.20). The probability that a zero-order cell will be fragmented into eight zero-order elements is $f$. The fragments with dimensions $h/2$ become first-order cells; each of these have a probability of being fragmented into eight second-order elements with dimensions $h/4$.

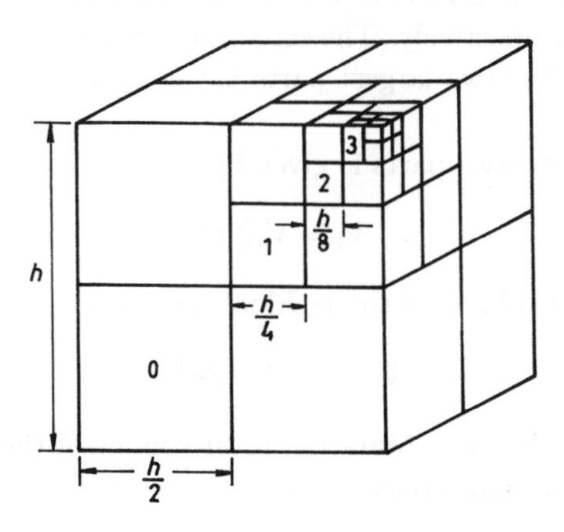


Fig. 4.20. Illustration of the renormalization group approach to fragmentation. (Following Turcotte, 1986a)

The process is repeated to higher orders. The basic structure is a fractal with fractal dimension

$$D = 3\ln(2f)/\ln 2\,. \qquad (4.16.9)$$

Although this model is very idealized and non-unique, it illustrates the basic principles of how scale-invariant fragmentation leads to a fractal distribution. It also illustrates the principle of renormalization. The division into eight fragments is an arbitrary choice; however, other choices such as the division into $m$ fragments will give the same result, where $m$ is an integer (1, 2, 3 ... etc.) without destroying the self-similarity property. This model is deterministic rather than statistical. This model relates the probability of fragmentation $f$ to the fractal dimension $D$ but does not place constraints on the value of the fractal dimension. The allowed range of $f$ is $1/2 < f < 1$ and the equivalent range of $D$ is $0 < D < 3$. Since the ratios $1/r$ of the above model should be positive integers, the distributions of fragments are discrete. However, actual distributions of fragments are continuous. It would be worthwhile to find a model for which the allowed values of the ratios $1/r$ may not necessarily be limited to integers. Jiang and Lung (1995) proposed two models, the scaling ratios of which may change continuously and which do not place constraints on the value of the fractal dimension. This was not paid attention to previously (see Turcotte, 1992).

### 4.16.1. *Equal-Ratio-Edged Orthorhombic Fragments Model* (*ERE*)

Suppose we have an equal-ratio-edged orthorhombic grain with lengths of edges being 1, $r^{-1}$ and $r$ respectively. The zero order grain may be divided into three parts (Fig. 4.21). The two largest parts have edges with lengths $1, r$, and $r^2$ respectively.

The volume of the two parts is given by

$$V_1 = V_0(2r^3)$$

where $V_0 = 1\cdot$. The volume of the residual is given by

$$V_r = V_0(1 - 2r^3)\,.$$

Repeating this process, we obtain the total volume of the second order cells, i.e. cutting into three parts once more:

$$V_2 = V_0(2r^3)^2\,.$$

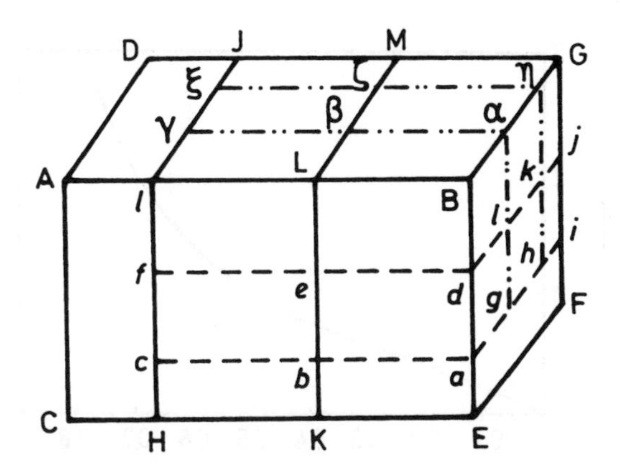


Fig. 4.21. The ERE model. (a) The initial grain is ABCDEFG, AB $= r^{-1}$, BE $= 1$, BG $= r$ and $V_0 = 1$; (b) First generation: BE $= 1$, BG $= r$, BL $= r^2$, and $V_1 = V_0(2r^3)$; (c) Second generation: BG $= r$, BL $= r^2$, Bd $= r^3$ and $V_2 = V_0(2r^3)^2$.

For the $n$th order cells, we have

$$V_n = V_0(2r^3)^n .$$

Then, according to the definition of fractal dimension by Mandelbrot (1982)

$$\frac{V_n}{V_0} = (2r^3)^n \equiv (r^{3-D})^n$$

$$2 = r^{-D} \tag{4.16.10}$$

$$D = \frac{\ln 2}{\ln(\frac{1}{r})} . \tag{4.16.11}$$

According to this model, the value of $1/r$ can be changed continuously. The allowed range of $r$ is $0 < r < (\frac{1}{2})^{1/3} = 0.7937$ and the range of $D$ is $0 < D < 3$ (Fig. 4.22).

The geometrical shapes of the cells in different orders of generations are similar. The ratio of the sides are $1 : r : r^2$, and each one is scaling with ratio $r$.

For each generation,

$$V_{n+1}(rl) = r^{3-D}V_n(l) . \tag{4.16.12}$$

Equation (4.16.12) satisfies the scaling form (Feder, 1988)

$$f(\lambda t) = \lambda^\alpha f(t) \tag{4.16.13}$$

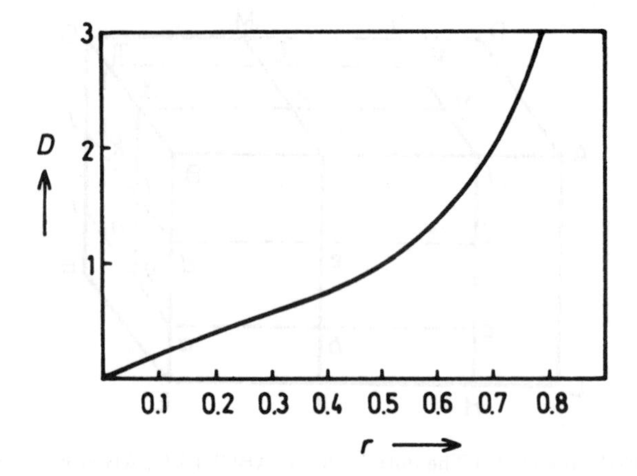


Fig. 4.22. The relationship of $D$ with $r$ in ERE model.

where $\lambda$ in Eq. (4.16.13) and $r$ in Eq. (4.16.12) are scaling factors and $\alpha$ and $3-D$ are the powers of the function, in Eq. (4.16.13) and Eq. (4.16.12) respectively. As with other fractals, the ERE model has a nice scaling symmetry.

The residual volume of the $n$th order cells is given by

$$\frac{V_{n,R}}{V_0} = 1 - (2r^3)^n \,. \tag{4.16.14}$$

Equation (4.16.14) is not a power-law (fractal) relationship. However, the distribution in one generation of the residual volume is given by

$$\frac{(V_{n+1,R} - V_{n,R})}{V_0} = (2r^3)^n(1 - 2r^3) = \text{const.}\ (2r^3)^n \,. \tag{4.16.15}$$

Equation (4.16.15) is a power-law relation. The fractal dimension of the density of residual volume is given by

$$\frac{(V_{n+1,R} - V_{n,R})}{(V_{n,R} - V_{n-1,R})} = 2r^3 \equiv r^{3-D'} \,. \tag{4.16.16}$$

Then

$$D' = \ln 2/\ln(1/r) = D \,. \tag{4.16.17}$$

The fractal dimension, $D'$ of the distribution of the residual volume is equal to that of the original fractal object.

4.16.2. *Volume-scaling Fragmentation* (*VSF*)

The ERE model has strict limitation on geometrical shapes and this may not be exactly consistent with practical fragments. Therefore a more generalized model of voluminal self-similarity is proposed below, this model having no limitation on geometrical shapes. Assuming the size of a zero order grain is $V_0$, it is divided into two unequal first order grains, $rV$ and $frV$, where $r$ and $f$ are the scaling factor and the ratio of the two first order grain sizes respectively. The values of $r$ and $f$ are smaller than unity. Repeating this process, we obtain the fractal (Fig. 4.23).

The volume of the first order cells is

$$V_F = r^E(1+f)V_0$$

where $E$ is the Euclidean dimension. The residual volume is given by

$$V_R = V_0 - r^E(1+f)V_0 = [1 - r^E(1+f)]V_0 \geq 0 \qquad (4.16.18)$$

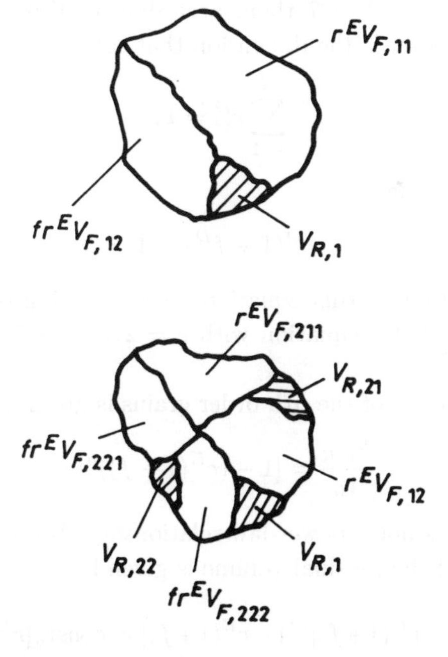


Fig. 4.23. Schematic figure for VSF model.

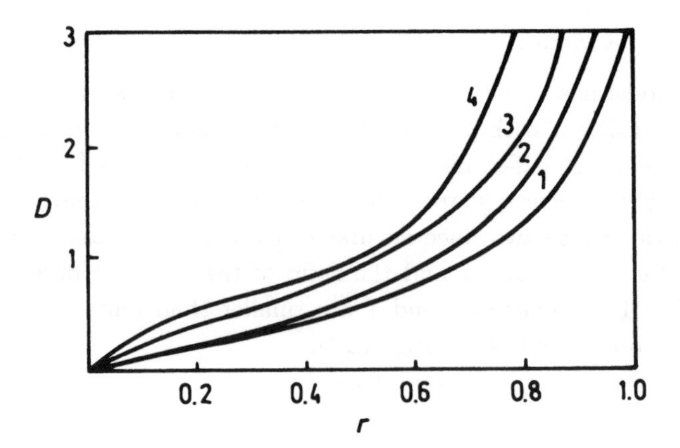


Fig. 4.24. The relationship of $D$ with $r$ in VSF model.

and $V_R$ should be not negative. Therefore, the necessary condition for $r$ is

$$r^{-E} \leq 1 + f\,. \tag{4.16.19}$$

If $f = 1$, $E = 3$, $r \leq 0.7937$ then, according to Feder (1988), the fractal dimension for such a set is the dimension that satisfies

$$\sum_{i=1}^{N} r_i^D = 1\,. \tag{4.16.20}$$

In our case, $N = 2$. Then,

$$r^D(1 + f^D) = 1\,, \tag{4.16.21}$$

and a special solution of this equation with $f = 1$ gives Eq. (4.16.3). A numerical solution of this equation with $r = 2/5$, $f = 5/8$ gives $D = 0.6110$ (Fig. 4.24).

The residual volume of the $n$th order grains is given by

$$\frac{V_{n,R}}{V_0} = [1 - (r^E(1 + f))^n]\,. \tag{4.16.22}$$

Equation (4.16.22) is not a power-law relationship. However, the distribution in one generation of the residual volume is given by

$$\frac{(V_{n+1,R} - V_{n,R})}{V_0} = [r^E(1+f)]^n[1-r^E(1+f)] = \text{const.}\ [r^E(1+f)]^n\,. \tag{4.16.23}$$

Equation (4.16.23) is a power law relationship: it is scaling with unequal ratios as the original fractal. The fractal dimension can be solved numerically as Eq. (4.16.21), and is equal to the original fractal scaling with unequal ratios.

Previously, Mandelbrot (1982) offered a tentative definition of fractal: "*A fractal is by definition a set for which the Hausdorff-Besicovitch dimension D strictly exceeds the topological dimension D*." According to this definition, the VSF model is a fractal, because its Hausdorff-Besicovitch dimension is larger than the topological dimension, which is equal to zero. However, Mandelbrot (1986) proposed another definition: "*A fractal is a shape made of parts similar to the whole in some way*."(Feder, 1988). We see from above that the VSF model satisfies the condition of this definition only in a loose sense. It seems that the definition of 1982 has more content than that of 1986. It should be pointed out that a neat and complete characterization of fractals is still lacking (Mandelbrot, 1987, see also Feder, 1988), but as one works through the various ways of constructing fractal structures one will get a better feeling for what the concept means (Feder, 1988; Lung, 1992).

For many years, it has been the aim of engineers to develop quantitative fractography. A parameter capable of continuous change for characterization of the fractured surfaces rather than discrete, qualitative and empirical inspection has been an objective for a long time. Now, fractal dimension is a possible candidate for this purpose. However, there are some problems on the fractal description of fractures, such as the method of experimental measurements, the multirange scaling complexity in materials and self-similarity or self-affinity properties. In spite of these problems requiring solution, the parameter, fractal dimension, possibly can be used in some limited and empirical cases, such as in the case when only one factor is changed in the process. For example, if only one element is changed in the composition of alloy design, only one elementary process in phase transformation is related in the heat treatment, or only one mechanism for degradation of materials is responsible in service failure, then the fractal dimension may provide information on continuous changes of structure and properties of materials.

The questions of self-similarity and self-affinity properties on fractured surfaces are still open. Which fracture mechanism (sudden break, stress-corrosion weathering processes, etc.) produces what kind of fractal structure is not known at the time of writing. We also do not know which materials (bcc and fcc metals, rocks and woods and etc.) create self-affine fractured surfaces and which do not. The difference between macro- and micromechanism of fracture is not clear. It will be very worthwhile to make experimental studies which compare these two descriptions along the lines of Sec. 4.13.

## 4.17. *Phenomenological Relation Between Fractal Dimension of Fractured Surfaces and Fracture Toughness of Materials*

Since Mandelbrot *et al.* (1984) showed that fractured surfaces are fractals in nature and that the fractal dimensions of the surfaces correlate well with the toughness of the material, many authors found that the fractal dimension depends on the fracture properties of materials, but the values of it seem in a narrow range for measurements with a resolution down to the micron scale. Some authors found that the roughness exponent could be universal (independent of the material for a range of materials). Some questions which have arisen are: Is the roughness exponent universal? What is the relationship between fracture properties and fractal structures ($D$ or $H$)? In this section, we shall discuss the universality and specificity of fractures and then, subsequently, discuss the relationship of fractal dimensions to fracture properties.

### 4.17.1. *Universality and Specificity of Fractures*

According to our concept of multirange fractals in materials, introduced as above, the fractal dimension of fractured surfaces measured is the total contribution of many elementary processes. Some processes, we think, are less dependent on the material, but some other processes are materials dependent. They superpose in the material. If one less materials dependent mechanism, say instability of the dynamical crack propagation, (Abraham *et al.*, 1994), complex mode crack propagation (Thomson, 1986; Lung, 1980), etc., produces fractal dimension $D_u$, and another materials dependent mechanism, say transgranular cracking, produces fractal dimension $D_s$; the effective fractal dimension $D_{\text{eff}}$ (see Sec. 4.14) synthesized by them would be $D_u < D_{\text{eff}} < D_s$ if both exist in the same range of scale and if $D_u < D_s$. The concrete value of $D_{\text{eff}}$ depends on the fraction of population of various mechanisms in the material. Then, the differences of the effective fractal dimensions in various materials might be reduced by this effect. It may be one of the reasons why the values of $D_{\text{eff}}$ measured seem always to lie within a narrow range. Secondly, as we have pointed out in Sec. 4.13, the measurement of roughness exponent is less sensitive than that of fractal dimension. Using the $H$ parameter as the roughness index the differences among various materials are less in evidence. Our previous works also have shown systematic differences of fractal dimensions for various materials as Milman *et al.* reported (Milman *et al.*, 1994). However, it appears that there are some less materials dependent mechanisms as Hansen *et al.* (1993) have pointed out in their investigations. These mechanisms, which we only mean less dependent on materials, may be multiple. The effective fractal dimension measured seems to be the synthesis of various elementary fractal structures including material dependent and less dependent ('universal') fractals. It seems impossible that the fractal dimension of fractured surfaces is a universal constant, though we believe there is not by any means a real conflict between current viewpoints.*

### 4.17.2. *Linear Elastic Fracture*

A model based on linear elastic fracture mechanics theory has been proposed (Lung, 1986). In Irwin's approach, the critical strain energy release rate, i.e., the critical crack extension force, $G_{Ic}$ may be written as $G_{Ic}(D = 1, \varepsilon_n = \varepsilon_0 = 1)$, if there is no fractal or the fractured surface is smooth:

$$G_{Ic}(1,1) = 2\gamma_s \quad \text{(for brittle fracture)}$$

$$G_{Ic}(1,1) = 2(\gamma_s + \gamma_p) \quad \text{(for quasi-brittle fracture)}, \tag{4.17.1}$$

where $\gamma_s$ is the specific surface energy and $\gamma_p$ represents the energy expended in the plastic zone necessary to produce unstable crack propagation at the crack tip. In the quasi-brittle case (or the small scale yielding case) we assume that the plastic zone at the crack tip is very small relative to the crack length and the thickness of plastic deformation is very thin.

Owing to the crack propagation along a zigzag line (Lung, 1986; Lung and Mu, 1988) the true areas of fractured surfaces (or lengths of crack lines in 2D system) are actually larger than the data obtained by macroscopic measurements (Fig. 4.25).

The area of the fractured surface per unit thickness of specimen would be $[\frac{\ln(\eta)}{L_0}] \cdot 1$. (In fracture mechanics, we always simplify the crack as a line in a two-dimensional system; say, in plane stress or plane strain cases.)

*The work of J. M. Lopez and J. Schmittbuhl (Phys. Rev. *E57*, 6405, 1998) confirms again that, in general, crack surfaces exhibit self-affine scaling properties in a wide range of length scales. These workers also suggest that fracture surfaces may have anomalous dynamic scaling properties similar to what can occur in models of kinetic roughening.

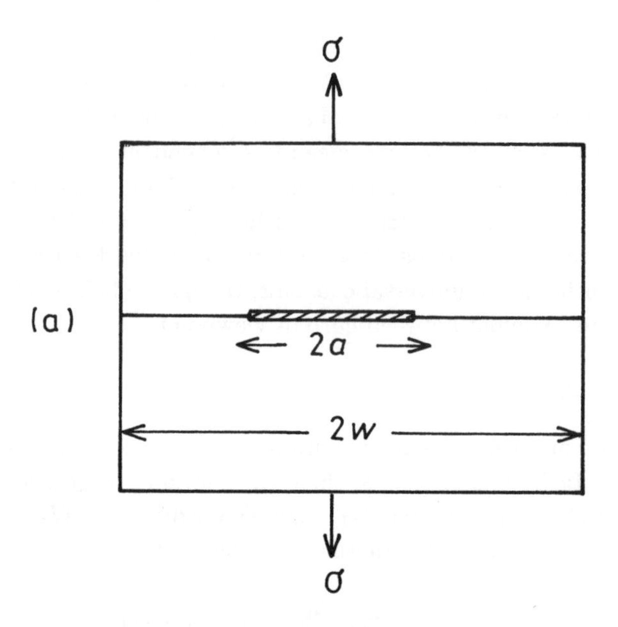


Fig. 4.25. (a) Ideal brittle fracture in glass.

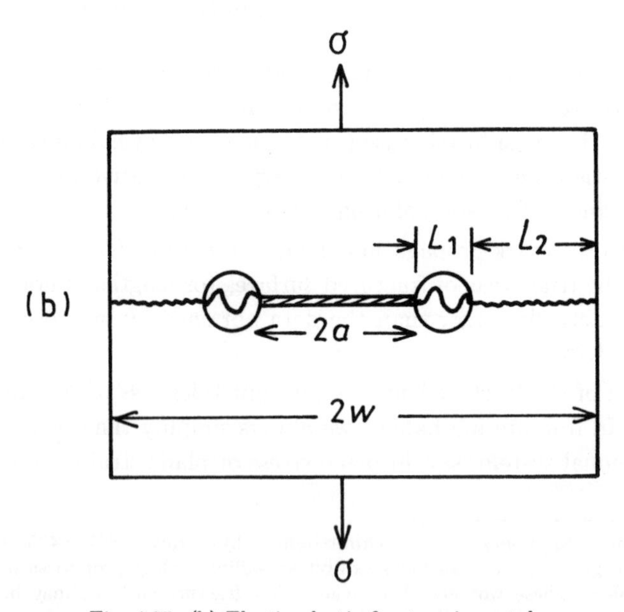


Fig. 4.25. (b) Elastic plastic fracture in metal.

Then, instead of Eq. (4.17.1), we have

$$G_{Ic}(D,\varepsilon_n) = G_{Ic}(1,1)\left(\frac{L_n(\eta)}{L_0}\right)$$
$$G_{Ic}(D,\varepsilon_n) = 2\gamma_s\left(\frac{L_n(\eta)}{L_0}\right). \tag{4.17.2}$$

The total length of a Koch curve is given by

$$\frac{L_n(\eta)}{L_0} = \left(\frac{\eta}{L_0}\right)^{(1-D)}. \tag{4.17.3}$$

From Eqs. (4.17.2) and (4.17.3)

$$G_{Ic}(D,\varepsilon_n) = G_{Ic}(1,1)\left(\frac{\eta}{L_0}\right)^{(1-D)} = 2\gamma_s\left(\frac{\eta}{L_0}\right)^{1-D} \tag{4.17.4}$$

where $D$ is the fractal dimension of the fractured surface. Equation (4.17.4) can be written as

$$\ln G_{Ic}(D,\varepsilon_n) = \ln G_{Ic}(1,1) + (1-D)\ln\left(\frac{\eta}{L_0}\right) \tag{4.17.5}$$

From Eq. (4.17.4), we may see that $G_{Ic}(D,\varepsilon_n)$ and $D$ are positive correlation, due to

$$\left(\frac{\partial G_I c(D,\varepsilon_n)}{\partial D}\right)_{\gamma_s,\varepsilon_n} > 0 \tag{4.17.6}$$

since

$$\ln\left(\frac{\eta}{L_0}\right) < 0. \tag{4.17.7}$$

It means that the logarithm value of fractal critical crack extension force or fracture toughness, is a function of $D$ (fractal dimension) and $\varepsilon_n$ (the lower bound of the fractal, or the length of the smallest crack step). It is positively correlated with $D$, and is in linear relationship with the fractal dimension of the fractured surface *provided* $G_{Ic}(1,1)$ *(or* $\gamma_s$*) and* $\frac{\eta}{L_0}$ *(or* $\varepsilon_n$*) are constants*. This relation, based on linear elastic fracture mechanics, holds in many experimental measurements for brittle materials (see Mu and Lung, 1988; Lung, 1992; and also for the positive correlation, Williford,1988; Milman *et al.*, 1994). In fact, $\ln \gamma_s$ does not change largely in many brittle materials. The linear relationship of Eq. (4.17.5) is approximately true in many brittle fracture cases.

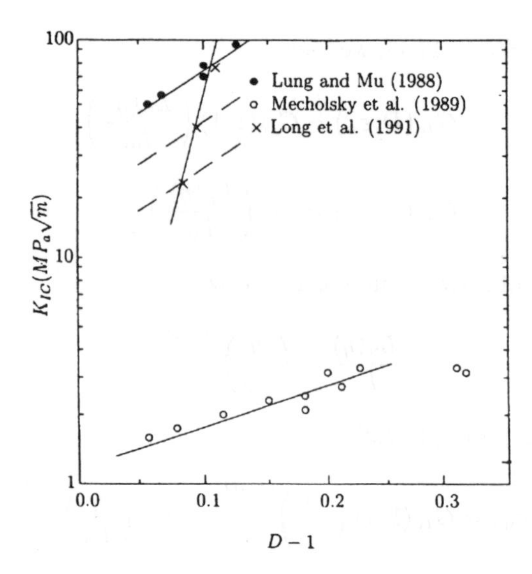


Fig. 4.26 The relationship of $K_{Ic}$ with $D$ of fractured surfaces. (Lung, 1998)

For example, Fig. 4.26 shows a good linear relation using a double logarithmic plot (Lung, 1998).*

---

*Two straight lines in Fig. 4.26 express the data of fracture toughness and hydrogen cracking experiments of 30CrMnSi2A high strength steels measured by our group. The lowest straight line expresses the data performed by Mecholsky *et al.* on the fracture toughnes of ceramic materials.

Three conclusions may be drawn:

(1) Three straight lines in Fig. 4.26 show the reasonableness of Eq. (4.17.5).

(2) The values of fractal dimension and of fracture toughness of 30CrMnSi2A steels are the same at the final critical fracture conditions in spite of their different loading modes.

(3) The data of fracture toughness of ceramic materials were lower than that of high strength steels. This may be due to the low values of $G_{Ic}(1,1)$ in ceramic materials.

However, three points should be noted:

(1) Eq. (4.17.4) was derived from linear elastic fracture mechanics; therefore, it is true only when the material is brittle or in small scale yielding case approximately.

(2) In Fig. 4.26, we may find that for one fractal dinension there are three values of fracture toughness in the three straight lines. This shows that we can hardly determine the fracture toughness by one parameter, fractal dimension, only. Dimensions, including fractal dimensions cannot subsume all the effects in complicated materials. For example, a one-dimensional steel wire is much stronger than a one-dimensional thread, even though their dimensions are the same.

(3) All experiments reported positive correlation between the fractal dimension and fracture toughness of materials.

In short, if one parameter is changed in one kind of material, say testing temperature, condition of heat treatment and etc., the relationship of Eq. (4.17.4) holds quite well if no abrupt structural changes occur.

Milman *et al.* (1994) fitted data by Long *et al.* (1992) and found that the empirical relation

$$K_{Ic} \propto K_0 \exp(\beta(D' - 1)) \tag{4.17.8}$$

with a positive coefficient $\beta \approx 30 \pm 6$ gave a good fit. This empirical relation is consistent with Eq. (4.17.5 ) formally, if $\beta = -\ln \varepsilon_n$. However, in this case $\varepsilon_n \approx 10^{-13}$. It is smaller than the inter-atomic distance in crystals! It seems unreasonable. However, it can be explained as follows: From Eqs. (4.17.4), $G_{Ic}$ is determined not only by $D$ but the specific surface energy also. In Long *et al*'s experiment (1991), the larger the $K_I$ at the crack tip, the lower the critical hydrogen concentration needed to cause crack propagation and the more ductile the materials at the crack tip as well. Therefore, the energy ($\gamma_s$ in Eqs. (4.17.4)) needed to form the new fracture surface is also higher. The dotted lines in Fig. 4.26 would be expected if another specimen under tensile test with the same content of hydrogen atoms as the local content at the crack tip in the slow crack propagation specimen under cantilever bending test. This explanation was verified by the fact that the fractal dimension reported by the authors agree well with the results obtained for the same steel under different fracture conditions (Mu *et al.*, 1993; Long *et al.*, 1991). The effect of multirange fractals (see Sec. 4.14) might be another physical cause to explain the empirical relation. Unfortunately, previous measurements have been performed only in a narrow range of micron scale. The satisfactory procedure is to measure values in all ranges of scale. Values of $D_i$ (or $H_i$) could be obtained by Eq. (4.14.6). If one can find a function $F(D_i)$ and the relationship of $F$ with fracture property, it would be reasonable. At the time of writing, the situation resembles the Chinese saying: "*we have seen only the individual tree instead of the whole forest.*" Fractal dimensions measured on fractured surfaces in atomic scale (Milman *et al.*, 1994) represents a worthwhile research area.

The difference between the fracture processes on nanometer and mesoscopic scales may be the effects of long-range elastic fields. It is well known that a macro plane strain brittle fracture may occur in low carbon steels though they are ductile from microscopic investigations on slip lines near the crack tip. A rough surface on an atomic scale may appear smooth on the micron scale. Indeed, the fracture processes in the two regimes are fundamentally different (Milman *et al.*, 1994; Hansen *et al.*, 1993). At the end of this chapter, we may see that the $H$ value of the roughness caused by complex mode fracture will be not lower than 0.95. The $H$ value of another model of continuum media proposed by Bai *et al.*, is about 0.9. Therefore, the synthesized fractal dimension including the contribution of these low fractal dimensions could not be high.

Some results exist which seem to contradict the above relationship. Some experimental data showed that the correlation between $D$ and toughness (dynamic tear energy, fracture toughness etc.) is a negative one. We found that many authors used the perimeter-area method inadequately (Lung and Mu, 1988; Lung, 1992; Milman *et al.*, 1994; Jiang *et al.*, 1994 and Shi *et al.*, 1996).

Another problem is that the measured values of $G_{Ic}$ may be different for various materials though they have the same fractal dimension value of fractured surfaces. However, the surface energies are different for various materials. In Eq. (4.17.5), $G_{Ic}$ is determined by the surface energy and the fractal dimension, not the latter only. It seems reasonable that different materials (ceramics and ultra high strength steels) have different $G_{Ic}$ even with the same values of $D_F$ (see Sec. 4.17.3). For the same kind of materials, $\gamma_s$ does not change too much, even in the case of a small change of structure. Because $\gamma_s$ is insensitive to the structure, it may be considered to be approximately constant; then, the linear relation between $\ln G_{Ic}$ and $D_F$ (Eq. 4.17.5) holds.

### 4.17.3. *Quasi-brittle Fracture*

It is true to say that at the time of writing it has not been proved possible to clearly understand the correlation between fractal dimension and ductile properties.

Williford (1988) analysed the collected data of positive and negative correlation between toughness and $D$ and plotted an analogy to the dimension spectrum as a multifractal (Fig. 4.27).

In his investigation, on the left side of Fig. 4.27, $D-2$ for very brittle ceramics is low because their microcracks are flat. On the right side, $D-2$ is also low when damage is dominated by dislocation processes which also have a low dimension. The central position is the transition part. The multifractal concept for this problem given by Williford, aroused the interest of some scientists as giving further insight into the correlation between the fracture toughness and fractal dimension. In the limiting case of very ductile material, he argued that the large energy given is the work to ideally shear (crack) a very ductile material into halves by passing a large number of dislocations

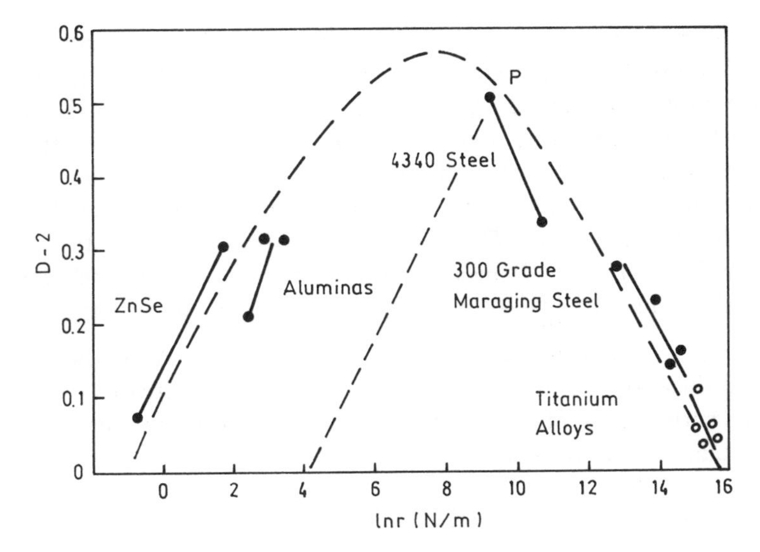


Fig. 4.27. The relationship of $D$ with fracture energy of various metals. (After Williford, 1988).

through the same plane, thus producing an atomically smooth 'fracture' surface with very low surface roughness, $R$ and $D \approx 2$. However, two problems have arisen: (1) usually, many measurements were made in the range of scale of microns. The atomically rough 'fracture' surface would be seen to be flat in the scale of microns, and an atomically smooth 'fracture' surface may be rough in microns due to the random orientations of the grains or of fracture crystal planes. Moreover, later studies by Milman *et al.* (1994) showed that the values of $H$ are usually lower (more rough) in the nanometer scale. (2) The analogy to a multifractal seems not to have a clear physical interpretation.

Many experiments reported negative correlations of fractal dimensions with fracture properties of materials. Most of them seem due to the SIM having been used. However, even the vertical section method having been used in Pande *et al.*'s experiments, the negative correlation still appeared and a satisfactory explanation has not yet been provided.

To find another way to explain the experimental results, based on fracture mechanics, the relationship of $D$ with fracture toughness and the concept of multirange fractals proposed by Lung (1986, 1993) can be used as a starting point for the discussion. From Eq. (4.17.5), in the case of elastic-plastic fracture, $G_{Ic}(1,1)$ loses it rigorous meaning. We should use $J_{Ic}$ or the effective specific surface energy $(2\gamma_p)$ as proposed by Orowan (Lung, 1992; Mu and Lung, 1988; Lung, 1986), which is quite sensitive to the structure of materials. It changes substantially from one structure to another even in one kind of material. For discussing the relationship of fracture toughness with fractal dimension, the $G_{Ic}(D,\varepsilon_n)$ value (or other toughness data) should be normalized by $G_{Ic}(1,1)$. It means that only discussion on the relationship of the relative value of $\frac{G_{Ic}(D,\varepsilon_n)}{G_{Ic}(1,1)}$ with $D_F$ is reasonable. According to Tetelman (1963), his $\gamma_m$ (see Chap. 1) has a factor of $N_0^{\frac{3}{2}}$. Here $N_0$ is the number of mobile dislocations in unit volume. The value of $\gamma_m$ increases more rapidly than the surface (2D) effect with the increase of ductility in ductile materials. Also in Lung and Gao's study (1985),

$$G_{Ic}^p = 2\gamma_p = \frac{W_i}{L\Delta}$$

where $L\Delta \times 1$ is the smooth cracked surface of the plastic zone. The dimension of $L\Delta$ is 2, but the dimension of $W_i$ is 3. It is a volume (3D) effect. The fractal dimension of fractured surfaces is only between 2 and 3. When the material becomes more ductile, the volume of the plastic zone increases more rapidly than the area of the fractal fractured surface. Then, $\gamma_p$ changes more rapidly than the enlargement of the fractal area of fractured surfaces. Unlike the brittle fracture case, $\ln \gamma_p$ cannot be approximated as a constant. Then, the value of the left side in Eq. (4.17.6) does not exist in reality and the inequality in Eq. (4.17.6) may fail. According to the concept of multirange fractals proposed by one of the present authors (Lung, 1993, 1994, 1995), a crack line can be considered as composed of several Koch lines in different ranges of scale. We assume that fractals composed of dislocation configurations have no overlapping region with fractals composed of trans or intergranular cracks. The crack lines created by dislocation processes may be seen as a flat line in the larger range of micron scale. The Koch lines (crack) created by intergranular cracking may be seen as a flat line over a larger range of millimeter scale. The schematic figure is shown in Fig. 4.28. This figure was designed by the late Dr. Mike Ohr for the 2nd International Conference on the Fundamentals of Fracture, 1985. He was successful in expressing this natural phenomenon though he was not familiar with the concept of fractals.

Suppose we have two nonoverlapping fractals in two different ranges of scale (atomic and metallographic). The total length of these multirange fractals is (Lung, 1993, 1994, 1996)

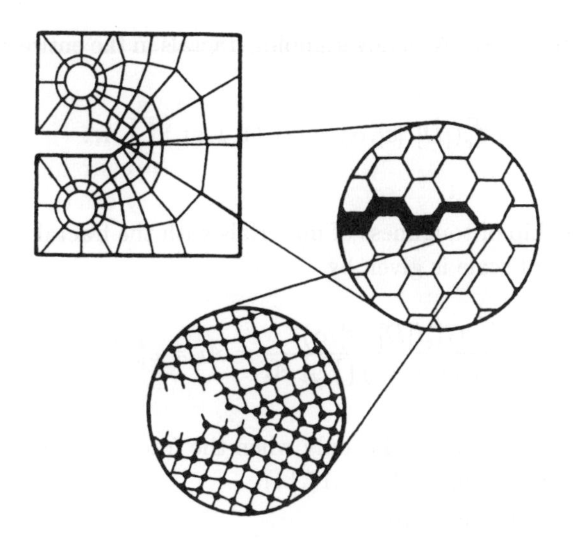

Fig. 4.28. The rough crack line looks relatively flat on a macroscopic range of scale.

$$L^F(\varepsilon) = \prod_{i=1}^{2} (\varepsilon_i/\varepsilon_{i-1})^{1-D_i} . \tag{4.17.9}$$

Then

$$G^F_{1c}([D], [\varepsilon_n], \alpha) = G_{1c}(1, 1, \text{at.}) A_1(\sim nm) A_2(\sim \mu m) \tag{4.17.10}$$

where $G_{1c}(1, 1, \text{at.}) = 2\gamma_s$, the specific surface energy of the fractured material, which is determined by the electronic structure or interatomic forces at the surface, $\alpha$ indicates the range of scale and $A_i = (\varepsilon_i/\varepsilon_{i-1})^{1-D_i}$. [ ] indicates that all $D_i$'s in various ranges of scale are included.

In the scale range of metallographic grains

$$G_{1c}(1, 1, \text{gr.}) = G_{1c}(1, 1, \text{at.}) A_1(nm) = 2\gamma_p \tag{4.17.11}$$

where $2\gamma_p$ is the energy expended in the plastic work necessary to produce unstable crack propagation.

Again, in a larger scale range of continuum media

$$\begin{aligned} G_{1c}(1, 1, \text{cont.}) &= G_{1c}(1, 1, \text{gr.}) A_2(\mu m) \\ &= G_{1c}(1, 1, \text{at.}) A_1(nm) A_2(\mu m) . \end{aligned} \tag{4.17.12}$$

In general, if we have $N$ nonoverlapping fractals in the entire ranges of scale

$$G_{1c}^F([D], [\varepsilon], \alpha) = G_{1c}(1, 1, \text{at.}) \prod_{i=1}^{n} A_i \,. \tag{4.17.13}$$

The relationship of toughness of materials with the fractal dimension measured in $k$ range of scale is given by

$$\frac{G_{1c}^F([D], [\varepsilon], \alpha)}{G_{1c}(1, 1, \text{at.}) \prod_{i \neq k}^{n} A_i} = A_k = \varepsilon_k^{1-D_k} \,. \tag{4.17.14}$$

The fractal relationship holds only when the fracture toughness is normalized to a value according to Eq. (4.17.14)

In the quasi-brittle fracture case

$$G_{1c} \approx 2\gamma_p \varepsilon_1^{(1-D_1)} \tag{4.17.15}$$

$$\ln G_{1c}^* = \ln G_{1c} - \ln(2\gamma_p) = (1 - D) \ln \varepsilon_1 \,. \tag{4.17.16}$$

At the time of writing, we still lack knowledge on $\gamma_p$ and its changes in ductile materials. If we treat data cited from Williford's figure (1988) along the line of the above thinking we may find some qualitative relationship for

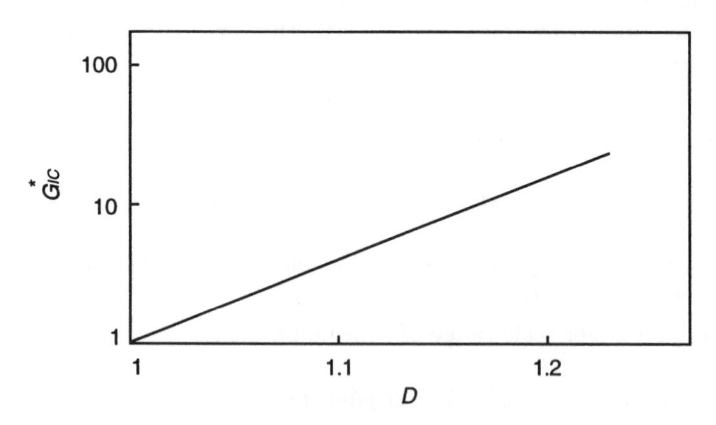


Fig. 4.29. Schematic figure $D$ of $G_{Ic}^* - D$ relationship.

toughness and $D$ in micron range of scale for quasi-brittle materials though data are not from rigorous fracture mechanics tests. However, at present we cannot. Assuming that ZnSe is perfect brittle and all materials obey the same relationship as ZnSe, we may draw a line parallel to ZnSe (PA). Subtractions of $\ln\gamma$ values may give us a rough impression on $\gamma_p$ of various materials. The schematic figure showing the relation between $G^*_{Ic}$ and $D$ would be Fig. 4.29. In this figure, there would be two features common to all materials:

(i) Positive fractal correlation for all materials might be obtained.
(ii) Data for all materials would pass through the point (1,1).

It seems reasonable that the straight lines pass through the point (1,1) in the figure. We do not know the change of $\ln\,\varepsilon_n$ in various materials. Here, we assume that the logarithmic value does not change very much. Figure 4.29 shows our analysis to be qualitatively reasonable though not of course rigorous. In this treatment, we have separated the effect of fractal surfaces and that of other processes, say plastic deformation etc. *Dimensions, including fractal dimensions cannot subsume all the effects in complicated materials*. For example, a one-dimensional steel wire is much stronger than a one-dimensional thread, even though their dimensions are the same.

Furthermore, from Eq. (4.17.16),

$$\left[\frac{\partial \ln G^*_{Ic}(D,\varepsilon_n)}{\partial D}\right]_{\varepsilon_n} = -\ln\varepsilon_n > 0 \tag{4.17.17}$$

$$\left[\frac{\partial \ln G^*_{Ic}(D,\varepsilon_n)}{\partial \ln\varepsilon_n}\right]_{D} = 1 - D < 0 \tag{4.17.18}$$

$$\left[\frac{\partial(\ln\varepsilon_n)}{\partial(1-D)}\right]_{G^*_{Ic}} = -\frac{\ln\varepsilon_n}{(1-D)} < 0 \tag{4.17.19}$$

$$\frac{d\ln G^*_{Ic}(D,\varepsilon_n)}{dD} = -\ln\varepsilon_n + (1-D)\frac{d(\ln\varepsilon_n)}{dD} \tag{4.17.20}$$

$$D = \frac{\ln N}{\ln\frac{1}{r}} = \frac{-n\ln N}{\ln\varepsilon_n}\,. \tag{4.17.21}$$

Because $n$ and $N$ are variable in materials, we cannot say that the differential in the second term of Eq. (4.17.20) is positive or negative. However, if $n$ and $N$ are constants approximately, the differential would be negative and then the value of the expression in Eq. (4.17.20) would be positive ($\varepsilon_n < 0$). A positive correlation between $G^*_{Ic}$ and D would still be expected.

In summary, the situation regarding ductile fracture in materials is very complicated. It is an irreversible, dissipative process. More experimental measurements and investigations are needed. Some experiments related to this problem have already been performed (Williford, 1990; Gong and Lai, 1993).

Moreover, according to the concept of multirange fractals in materials, the fractal behaviour of a fractured surface is the total contribution of many elementary processes. Every elementary microstructure contributes its fractal or non-fractal behaviour at a certain range of scale. Even in the same range, there are perhaps several elementary physical processes superposed on one another or leading to multifractal. The relationship between the fractal dimension measured and the property of materials could be complicated. We should take care to distinguish which mechanism plays the decisive role in the fracture process and to measure the fractal dimension in the same range of scale as the mechanism occurred. Then, the main factor may emerge.

For establishment of the relationship of fractal structure with property, *it is important to factorize out the effect of fractal structure from other physical causes and separate the appropriate range of scale from multirange fractals.*

The universality and specificity of the fractal dimension of fractures is an interesting problem. More experimental measurements and theoretical analyses are needed. In general, we think that any universality exists in the concrete natural things on which the specificity appears. In other words, universality exists in specificity and manifests its existence through specificity. They are mutually correlated with each other. For example, the experimental crack propagation picture cited by Herrman (1990) is a specificity of one kind of material. Herrman's fractal crack growth simulation (universal) can compare qualitatively with it only when the parameter $\eta$ in Eq. (4.18.5) has a definite value which reflects the specificity of the material.

According to the idea of multirange fractals, if we find a way to separate the specific fractal dimension of an elementary process from the measured (or synthesized) fractal dimension, which narrows the differences among various materials, we may find the relationships of fractal dimensions to fracture properties of various materials more distinctly.

In failure analysis, it is important to estimate the degradation of materials promptly after some fractographic inspections. The knowledge of the relationship of fractal dimension to properties would be very helpful (Lung, 1986; Wang *et al.*, 1988; Milman *et al.*, 1994; Jaeger *et al.*, 1996 and etc.).

## 4.18. *Physical Sources of Fractal Surfaces*

Usually, crack propagation in metals is like a process in driven system explained by Schmittmann and Zia, 1995. This simply means that some "force" constantly pushes the system of interest away from thermal equilibrium. Interfaces, crystal lattice planes, corrosion and impurity atoms are sources of the "force". This "force" in the material and the external uniform applied force drive the crack propagation. The "force" greatly influences the formation of fractal structure.

The fracture property of materials is the synthesized phenomena of microstructures in different ranges of scale. Metallographic structure, lattice defects and alloy elements influence the properties on different levels of structure. A double logarithmic plot permits an analysis of the fractal character of a microstructural feature. This scale range of experimental fractal analysis should be consistent with the scale range of the microstructure which is related to the property of material. Many previous experiments were done in the micron range. It is well known that dislocation configurations and dynamics play a major role on plastic deformations in metals. Fractals in the nanometer range and atomic dimension should be focussed on. Only a few investigations have been made (see Milman *et al.*, 1994) at the time of writing.

### 4.18.1. *Intergranular Fracture Model*

Intergranular cracks may form a Koch curve in a statistical and approximate sense (Fig. 4.30).

One side of the grain boundary of a grain may be considered as the lower bound of a Koch curve. The crack propagates along a zigzag passage which, like the sea shore, possesses hierarchical structure. We do not know how many generations it may have. Experimental data (see Long *et al.*, 1992) indicate that the real crack length measured is dependent on the yardstick length and obeys a power-law relationship. In the analysis of the last section, there would be more than one fractal structure in the entire range of scale. But for the

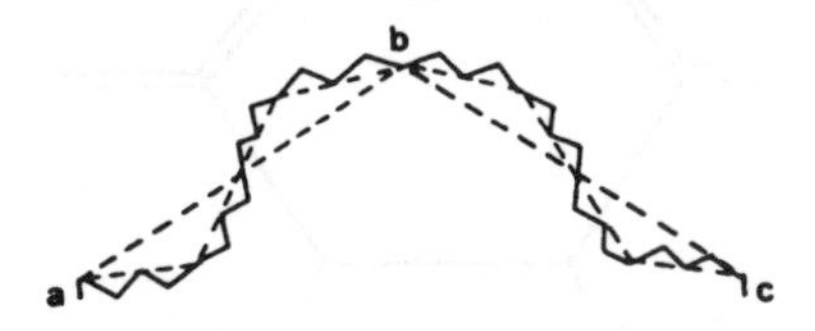


Fig. 4.30. Zigzag cracks formed in fractal modelled metals.

intergranular crack, it has its own upper and lower bound in the micron scale range. In principle, the upper bound also can be measured by experiments. Real grain boundaries have irregular shapes. In many cases, they are modelled as hexagonal shapes in a two-dimensional figure. Using this model figure, we may estimate the fractal dimension of intergranular cracks. The crack prefers to propagate along a weaker passage near to the general direction determined by macroscopic continuum fracture mechanics. Figure 4.31 is a schematic figure of an ideal process.

There are two forms of intergranular fracture (Figs. 4.31 and 4.32). Their fractal dimensions can be estimated by the definition

$$D = \frac{\ln N}{\ln \frac{1}{r}} \tag{4.18.1}$$

where $N = \frac{L_i}{\varepsilon_{0i}}$, $r = \frac{\varepsilon_{0i}}{L_{0i}}$

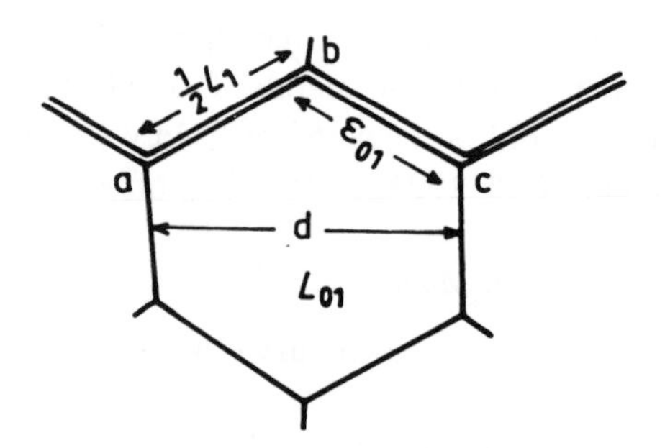


Fig. 4.31 Intergranular brittle fracture. (one case)

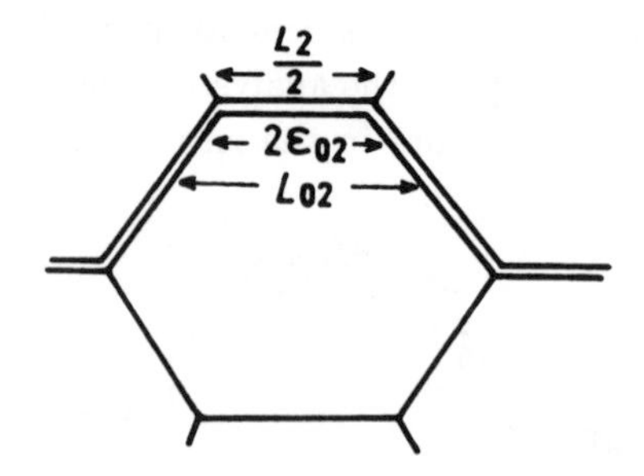


Fig. 4.32. Intergranular brittle fracture. (another case)

(i) $N = 2$, $r^{-1} = 1.732$, $D = 1.26$
(ii) $N = 4$, $r^{-1} = 3$, $D = 1.26$

Both fractal dimensions of these two forms are 1.26 but the grain sizes are different. The scaling ratios are also different:

$$G_c = 2\gamma_s \left(\frac{L_i}{L_{0i}}\right) = 2\gamma_s \left(\frac{1}{r}\right)^{(D-1)} \tag{4.18.2}$$

(i) $G_c = 1.73^{0.26}(2\gamma_s)$
(ii) $G_c = 3^{0.26}(2\gamma_s)$

We can see that case (i) consumes less energy than case (ii), and then it is preferable.

As the grain size becomes smaller and smaller, the true area of the fracture surface becomes larger and larger. As above, we have indicated that $L_0$ can be determined experimentally by Eq. (4.13.8) after $D_F$ has determined by the slope of the straight line in a double logarithmic plot. Theoretically, $L_0$ may be considered as the distance of one step of crack propagation from initial position to temporary arrest. For an intergranular crack, $D = 1.26$. This value seems universal for all intergranular cracking processes in materials due to the standard simple model. However, experiments of Long *et al.* (1992) showed that the measured fractal dimensions of intergranular cracks are lower than this value. The explanation is that this phenomenon is due to the random orientation of grains and that the crack prefers the weakest passage if the stress distribution is appropriate for crack propagation. Another reason may involve multirange fractals. As was discussed in Secs. 4.13 and 4.14, one fractal structure, superposed with another low $D_F$ fractal, may have the value of synthesized apparent fractal dimension lying between the two. Then the fractal dimension measured will appear to be lower.

For brittle materials, $L_0$ is larger. Comparing materials with the same grain size, the relative lower bound of brittle materials, $\frac{\eta}{L_0}$, is smaller; then, the effect of increase of the fracture toughness due to the fractal structure is larger than that of ductile materials. This can be seen also in the relation

$$G_c = 2\gamma_s \left(\frac{\eta}{L_0}\right)^{1-D} .$$

In this equation, $(\frac{\partial G_c}{\partial L_0})_{\eta,D} > 0$. This means that longer value of $L_0$ (or the more brittle the material), increases the effect of fracture on $G_c$ value. This effect is in contrast with that of $D_F$. This effect competes with that of $D_F$.

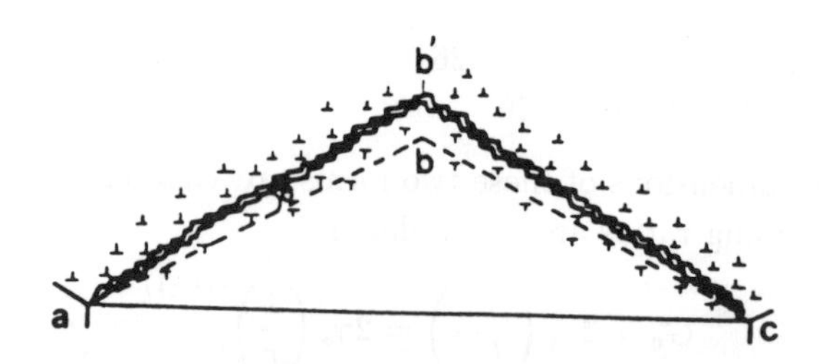


Fig. 4.33. The additional angle formed by plastic deformation in the grain.

However, if the number of generations for the fractal is limited or fixed, this effect is not important because it will have no effect when $L_0$ is so large that the number of generations lies outside the lower limit. In brittle materials, this effect was not found. It seems that the effect of $D_F$ plays the main role.

In the intergranular cracking mechanism, the fractal dimension 1.26 ($D - 1 = 0.26$) is not large enough to raise the fracture toughness. Plastic deformations in the grain would result in the grain boundaries having an additional angle (Fig. 4.33).

In this case, $N = 2$; $r = [2\cos(30° + \theta)]^{-1}$

$$D = \frac{\ln 2}{\ln[2\cos(30° + \theta)]}\,. \tag{4.18.3}$$

The value of $\theta$ can be estimated as follows

$$\theta = \frac{(\rho b L)}{L} = \rho b \tag{4.18.4}$$

where $\rho$ is the linear density of mobile dislocations. Typical values of the total linear density of dislocations range from $10^6 - 10^7$/cm for cold worked crystals to $10^3$/cm for annealed crystals. With $b \approx 3 \times 10^{-8}$ cm, the range of $\theta$ in Eq. (4.18.4) is from $3 \times 10^{-5}$ (rad.) to 0.03–0.3 (rad.) or (1.7° – 17°). Then, the fractal dimensions range from 1.26 to 2.23.

Taking $D = 2.23$, we have

$$G_c \approx (2\gamma_s)\left(\frac{d}{L_0}\right)^{-1.23}.$$

If $\frac{d}{L_0} = 10^{-4}$

$$G_c \approx 2\gamma_s \times 8.3 \times 10^4\,.$$

The critical crack extension force estimated by this fractal model would rise rapidly with the decrease of grain size. However, in practical cases, so many mobile dislocations in the grain would change the mechanism from intergranular to transgranular cracking. On the other hand, if the grain size decreases to less than micron or nanometer scales, the deformation and fracture mechanism would change back to be intergranular again. Anyhow, in that case, Hall-Petch's $d^{-\frac{1}{2}}$ relation will not hold.

Equation (4.18.3) shows that if there is an additional angle, the fractal dimension rises. This is consistent with many experimental results that the fractal dimension has a positive correlation with the fracture toughness of brittle materials.

### 4.18.2. *Transgranular Fracture Model*

Xie (1989, 1993) proposed a fractal model for transgranular brittle fracture (Fig. 4.34).

$$N = 3\,, \quad r = \frac{1}{2.236}$$

and

$$D = \frac{\ln 3}{\ln 2.236} = 1.365\,.$$

In general, it consumes more energy than intergranular cracking.

### 4.18.3. *Changes in Fracture Mechanism*

The fractal dimension of different parts of a fractured surface of 30CrMnSiNi2A steel formed by slow stable crack propagation induced by the combined effect

Fig. 4.34. (a) Cleavage step in the crystallographic plane of marble (three point bending loading). Arrows indicate the direction of crack propagation (scanning electron micrograph ×1000).

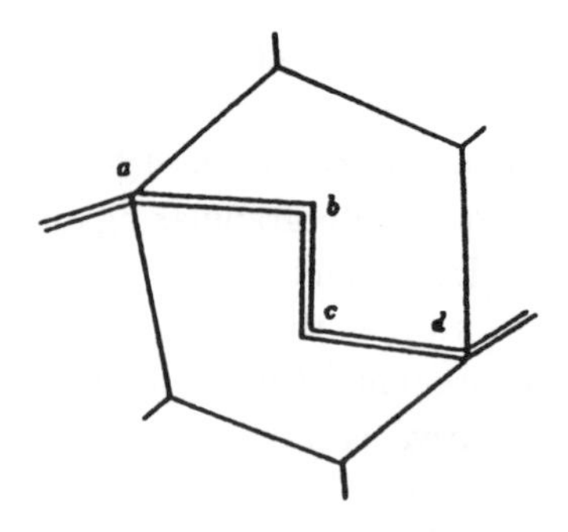


Fig. 4.34. (b) A fractal model of a transgranular brittle fracture.

of hydrogen and static bending moment was determined by using the method of fracture profile analysis (Long *et al.*, 1992). The results showed that the fractal dimension measured increases with increase of the relative fraction of transgranular fracture compared to intergranular fracture as the crack propagates. As we have discussed in Sec. 4.14, the fractal dimension measured is actually an effective value, superposed by two fractal systems (inter- and trans-granular fractal cracks). There seems not to be one fractal system of multiscaling.

It is well known that the crack propagation in a sample of high strength steel containing hydrogen results from the fact that the aggregation of the hydrogen atoms at the crack tip will occur under the influence of a stress gradient, and the local part of the material there becomes brittle. The smaller the $K_1$ at the crack tip, the higher the critical hydrogen concentration needed to cause crack propagation and the more brittle the local material at the crack tip as well. Therefore, the energy needed to form the new fracture surface is also smaller. As $K_1$ increases, the critical hydrogen concentration needed for cracking decreases, and the toughness of the local part of the material at the crack tip retained a higher value; then the energy needed to form the new fracture surface and the fraction of transgranular fracture also increase. Thereby, the effective roughness of the fracture surface formed at locally higher $K_1$ is larger than that formed at locally lower $K_1$. Consequently, the local density and the depth of the valleys and peaks on the profiles increase as $K_1$ increases, and then the effective fractal dimension as measured increases.

### 4.18.4. *Fractal Crack Growth*

Herrmann (1990) described the fracture growth as a moving boundary problem similar to dielectric breakdown or viscous fingering. He modelled this process on a square lattice (not crystal lattice or atom lattice) by computer simulation and found that the patterns of cracks can be fractal, even without including noise, due only to the interplay of anisotropy and memory. His model is similar to a stable crack propagation. The shapes of the cracks can be compared to the ones found experimentally for stress corrosion.

The important point is that he assumed a beam model in which the probability $p$ defined by

$$p = (f^2 + r. \max(|m_1||m_2|)^\eta \tag{4.18.5}$$

where $f$ is the traction force applied on the beam and $m_1$ and $m_2$ are the moments that are acting at the two ends of the beam; this $p$ determines if the beam will be broken. Each time a beam is broken, the shape of the crack and consequently the boundary conditions of the equation of motion changes and one has discretized the equation again to know which beam to break next. If in Eq. (4.18.5) an exponential instead of a power law was used, the structures seem to be dense. The form of Eq. (4.18.5) is empirical. For $\eta = 1$ this growth law is inspired by the von Mises yielding criterion, but it is not possible to derive it from first principles. The power-law seems reasonable if we examine the crack growth under alternating stress (fatigue).

$$\frac{da}{dN} \sim \Delta K^m, \quad (m \sim 3 - 4) \tag{4.18.6}$$

where $a$ is the crack length, $N$ is the number of cycles of the alternating stress and $\Delta K$ is the difference of upper and lower limits of the alternating stress intensity factor. The open problem that remains is to find the relationship between Eq. (4.18.5) and many physical mechanisms of fracture. Herrmann (1990) found some universal features due to mechanical instabilities leading to fracture. However, it seems that the specificities of materials are also subsumed in the equation and parameters of Eq. (4.18.5). In his computer simulation, the fractal dimension $D$ of the crack lines depends on the parameter $\eta$. $D$ decreases as $\eta$ decreases. We know that the material fractures at higher strength when $\eta$ is smaller. This result demonstrates that lower fractal dimension corresponds to more brittle fracture. This is consistent with the above discussion for brittle fractures.

Moreover, Herrmann (1990) used an equal size lattice, or equivalently an assumption for the lengths of steps of crack propagation being equal. However, this is not the case in reality. According to the experiments of Long *et al.* (1992), the steps would be smaller and smaller as more ductile transgranular cracking is activated and hence leading to smaller $L_0$ (see also Lung and Mu, 1996).

### 4.18.5. *Dynamic Instability*

Measurements by Fineberg *et al.* (1991) indicated that, at least in plastic material, the limiting fracture speed is significantly less than the Rayleigh velocity,* and the approach to this limiting speed is accompanied by the onset of dynamic instability. This has also been seen experimentally by Yuse *et al.* (1993). The velocity-limiting instability involves oscillations in the direction of crack growth.

Using the molecular dynamics method, Abraham *et al.* (1994) found that the crack tip initially propagates straight, then the onset of crack instability begins as a roughening of the created surfaces which eventually results in the zigzag tip motion at 30° from the mean crack direction. However, the roughness occurring on scales of tens of nanometers might look like a straight line on scales of microns and minimeters.

The instability roughens the surfaces, and the roughening of the surface limits the velocity of crack propagation. Xie (1994, 1995) studied crack tip motion along a fractal crack trace. A formula was derived to describe the effects of fractal crack propagation on the dynamic stress intensity factor and on crack velocity. His calculation showed that the dynamic stress intensity factor and apparent crack velocity are strongly affected by the microstructure parameter (grain size), fractal dimension, $D$, and fractal kinking angle of crack extension path. He reported that his calculation is in good agreement with experimental results. Here, Xie's fractal crack path is created not only by dynamic instability but also includes the microstructure factors.

Dynamic instability is universal; it is not only a mechanical process (Fineberg, 1991; Yuse, 1993), but is also related to energy dissipated on the fracture surface which may be converted to heat. Langer (1993) proposed a dynamic model of onset and propagation of fracture. In his model, the stress acting on the fracture surface includes a dissipative term and exhibits a dissipation-dependent effective threshold for fracture. His calculation showed that the crack creeps very slowly at external stress just above the Griffith threshold, and makes an abrupt transition to propagation at roughly the

---

*See the book *Dislocation Based Fracture Mechanics* by J. Weertman, (World Scientific, Singapore, 1996) or *Fracture of Brittle Solids*, 2nd Edition: (Cambridge University Press UK).

Rayleigh wave speed at higher stresses. When heating due to dissipation is taken into account, the model may exhibit a maximum in the crack propagation speed as a function of applied stress. In addition, the decrease in velocity at large applied stress might be an indication of some sort of dynamic instability (Ching *et al.*, 1996). Summing up Langer's analysis, Herrmann's computer simulation, and many experiments on stress corrosion cracking processes, dynamic instability is one of the main sources for fractal structures of fractured surfaces in materials.

### 4.18.6. *Evolution Induced Catastrophic Model*

A model of evolution induced catastrophe (EIC) was proposed by Bai *et al.* (1994). This model assumed parallel microcracks nucleated randomly with a certain size distribution. Provided a crack of length $c$ satisfies the coalescence condition, it will coalesce with its neighbor. The coalescence will continue on a greater and greater scale until complete fracture occurs. This non-local coalescence condition together with continuing crack nucleation will lead to a cascade of coalescence of microcracks. This model may predict the dependence of the fractal dimension $D_F$ on the interaction parameter, $L_c$, a normalized critical distance between two microcracks. In a two-dimensional system, the variation could be estimated from

$$D = \frac{\ln(2 + \frac{L_c}{2})}{\ln(2 + \frac{L_c}{\pi})}. \tag{4.18.7}$$

This formula demonstrates a unified spectrum $D(L_c)$ similar to Williford's qualitatively; but, as reported by Bai *et al.*, has a peak value much higher than that obtained in their computer simulation and much lower than that predicted by Williford's multifractal model (1988).

### 4.18.7. *Oscillatory Propagation of a Slant Crack*

Thomson (1986) analyzed the stress field at the crack tip. In the combined mode II loading case, there is always a branching force to make the crack change the direction of propagation. We know that pure mode I, II or III cases are rare. Even in pure mode I loading, the direction of the maximum stress, analysed for an isotropic medium, is not always consistent with the direction of the easiest glide plane in crystals. The angle between these two directions leads to a complex mode of cracking. This slant crack induces oscillations in the direction of crack propagation.

In general, defects may be considered as singularities in an elastic field (Eshelby, 1956). Eshelby's elastic energy momentum tensor theory in continuum media provided a powerful tool for understanding the motion of a singularity in an elastic field. Now, we consider the crack tip as a singularity in the elastic field. For uniaxial applied stress, a slant crack with an angle $\beta$ to the vertical loading will crack along the direction with an angle $\alpha_0$ to the horizontal axis. The relation between $\alpha_0$ and $\beta$ may be written as (Lung, 1980),

$$\alpha_0 = \text{arc } tg \left[1 - \frac{(1-2\nu)}{2(1-\nu)} \sin 2\beta\right] . \tag{4.18.8}$$

The cracking angle $\alpha_0$ has a relation to the fracture angle $\theta_0$ in traditional fracture mechanics (Fig. 3.17)

$$\alpha_0 = \theta_0 - \beta + \frac{\pi}{2} . \tag{4.18.9}$$

Figure 3.19 represents the theoretical curves calculated from Eq. (4.18.8), in comparison with the finite element method, strain energy density method and maximum stress method. Five groups of experimental data are also represented. In spite of a larger difference with other theoretical curves below $\beta = 40°$, the qualitative forms of the calculated curves and the maximum cracking angle are nearly the same.

Figure 4.35 represents the evolution of a slant crack. The curves are calculated according to Eq. (4.18.8). O′B and O′B′ are straight lines for: $\alpha_0 + \beta - \frac{\pi}{2} = 0$ and $-\alpha_0 + \beta - \frac{\pi}{2} = 0$. A slant crack at an initial angle $\beta$ point $A_1(\beta, 0)$ propagates to $A_2(\beta, \alpha_{01})$; and then from $A_2$ as a new slant crack of angle $\alpha_{01}$, at the new starting position, $A_3(\frac{\pi}{2} - \alpha_{01}, \alpha_{01})$, propagates to $A_4(\frac{\pi}{2} - \alpha_{01}, \alpha_{02})$, and so on. It will approach $O'(\frac{\pi}{2}, 0)$, the fixed point finally. This oscillatory propagation is deterministic. However, the length of each cracking step is stochastic. The distribution of the initial slant angle is random. Looking at the power-law relationship for the crack growth process under alternating stress, (Eq. 4.18.6), it is not unreasonable to assume that this kind of oscillatory propagation path is like a fractal curve approximately. The fractal dimension estimated is about 1.05 ($H = 0.95$). This might be also one of the sources for fractal formation in the micron range.

Fractals in atomic range of scales have not been much studied to date. We know that kinks of dislocation lines, clusters of impurity atoms and vacancies, and localized electrons can have marked influences on mechanical properties and they might form fractal structures in some cases. One needs further experimental investigations (Milman *et al.*, 1994). Theoretical investigations are also required concerning atomic scale fracture.

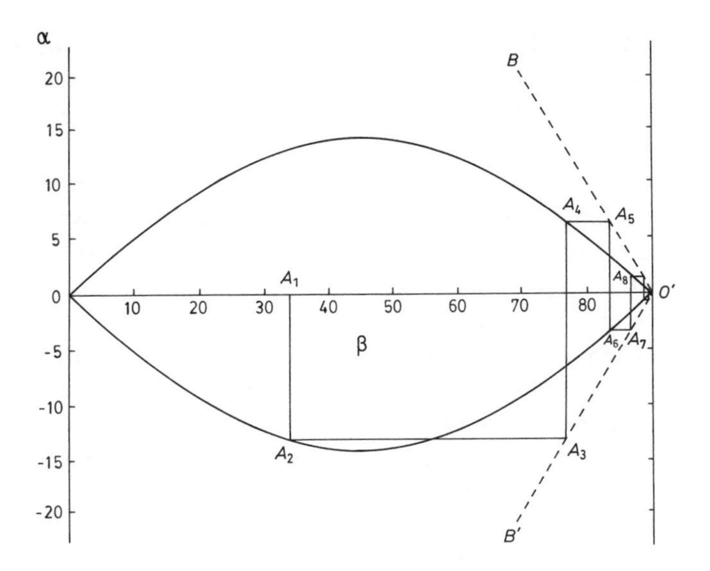


Fig. 4.35. The evolution of a slant crack from an initial angle approaching a fixed point finally.

For fractal studies, there are three fields worthy of future research: (1) Fractal description of fractures, (2) The relationship of fractal dimension with fracture properties of materials, and (3) The physical sources of the formation of fractal structures. At the time of writing, many works have been carried out on (1) and (2).

Fractals are the geometry of nature; one must still understand how nature produces them. The origin of fractals is a dynamical problem. Real systems are dissipative, that is they have friction, and rarely go to their ground state, unlike the ideal models usually discussed in a first year course of physics. A typical example is a simple pile of sand. Adding sand slowly to a flat pile will result only in some local rearrangement of particles. The individual grains, or degrees of freedom, do not interact over large distances. Continuing the process will result in the slope increasing to a critical value where an additional grain of sand gives rise to avalanches of any size, from a single grain falling up to the full size of the sand pile. The pile can no longer be described in terms of many local degrees of freedom, but only a whole description in terms of one sandpile will do. The distribution of avalanches follows a power law. "Self-Organized Criticality" (SOC) refers to this tendency of large dissipative systems to drive themselves to a critical state with a wide range of length and time scales (see Bak and Chen, 1989; Bak and Creutz, 1993). The idea provides a unifying concept for large scale behaviour in systems with many degrees of freedom. Fracture in materials seems very like SOC. Continuing the process of loading will result in a critical state, fracture. SOC can be expected to be one of the best theoretical models through which to understand the physical origin of formation of fractal structure of fractured surfaces in materials.

Subsequent advances in the understanding of the dynamics of interfaces and lines pinned by quenched random impurities might shed light on the problem of crack surfaces and fronts (Narayan and Fisher, 1993; Ertas and Karder, 1994; Ramanathan *et al.*, 1997a; 1997b) since a crack front traverses the sample. Depinning of interfaces (or lines) is a non-equilibrium critical phenomenon involving an external force and a pinning potential (Karder, 1990). When the force is weak the system is stationary, trapped in a metastable state. Beyond a threshold force the metastable state disappears and the system starts to move. Before the motion settles to a uniform velocity due to viscous force, the velocity of the point close to threshold behaves as $v \propto (F - F_c)^\beta$. A self-affine rough surface (or line) would form as the front of the crack propagates in the material.

However, as in Section 4.18, we have explained, mechanical properties are structurally sensitive. Fracture in real materials is a process also driven by internal forces in the driven media. Consideration of these factors would be helpful.

In summary, fractal dimension provides a continuous parameter for quantitative fractography; however, the following points should be noted:

(1) This method is suitable for materials with continuous changes of structure rather than abrupt changes.
(2) Selection of a nice method for fractal dimension measurement is important.
(3) Plane strain fracture toughness is a useful criterion for toughness of materials other than impact test and etc.
(4) It is recommended to use a standard specimen for comparison with mechanical parts in service.

# Chapter 5

# Elastic Moduli and More General Phonon Properties

## 5.1. *Outline*

A major theme of this Chapter will be the phonon theory of the cleavage force between half-planes of a metal, and its relation to surface energy. Propagation of screw dislocations, treated by lattice dynamics, with emission of phonons, will also be referred to, as will some approximate relations between elastic constants in hexagonal-close-packed metals.

## 5.2. *Force between Half-Planes of a Metal*

The advent of the atomic force microscope means that the force $F(z)$ between the half-planes of a metal as a function of separation $z$ becomes an observable. Of course, one of the metal surfaces is the tip; therefore size effects will have to be considered also.*

Let us note first the early work of Friedel (1976). In his study, all attention was focused on the small-$z$ form of the force $F(z)$, namely

$$F(z) = Az : \text{small } z\,. \tag{5.2.1}$$

Roughly, following Friedel (1976), the cleavage force $F(z)$ per unit area is

$$\left.\begin{aligned} F &= E(z/a)\,, & z \le a \\ &= \quad 0\,, & z > a \end{aligned}\right\}. \tag{5.2.2}$$

*A simple model relevant to such size effects has been worked out by M. Razavy, N. H. March, B. V. Paranjape, Phys. Rev. *B54*, 4492 (1996): see also N. H. March, Int. J. Quantum Chem: Proc. Sanibel Symp. 1997: in press.

Here $E$ is an appropriate elastic modulus, $a$ is the lattice interplanar spacing while $z$ is the separation of the two half metal crystals during the cleavage.

But, in contrast to the above emphasis on the short-range form of $F(z)$, Lifshitz (1956) has shown that when the two halves of the crystal are very far apart, an attractive polarization force per unit area exists, having the form

$$F = C/z^3 : \text{large } z\,. \tag{5.2.3}$$

The constant $C$ can be related to the frequency-dependent dielectric function $\epsilon(\omega)$ (see, for example Kohn and Yaniv, 1979). We note here that for a jellium-type of metal, with plasma frequency

$$\omega_p = \left(\frac{4\pi ne^2}{m}\right)^{1/2}, \tag{5.2.4}$$

$n$ being the electron density and $m$ the electron mass, then $C$ in Eq. (5.2.3) is given by

$$C = 1.79 \times 10^{-3}\hbar\omega_p\,. \tag{5.2.5}$$

### 5.2.1. *Relation of Cleavage Force to Surface Energy of Metal*

Following Kohn and Yaniv (1979), we consider below only metal crystals with one atom per unit cell. These workers consider then the situation in which equal and opposite forces are applied to all atoms on two adjacent planes. In Fig. 5.1, taken from Kohn and Yaniv (1979), reversible cleavage of the crystal is depicted. Fig. 5.1(a) shows the uncleaved crystal, with all interplanar spacings parallel to the cleavage plane (the dashed line) equal to $a$. The crystal during cleavage is depicted in Fig. 5.1(b), having a spacing of $a + z$ between the two outermost planes. Relaxation of the spacings between

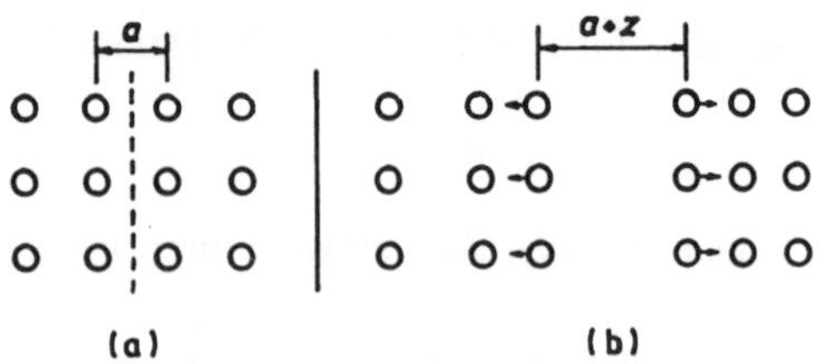

Fig. 5.1. Reversible cleavage of the crystal. (a) Shows the uncleaved crystal, with all interplanar spacings parallel to the cleavage plane (shown dashed) equal to $a$. (b) Shows the crystal during cleavage, with a spacing of $a + z$ between the two outermost planes, and naturally relaxed spacings between the other planes.

the other planes is accounted for in the treatment of Kohn and Yaniv (1979), their work transcending in this respect the earlier important study of Zaremba (1977) who made the assumption of a rigid lattice.

Including these relaxation effects, the surface energy $\sigma$ per unit surface area is equal to one-half the total cleavage work per unit cross-sectional area. Hence one can write (the factor of 1/2 below arises from the two surfaces formed during cleavage)

$$\sigma = \frac{1}{2}\int_0^\infty F(z)dz\,. \tag{5.2.6}$$

While Eq. (5.2.6) is formally exact, if we insert the approximate form (5.2.2) for $F(z)$ we obtain $\sigma \sim (1/4)Ea$, which estimate, of course, focusses all attention on the short-range form of $F(z)$. Alternatively one might try using the longer-range form (5.2.3) for $z \geq z_0$ and putting $F(z) = 0$ for $z \leq z_0$, yielding $\sigma = C/4z_0^2$ (Kohn and Yaniv, 1979). In this form, the theory is related to the ideas of Schmidt and Lucas (1972) and of Craig (1972). These workers presented arguments for choosing $z_0 = 0.33v_F/\omega_p$ for metals, where $v_F$ is the Fermi velocity. Making this particular choice leads to semiquantitative agreement with experimental surface energies.

### 5.2.2. *Kohn-Yaniv Interpolation Formula for Cleavage Force*

Kohn and Yaniv (1979) write

$$F(z) = dU(z)/dz \tag{5.2.7}$$

and then note that the limiting forms (5.2.1) and (5.2.3) are embraced by the choice

$$U = -\frac{1}{2}\left(\frac{C}{d^2+z^2}\right) : d^2 = (C/A)^{1/2}\,. \tag{5.2.8}$$

Inserting Eqs. (5.2.7) and (5.2.8) into Eq. (5.2.6) readily yields the Kohn-Yaniv result

$$\sigma = -\frac{1}{2}U(0) = \frac{1}{4}(AC)^{1/2}\,. \tag{5.2.9}$$

As already mentioned, the constant $A$ was calculated by Zaremba (1977) in terms of the phonon spectrum of the crystal, by making a rigid lattice assumption. Kohn and Yaniv (1979) allow for atomic relaxation as a result of cleavage. They calculate the constant $A$ in Eq. (5.2.1) as

$$A = \frac{1}{4}\rho a\omega_0^2 \tag{5.2.10}$$

where $\rho$ is the density, $a$ the equilibrium interplanar spacing parallel to the cleavage plane, while $\omega_0^2$ is expressed in terms of the frequencies $\omega(q)$ of the longitudinal phonons perpendicular to the cleavage plane (see Appendix 5.1 for the detailed relation between $\omega_0^2$ and $\omega^2(q)$).

### 5.2.3. *Universal Model for Cleavage Force*

Besides the interpolation formulae (5.2.7) and (5.2.8) for the cleavage force, Kohn and Yaniv (1979) propose a 'universal model'. To construct this, they scale the force $F$ in units of $(A^3C)^{1/4}$ and the distance $z$ in units of $(C/A)^{1/4}$:

$$F(z) = (A^3C)^{1/4} f[z/(C/A)^{1/4}] \tag{5.2.11}$$

so that $f(z) = z$ for $z \ll 1$ and $f(z) = 1/z^3$ for $z \gg 1$.

The surface energy $\sigma$ is then given by inserting Eq. (5.2.11) into Eq. (5.2.6) as

$$\sigma = \alpha(AC)^{1/2} \tag{5.2.12}$$

where A is given by Eq. (5.2.10), $C$ by the Lifshitz theory already referred to, and

$$\alpha = \frac{1}{2}\int_0^\infty f(z)dz \tag{5.2.13}$$

which Kohn and Yaniv (1979) argue will be characteristic of each material and each crystal face. These workers note that $f(z)$ has the same form for both small and large $z$ for all materials and, in the absence of additional information, they made the hypothesis that $f(z)$ is a universal function for all materials.

Kohn and Yaniv (1979) then compare the prediction (5.2.12) with experiment for the (110) faces of 10 cubic metals for which reliable experimental surface energies exist. They plot $\sigma_{\text{expt}}$ vs $(AC)^{1/2}$ and find a reasonable straight line passing through the origin (see Fig. 5.2). A least-squares fit led them to the value

$$\alpha = 0.476\,, \tag{5.2.14}$$

which is a little less than twice the value obtained with the simple interpolation form in Eqs. (5.2.7)–(5.2.9).

Further work, both experiment and theory, the former using the atomic force microscope, is clearly needed to find the detailed form of the cleavage force $F(z)$ for a variety of metal crystals, and hence to test the validity of the scaling

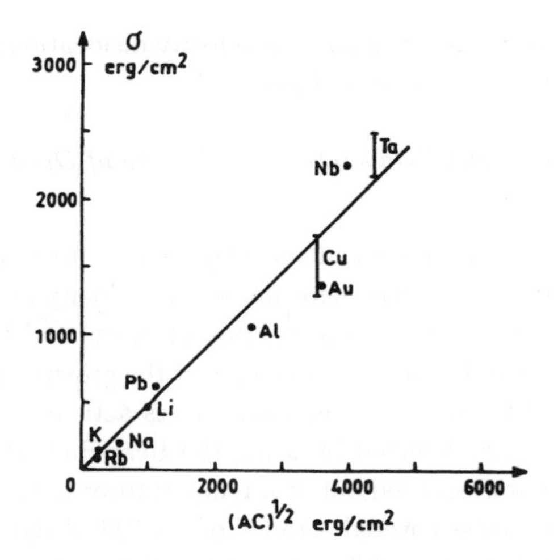


Fig. 5.2. Experimental surface energies vs $(AC)^{1/2}$. The straight line represents the best fit with $\alpha^* = 0.476$. After Kohn and Yaniv (1979).

hypothesis of Kohn and Yaniv (1979) in more detail. We finally note that March and Paranjape (1984) have calculated $A$ from electron theory using the jellium model plus solution of linearized Thomas-Fermi equation.*

Their work has been generalized subsequently by Heinrichs (1986) and related to the study of Budd and Vannimenus (1973).

### 5.2.4. *Emission of Phonons by Moving Dislocations*

Atkinson and Cabrera (1965) have used the Frenkel-Kontorova model to discuss the acoustic emission from a (one-dimensional) dislocation. Subsequently, Celli and Flytzanis (1970) have treated the problem of the motion of a screw dislocation in a crystal from the standpoint of lattice dynamics. For a specific lattice with piecewise linear nearest-neighbour interactions, the external stress necessary to maintain the dislocation in uniform motion is calculated as a function of the dislocation velocity. This stress turns out to have a minimum at one-half of the velocity of sound and motion still proves possible at hypersonic velocities. Since the calculations were appropriate at absolute zero of temperature, the only dissipative mechanism that is operative is the emission of phonons by the core of the moving dislocation. Celli and Flytzanis (1970)

---

*For subsequent work, see both references added in footnote on p. 1934.

stress that the theory is relevant to high-velocity dislocation motion,* which is not of the usual thermally activated kind.

### 5.2.5. *Effect of Stress and Temperature on Velocity of Dislocations in Fe Single Crystals*

In relation to thermal activation, it is relevant here to refer to the work of Turner and Vreeland (1970) who determined the velocity of edge oriented dislocations in Fe single crystals, as a function of stress and temperature. The velocities were extracted from measurements of the growth of slip bands which had been subjected to constant amplitude stress activated by a pulsed technique. Slip bands were detected by using the Berg-Barrett X-ray technique. The measurements were carried out at 4 temperatures in the range 77–373°K. The resolved shear stress covered a range of 10–500 M dyn/cm$^2$, while measured velocities lay between $10^{-6}$ to 1 cm/sec. Turner and Vreeland observed a strong temperature dependence of the dislocation velocity. They concluded that their results can only be correlated with theories with a single thermally activated mechanism if a substantial entropy of activation exists. They discuss alternative explanations of their experimental findings in term of multiple processes and differences in slip band structure.

### 5.2.6. *Pair Potential: Uses and Limitations*

Simulations using empirical interatomic potentials can often supply efficient and usually inexpensive routes for studying ionic structure and dynamics in metallic systems. For a long time, pair potentials were used very extensively in such simulation studies. They can reproduce usefully total energies for many systems. But when one turns to elastic properties, deficiencies emerge (e.g. their instability to reproduce the so-called Cauchy discrepancy; see for instance Johnson (1972)). This situation can be remedied by the addition to the pair potential contribution of a volume-dependent, structure independent energy (the reasons being set out in Chaps. 6 and 7 below). But in specific examples, such as fracture surfaces, where the volume is ambiguous, pair potential models need transcending. A further difficulty, in the (simplest) pair potential scheme comes up in the determination of the vacancy formation energy $E_v$ (compare Johnson, 1987). It is found empirically that this energy $E_v$ is typically about 1/3 of the cohesive energy. In contrast, the straightforward

*The Celli-Flytzanis lattice dynamical model is discussed in some detail in Chapter 10.

pair potential models predict that, excluding the contribution from relaxation which is modest in close-packed metals, these two energies are equal. These limitations of the simple pair potential approximation have been addressed by the development of empirical many-body potentials which is the major theme of Chap. 8 below.

### 5.2.7. *Some Useful Combinations of Elastic Constants for Hexagonal Close Packed Metals*

We wish to include in this chapter related to phonons and elastic constants some considerations relating particularly to the latter in the specific case of hexagonal close-packed (hcp) metals.

Early work by Czachor (1965) gave the elastic constants for hcp crystals in terms of the components of the dynamical matrix. Subsequently, expressions for the elastic moduli were obtained in the case of a pair interatomic potential (with third-neighbours interaction range) by Trott and Heald (1971). Both the above approaches depend on the method of long waves (see Born and Huang (1956)), in which the elastic properties emerge as a by-product of the dynamical matrix in the long wavelength limit.

Later work by Martin (1975) used the method of uniform deformation as applied to general monoatomic lattices. The work of Pasianot and Savino (1992) on which the remainder of this section is largely based also utilizes this method of uniform deformation applied to the hcp lattice (see also the earlier treatment for the cubic metals by Pasianot *et al.*, 1991). Their work is within the context of a specific type of many-body interatomic potential — this aspect will be covered in Chap. 8.

However, within such an admittedly constrained framework, Pasianot and Savino (1992) obtain the following restriction on the elastic constants of an hcp structure:

$$\frac{1}{2}(3c_{12} - c_{11}) > c_{13} - c_{44} \,. \qquad (5.2.15)$$

Table 5.1, taken from Pasianot and Savino (1992), collects the experimental values for the two sides of the inequality (5.2.15) for around 10 hcp metals. Zn and Cd do not satisfy the inequality (5.2.15) and Zr is a marginal case, which will be discussed further in Chap. 8.

In connection with the same type of many-body interatomic potential, various authors (Daw, 1989; Carlsson, 1990) have emphasized that one should

Table 5.1. Experimental values for hcp metals of the two sides of inequality (5.2.14) (taken from Pasianot and Savino, 1992). $c_{ij}$ are given in eV/Å$^3$.

| | $\frac{1}{2}(3c_{12} - c_{11})$ | $(c_{13} - c_{44})$ | $c/a$ ratio |
|---|---|---|---|
| Be | −0.67 | −0.97 | 1.568 |
| Mg | 0.047 | 0.025 | 1.624 |
| Zn | −0.20 | 0.06 | 1.856 |
| Cd | −0.028 | 0.106 | 1.886 |
| Y | 0.012 | −0.05 | 1.572 |
| Hf | 0.11 | 0.037 | 1.581 |
| Ti | 0.26 | 0.11 | 1.588 |
| Zr | 0.14 | 0.18 | 1.593 |
| Sc | 0.09 | 0.01 | 1.594 |
| Tl | 0.22 | 0.13 | 1.598 |
| Co | 0.55 | 0.12 | 1.623 |

have (in addition to the inequality (5.2.15)):

$$c_{13} - c_{44} > 0\,. \tag{5.2.16}$$

Table 5.1 shows that this is not the case for Be and Y, but all other hcp metals satisfy this inequality (5.2.16).

We shall take up these matters again, within the embedded atom and related methods, in Chap. 8.

## 5.3. *Empirical Relations between Elastic Moduli, Vacancy Formation Energy and Melting Temperature*

We shall conclude this Chapter by briefly mentioning some empirical relations which are established between elastic moduli, vacancy formation energy and melting temperature. Some insight into such relations can be gained from electron theory, as well as from pair potential arguments.

One of these empirical relations connects the Debye temperature, related to elastic properties (see Eq. (A6.15)) to the vacancy formation energy $E_v$ (Eq. (A.6.6)). In Appendix A5.2, a model is presented which shows that $E_v \propto B\Omega$ in some special situations, with $B$ the bulk modulus and $\Omega$ the atomic volume. This is equivalent to the empirical Mukherjee (1965) relation (A6.16). Finally, the established correlation between $E_v$ and thermal energy

$k_B T_m$ associated with the melting temperature is treated by a statistical mechanical model in Appendix A7.4.

# Chapter 6

# Elements of Electronic Structure Theory

## 6.1. *Free Electron Theory*

The phase space result for the electron density at the Fermi level is readily extended to yield the density of electrons $n_0(E)$ below energy $E$ as

$$n_0(E) = \frac{8\pi}{3h^3}(2mE)^{3/2}\,. \tag{6.1.1}$$

Thus, since states are doubly filled by electrons with opposed spins, the number of states/unit volume $N_0(E)dE$ lying between energy $E$ and $E + dE$ is given by the derivative of Eq. (6.1.1); to yield

$$N_0(E) = \frac{4\pi}{h^3}(2m)^{3/2}E^{1/2}\,. \tag{6.1.2}$$

Electrons can only undergo excitation into unoccupied states above the Fermi energy $E_F$ because of the Pauli Exclusion Principle. For thermal excitation, only electrons within $\sim k_BT$, the thermal energy corresponding to temperature $T$, can be excited across $E_F$. At room temperature, $k_BT \sim (1/40)$ eV, and thus these electrons constitute a small fraction $\sim k_BT/E_F$ of the total number of electrons in a simple metal like Cu or Al. The classical value of the heat capacity is thus reduced by this factor $k_BT/E_F$. Employing Fermi-Dirac statistics appropriate to almost degenerate electrons, rather than classical Maxwell-Boltzmann statistics, one obtains (see for example Mott and Jones (1936)) for the specific heat at constant volume at low $T$:

$$c_v = \frac{\pi^2}{2}k_B\frac{k_BT}{E_F}\,. \tag{6.1.3}$$

### 6.1.1. *Bulk Moduli of Metals Compared with Free Electron Model*

In the free electron model, the total energy $E$ of the metal is just the kinetic (mean Fermi) energy of the electrons. This energy, per electron is usefully expressed in terms of the mean interelectronic spacing $r_s$. This represents the radius of a sphere of volume equal to the volume per electron in the metal. If $r_s$ is given in units of the Bohr radius $a_0 = \hbar^2/me^2$, then

$$\frac{E}{NZ} = \frac{2 \cdot 21}{r_s^2}, \tag{6.1.4}$$

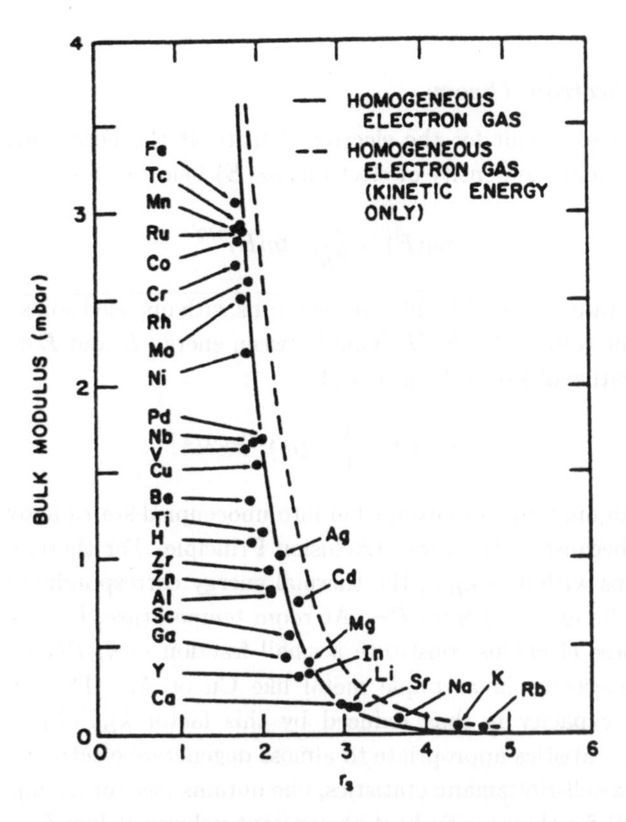


**Fig. 6.1. Comparison of bulk moduli of metals (obtained from band calculations) with that of a homogeneous electron gas having a density equal to that in the interstitial region between atoms in the metal reproduced from Fig. 18 of Alonso and March (1989) and also Williams *et al.* (1980).**

where $E/NZ$ is then in Rydbergs(Ry), $Z$ being the valence (1 for Cu: 3 for Al). The isothermal bulk modulus $B$ of the free-electron gas is then readily obtained as

$$B = \frac{0.586}{r_s^5}. \tag{6.1.5}$$

The simple result (6.1.5) is already in agreement with the experimental bulk modulus data for alkali metals just above their melting points (see Alonso and March, 1989). The agreement worsens as the valence $Z$ increases, though it is true that the prediction (6.1.5) gives the qualitative characteristic that the bulk modulus $B$ increases with decreasing interelectronic separation $r_s$. The dashed curve in Fig. 6.1 shows the prediction (6.1.5). The solid circles are obtained from energy band theory calculations (Williams *et al.*, 1980).

## 6.2. *Exchange and Correlation Interactions*

In the homogeneous electron gas, a refined energy calculation transcends the kinetic energy result (6.1.4) by adding exchange and correlation, as first discussed by Wigner and Seitz in pioneering work. The physical origin of these (potential) energy terms is easy to describe in words.

### 6.2.1. *Exchange Hole*

The concept of the Fermi hole around an electron one chooses to 'sit on', say with ↑ spin as it moves through the uniform electron gas (jellium model) goes back to Wigner and Seitz. In the Hartree-Fock approximation, there is no correlation between antiparallel electrons, while the Pauli Principle causes electrons of ↑ spin to be repelled from the ↑ spin electron chosen as origin. Then the uniform electron density $n_0$, relative to the origin and distance $r$ away becomes $n_0 g(r)$ where $g(r)$ is the electronic pair function. Evidently $g(r) \to 1$ as r $\to \infty$ and Wigner and Seitz (1934) obtained the Hartree-Fock result

$$g(r) = 1 - \frac{9}{2}\left(\frac{j_1(k_F r)}{k_F r}\right)^2 \tag{6.2.1}$$

where $j_1(x)$ is the first-order spherical Bessel function $[\sin x - x\cos x]/x^2$ and $k_F$ is the Fermi wave number, related to the Fermi momentum $p_F$ by $k_F = p_F/\hbar$.

It is easy to show that there is precisely a deficit of 1 electron as a result of this hole created around the electron at the origin. This result is not changed when Coulombic repulsion $e^2/r_{ij}$ between electrons $i$ and $j$ at separation $r_{ij}$

is included. In the paramagnetic electron gas considered above, $g(r = 0) = 1/2$, reflecting that parallel spins cannot sit at the origin because of the Pauli Principle but antiparallel spins are uncorrelated. When one allows for Coulomb repulsion, $g(r = 0) < 1/2$ and since the hole (now called exchange-correlation hole) must still contain precisely a deficit of one electron, the hole is deepened but is somewhat less spread out, to preserve this normalization.

Returning to the exchange hole $g(r)$ above, the mean potential energy per electron is simply exchange energy, as in the jellium model a neutralizing background of (non-responsive) positive charge is implicit, which cancels the classical Coulomb potential energy terms. It is then to be noted that the energy above is essentially the interaction of the Fermi hole electron density described by Eq. (6.2.1) with the electron, of charge $-e$, at the origin. Thus, one merely calculates the electrostatic potential at the origin due to the Fermi hole density $n_0[g(r) - 1]$ to obtain

$$\text{Exchange energy per electron} = \frac{e^2}{2} \int \frac{n_0[g(r) - 1]}{r} d\mathbf{r} \tag{6.2.2}$$

where the factor $1/2$ avoids double counting of the electron-electron interactions. It is readily shown, by using the above free-electron result locally, that the exchange energy per unit volume $\epsilon_x$ of an inhomogeneous electron gas of density $n(r)$ is

$$\epsilon_x(\mathbf{r}) = -c_x n(\mathbf{r})^{4/3} : c_x = \frac{3}{4} e^2 \left(\frac{3}{\pi}\right)^{1/3} . \tag{6.2.3}$$

From this follows the Dirac-Slater exchange potential $V_x(\mathbf{r}) \equiv \delta A / \delta n(\mathbf{r})$ where

$$A = \int \epsilon_x(\mathbf{r}) d\mathbf{r} \tag{6.2.4}$$

as

$$V_x(\mathbf{r}) = -\frac{4}{3} c_x n(r)^{1/3} . \tag{6.2.5}$$

This is the still widely used 'local density approximation' to the exchange potential entering the Schrödinger equations to calculate the Slater-Kohn-Sham orbitals in density functional theory.* Of course, $V_x(\mathbf{r})$ needs corrections for

*The important method of Car and Parrinello, which combines density functional theory and molecular dynamics is reviewed in a practically oriented article by Remler and Madden (1990). Later, E. Smargiassi and P. A. Madden (Phys. Rev. *B51*, 129, 1995) use the Car-Parrinello approach to calculate defect properties without solving the Slater-Kohn-Sham Eq. (6.2.6), but by treating electronic kinetic energy as a functional of electron density.

the 'correlation' contribution to the 'exchange-correlation' hole and this can be done, again in the local density approximation, by using the quantum Monte-Carlo simulations by Ceperley and Alder (1980). Their results have led Vosko *et al.* (1980), and Perdew and Zunger (1981) to propose sophisticated fitting formulae, the accuracy of which have been briefly considered by Herman and March (1984). This all leads, in turn to an exchange-correlation potential* $V_{xc}(\mathbf{r})$ to transcend Eq. (6.2.5). This then allows the one-electron Schrödinger equation

$$\nabla^2\phi_i + \frac{2m}{\hbar^2}[\epsilon_i - V_{\text{Hartree}}(\mathbf{r}) - V_{xc}(\mathbf{r})]\phi_i = 0 \tag{6.2.6}$$

to be solved for the Slater-Kohn-Sham orbitals $\phi_i$. This equation, if $V_{xc}(\mathbf{r})$ were known exactly (see Holas and March,1995) would lead, by construction in an $N$-electron problem, to the exact ground-state density

$$n(\mathbf{r}) = \sum_{i=1}^{N} \phi_i(\mathbf{r})\phi_i^*(\mathbf{r})\,. \tag{6.2.7}$$

This route, with approximations such as 'local density' detailed above, is now widely used in condensed phases, both solids that are crystalline or amorphous, and liquids. The electron theory pair potential for liquid Na just above the freezing point, discussed in Chap. 7, has been obtained starting from such Slater-Kohn-Sham equations for $\phi_i$ and then constructing the electron density $n$ by summing the squares of these wave functions out to the Fermi level.

## 6.3. *Bulk Modulus Including Exchange and Correlation*

Having given this brief introduction to the density functional theory of exchange and correlation in an inhomogeneous electron gas, let us return to jellium and correct the bulk modulus formula (6.1.5) for electron-electron interactions. If for the correlation part we use the Nozieres-Pines formula (Nozieres and Pines 1958) then the total energy (per electron) of the homogeneous electron gas becomes (in Ry)

$$\frac{E}{NZ} = \frac{2.21}{r_s^2} - \frac{0.916}{r_s} - (0.115 - 0.0313\ln r_s) \tag{6.3.1}$$

*For semiempirical forms of the exchange — correlation energy $E_{xc}[n]$ and the corresponding exchange — correlation potential $V_{xc}(\mathbf{r}) = \delta E_{xc}/\delta n(\mathbf{r})$, see N. C. Handy and D. J. Tozer, Molecular Physics *94*, 707 (1998).

where the second and third terms on the right-hand side are respectively (Dirac) exchange (see above) and correlation contributions. The corresponding bulk modulus $B$ is then given by Ling and Gelatt (1980)

$$B\Omega = Z[22.1 - 3.66r_s - 0.093r_s^2]/9r_s^2 \tag{6.3.2}$$

where $\Omega$ is the atomic volume. The interest in this result becomes clear when one applies it to the electron density ($n_b$ say) in the interstitial region between Wigner-Seitz spheres. The continuous curve of Fig. 6.1 demonstrates that the bulk modulus, even in transition metals, reflects the energy required to compress the interstitial electrons (see also Alonso and March, 1989).

## 6.4. *Structural Stability of Non-Transition Metals*

Following the work of Corless and March (1961) on the long range oscillating interaction between test charges (see also Appendix 7.1.) Worster and March (1964) pointed out that such non-monotonic interatomic potentials would have interesting implications regarding the equilibrium lattice structures of simple metals. Much quantitative work has been done subsequently, and we shall summarize in this section some of the results for nontransition metals.

As discussed later in this chapter, within second-order perturbation theory the division of the total energy of a metal into a volume-dependent, structure-independent term, plus a sum of two-body interactions, is well suited to a treatment of the structural stability of metal. An early review of the pseudopotential theory* of the crystal structures of non-transition metals, in which trends across the Periodic Table were discussed was that of Heine and Weaire (1970). Since then, a good deal of progress has been made in the explanation of structural stability, associated especially with the names of Cohen, Martin, Hafer and Moriarty (for detailed references, see the book by Alonso and March (1989).

The account below is based on the approach of Hafner and Heine (1983) and the review of this work given by Alonso and March (1989). The Hafner-Heine approach (compare also Corless and March, 1961; Worster and March, 1964) works in coordinate $\mathbf{r}$ space, and it demonstrates that the use of a simple empty-core pseudopotential is adequate to understand the structural trends. In essence, their work also shows that the structural trends arise from a

*For the reader requiring definitions and a little detail, Section 6.12 could be consulted at this point.

characteristic variation of the **r** space interaction potential with electron density and pseudopotential.

For reasons set out above, it is useful in discussing structural trends to analyze structural stability at constant volume. Then one needs only to appeal to the two-body part of the total energy. Then, for a typical **r** space interionic potential, the essential idea governing structure is easy to state; a closed-packed structure is favored if the nearest neighbors fall on the minimum of the potential. On the other hand, if the nearest neighbor in the fcc or bcp lattices fall on a peak, then the crystal may lower its energy by moving some neighbors closer and moving others further away.

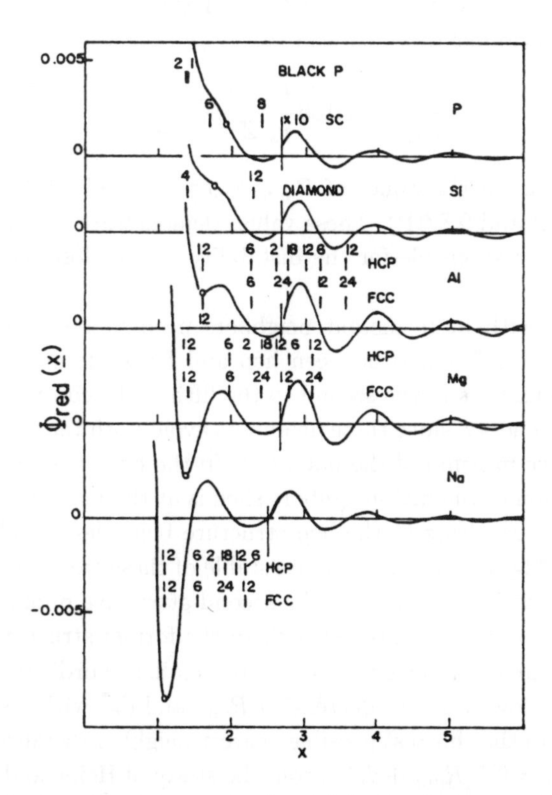


Fig. 6.2. Trends in the reduced interaction $\Phi_{\rm red}(x)$ for the elements in the Na period. The circle indicates where the nearest-neighbour distance for close packing falls on the interatomic potential. Vertical bars indicate nearest-neighbour distances and coordination numbers for the indicated structures. Redrawn after Hafner and Heine, 1983.

The trend across a row of the Periodic Table is a change from close-packed structures at the left to open structures with lower coordination numbers on the right. Figure 6.2 (Fig. 28 of Alonso and March, 1989) shows the interionic potential for 2nd row elements, taken from Hafner and Heine (1983). For Na, Mg and Al the nearest neighbors for a close-packed structure reside at a minimum of the potential but this is not so for Si or P. This then is the reason for the stability of the closed-packed structure in the first 3 elements referred to above, but not in the other 2 cases. This trend, it turns out, is mainly a consequence of the ratio of the radius of the Wigner-Seitz sphere $R_{ws}$ to the mean interelectronic spacing $r_s$ owing to the different valencies. The nearest-neighbor distance in a close-packed structure, $d_{cp}$ say, is related to the Wigner-Seitz radius by $d_{cp} = 1.809R_{ws} = 1.809r_s Z^{1/3}$. Alonso and March (1989) give the arguments which lead to†

$$\frac{R_{\min}}{d_{cp}} = 1.106\frac{R_c}{r_s}\frac{1}{Z^{1/3}} + \frac{1.824}{r_s Z^{\frac{1}{3}}} \tag{6.4.1}$$

from which the following values of $R_{\min}/d_{cp}(\equiv x)$ result; 1.00(Na), 0.98(Mg) 0.99(Al) 0.91(Si) and 0.87(P). These values demonstrate that the variation of $d_{cp}$ with $Z^{1/3}$ is responsible for the instability of the close-packed structures in Si and P.

Interactions with more distant neighbors need to be invoked to explain structural energy differences between hcp and fcc structures. In fcc Al, for example, the neighbors in the second to the fifth shells contribute little to the pair-interaction energy, since these neighbors avoid both the attractive minima and the repulsive maxima of the potential. In the hcp structure, on the other hand, the position of the fifth-neighbors shell is at the maximum of the second repulsive wiggle, resulting in the hcp structure being less stable than fcc. In the example of Mg, the hcp is the more stable of these two structures, because more neighbors sit in the region of the second attractive minimum and less in the region of the second repulsive wiggle in the former structure.

The trend already referred to above to smaller coordination numbers to the right of this row is due to decreasing $R_{\min}$ and $d_r$* with respect to $d_{cp}$. A rough estimate of the shortest possible nearest-neighbor distance in a distorted structure is $d_{nn} = 0.5(R_{\min} + d_r)$. From the study of Heine and Weaire (1970)

† $R_c$ is core radius of pseudopotential: $R_{ws} = 1.4R_c + 1.305$; see also Section 6.12 below.

* $d_r$ is a repulsive pseudoatom diameter (see Alonso and March, 1989). $R_{\min}$ is position of deep minimum in pair potential.

the coordination number $c$ can be estimated from the formula

$$\frac{d_{nn}}{d_{cp}} = \left(\frac{c+1}{13}\right)^{1/3} . \tag{6.4.2}$$

This leads to $c = 5.2$ for Si ($c = 4$ in the diamond structure and $c = 6$ in the high-pressure modification) and $c = 3.1$ for P ($c = 3$ is the experimental value).

### 6.4.1. *Trends Along Columns in Periodic Table*

The account below follows closely that of Alonso and March (1989). In the alkali and the alkaline-earth metals, the atomic volume increases quite rapidly with principal quantum number. This can be thought of, in OPW* or pseudopotential terms, as due to each new electronic shell being pushed outwards

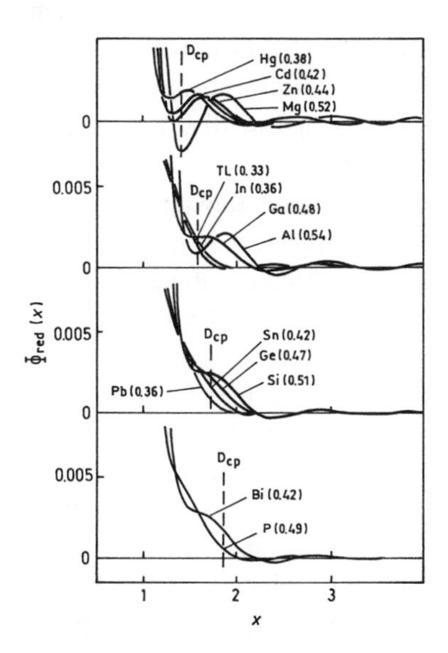


Fig. 6.3. Trends in the reduced interatomic interaction for the B-group elements. The electron-density parameter has been set constant for a given valence ($r_s = 2.56$ for $Z = 2$, $r_s = 2.20$ for $Z = 3$, $r_s = 2.05$ for $Z = 4$ and $Z = 5$). The ratio $R_c/r_s$ is given in parentheses for each case. The dashed line marks the nearest-neighbour distance $D_{cp}$ for close packing. Redrawn after Hafner and Heine, 1983.

---

*Orthogonalized plane wave (OPW) method works with a basis of plane waves orthogonalized to core states.

by orthogonalization to the core. The interionic potentials corresponding to these two groups are consistent with the existence of closed-packed structures in all these metals at some temperature and pressure.

The trend in Group IIB is a transition from hcp structure in Mg to distorted hexagonal structures in Zn and Cd and then to a unique lattice appropriate to Hg. Results of a simplified calculation (see Hafner and Heine) with a constant value of $r_s$ are shown in Fig. 6.3 (Fig. 29 of Alonso and March, p. 90, 1989) and suffice to explain this trend, which is due to the decrease of $d_r$ and $R_{\min}$ relative to $d_{cp}$ with decreasing $R_c/r_s$. Then $d_{cp}$ moves from a minimum to a maximum. The damping in the amplitude of the oscillations is also important. The distorted hexagonal structures of Zn and Cd are related to the ideal case by a shear distortion. The instability of the close-packed structure can be expected to manifest itself as a softening in its elastic shear moduli, and this is well accounted for by the elastic constants obtained from the interionic potentials (Hafner and Heine, 1983). The two polymorphic forms of Hg can be looked upon as tetragonal ($\beta Hg$) or rhombohedral ($\alpha Hg$) distortions of the fcc structure. The pseudopotential approach also predicts the instability of the fcc lattice against rhombohedral distortions and the metastability with respect to tetragonal distortions.

Similar physical effects discussed above for the divalent metals also explain the structural trends through the trivalent elements. Al is fcc since $d_{cp}$ is at the first minimum of the interatomic potential. The form of such a potential for Ga indicates that the close-packed structure is destabilized and a new (complex) structure results for Ga. Nevertheless, the rapid rise of the repulsive part of the potential prevents the formation of an open covalent structure with nearest-neighbor distances smaller than these found in the Ga lattice. Hafner and Heine (1983) have performed an analysis of the elastic constants which demonstrates that, as $R_c/r_s$ decreases further, a tendency appears for the close-packed structures to become stable against shear distortions. This situation is not yet completely reached in In, where the rhombohedral shear has been removed, resulting in tetragonal In, but finally, Tl is again a close-packed structure. To summarize, close-packed structures appear again in the heavy polyvalent metals and this can be related to the damping of the oscillations in the potential. For the metals In, Tl and Pb, the interionic potential is just of screened repulsive character and in this situation, a close-packed structure is favored.

The observed trends in the group with $Z = 4$ are consistent with the preceding discussion; covalent structures in the light elements and stabilization of the fcc structure for Pb. The same can be said about $Z = 5$. Comparison of the interionic potentials of P and Bi supports the trend of decreasing distortion and increasing coordination number, though Bi is not yet close-packed. Alonso and March also discuss the work of Pettifor and Ward (1984) but the interested reader is referred to these sources for details.

## 6.5. *Elastic Constants of Hexagonal Transition Metals*[*] *from Electronic Structure Calculations*

Fast *et al.* (1995) have calculated the elastic constants of the hexagonal 4d transition metals (Y, Zr, Tc and Ru) and also the 5d elements Re and Os by first-principles electronic structure theory.

We note first that for a hexagonal lattice, there are five independent elastic constants, usually denoted as $C_{11}, C_{12}, C_{13}, C_{33}$ and $C_{55}$ (see, e.g. Wallace, 1970). This is in contrast to just three independent elastic constants for cubic materials.

### 6.5.1. *Cauchy Relations for Central Forces*

For hexagonal materials, assuming central forces, the following relations hold between the above elastic constants (see Wallace, 1970):

$$C_{13} = C_{55}$$

$$C_{12} = C_{66} = \frac{1}{2}(C_{11} - C_{12}) .$$

Fast *et al.* (1995) have introduced normalized elastic constants $C'_{ij}$, defined as

$$C'_{ij} = C_{ij}/B$$

with $B$ the bulk modulus. This ratio is favourable for exposing trends, as division by the bulk modulus is essentially normalizing the interatomic forces with an average restoring force of the system.

In Fig. 6.4(a), (b) (Figs. 5(a) and 4(b) from Fast *et al.*, 1995), the experimental results of $C'_{ij}$ are plotted (closed symbols) together with the electronic structure results calculated by Fast *et al.* (open circles). These workers stress that the $C'_{11}$ and $C'_{33}$ renormalized elastic constants scatter around the value 1.8 for all transition metals, from both experiment and theory. Similarly, the normalized $C'_{12}, C'_{13}$ and $C'_{55}$ constants group around 0.6.

---

[*]See also Section 5.2.7 above.

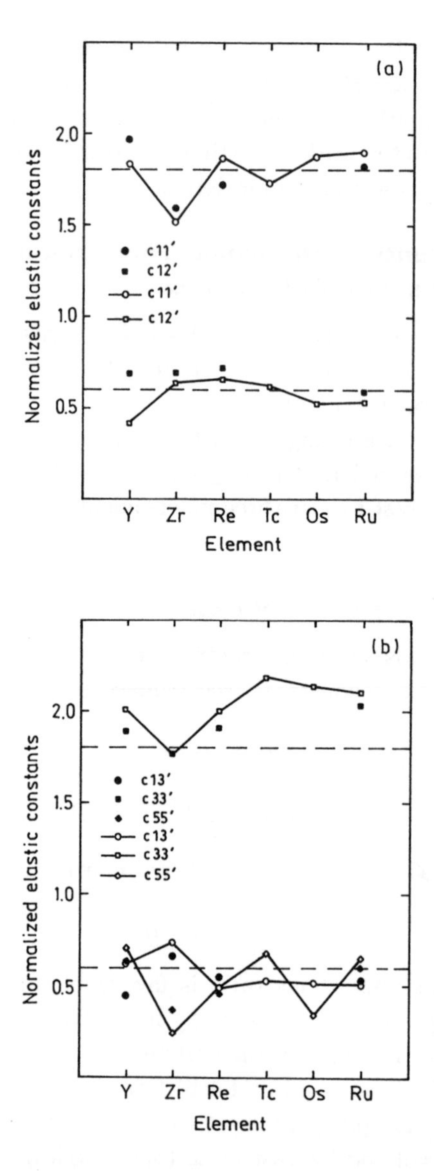


Fig. 6.4. Normalized elastic constants (see text), $C'_{ij} = C_{ij}/B$, for selected hexagonal transition metals. Closed symbols represent experimental data and open symbols represent theoretical data. Insert $C'_{11}$ and $C'_{12}$ are plotted in (a). $C'_{13}$, $C'_{33}$ and $C'_{55}$ are shown in (b). The dashed lines correspond to the values appropriate for an isotropic medium. (Reproduced with permission from Fast *et al.*, 1995.)

### 6.5.2. *Results for an Isotropic Medium*

Fast *et al.* (1995) next note that, for an isotropic medium, one has $C_{11} = C_{13}, C_{12} = C_{13} = C_{55}$. In this case, the Cauchy relation, $C_{12} = (C_{11} - C_{12})/2$ is valid and one can readily calculate the normalized elastic constants as

$$C'_{11} = 1.8,\; C'_{12} = 0.6 \quad \text{(isotropic material)}\,.$$

Figure 6.4(a), (b) therefore suggest that the hexagonal transition metals* are quite isotropic. This is in marked contrast to both bcc and fcc transition metals.

### 6.5.3. *Changes in Electronic Density of States for Various Strains*

Last *et al.* (1995) have studied the electronic density of states for hcp transition metals, for the variety of strains used in their calculations of elastic constants. Their main conclusion is that the density of states maintains its general shape for all types of distortions. This is again in marked contrast to the bands for bcc and fcc transition metals, where one observes a substantially larger change in the density of states when distorting the lattice.

The $c/a$ ratios have also been calculated for the hexagonal metals and the deviation from experiment is found to be $\sim 1\%$ at most (Last *et al.*, 1995). Their work, together with the earlier studies of Wills *et al.* (1992) and of Söderlind *et al.* (1993) give a rather complete theoretical description of the elastic constants of the transition metals. We discuss briefly, before returning to transition metals, to give a little detail on the potential calculations for simple metals.

## 6.6. *Energy of Simple Metals as Volume Term Plus Pair Potential Contribution*

The dispersion relation $\epsilon(\mathbf{k})$ of an electron in a crystal characterized by wave $\mathbf{k}$ can be calculated using pseudopotentials to second order. The result is

$$\epsilon(\mathbf{k}) = \frac{\hbar^2 k^2}{2m} + \langle \mathbf{k}|V_{ks}|\mathbf{k}\rangle + {\sum_{\mathbf{q}}}' \frac{\langle \mathbf{k}|V_{ks}|\mathbf{k}+\mathbf{q}\rangle\langle \mathbf{k}+\mathbf{q}|V_{ks}|\mathbf{k}\rangle}{(\hbar^2/2m)(k^2 - |\mathbf{k}+\mathbf{q}|^2)}\,. \tag{6.6.1}$$

*Relevant work here is that of D. J. Bacon and M. H. Liang (Phil. Mas. *A53*, 163, 1986) who enumerated the stacking faults in hcp metals. See also G. J. Ackland (Phil. Mas. *A66*, 917, 1992) and G. J. Ackland, S. J. Wooding and D. J. Bacon (Phil. Mas. *A71*, 553, 1995).

Here $V_{ps}$ is the total pseudopotential in the crystal, given by

$$V_{ps}(\mathbf{r}) = \sum v_{ps}(\mathbf{r} - \mathbf{R}_i) \tag{6.6.2}$$

where $v_{ps}$ is an individual pseudopotential centred at atomic position $\mathbf{R}_i$. The vectors $\mathbf{q}$ in Eq. (6.6.1) are reciprocal lattice vectors.

The matrix element of the pseudopotential $V_{ps}$ in Eq. (6.6.2) is easily cast into the form

$$\langle \mathbf{k} + \mathbf{q} | V_{ps} | \mathbf{k} \rangle = S(\mathbf{q}) r(\mathbf{q}) \tag{6.6.3}$$

where the atomic structure factor

$$S(\mathbf{q}) = \frac{1}{N} \sum_i \exp(-i\mathbf{q} \cdot \mathbf{R}_i) \tag{6.6.4}$$

depends only on the lattice, while the pseudopotential form factor

$$v(\mathbf{q}) = \frac{1}{\vartheta} \int \exp(-i\mathbf{q} \cdot \mathbf{r})\, v_{ps}(\mathbf{r}) d\mathbf{r} \tag{6.6.5}$$

characteristic the type of ion in the crystal. It depends on the ionic positions only through the atomic volume.

The sum of the final term in Eq. (6.1.1) over occupied state is referred to as the band-structure energy $E_{bs}$. One can cast this sum into the form

$$E_{bs} = \sum_{\mathbf{q}}{}' S^*(\mathbf{q}) S(\mathbf{q}) F(q) \tag{6.6.6}$$

where $F(q)$ is termed the energy-wave number characteristic. It involves the pseudopotential and the dielectric constant $\epsilon(q)$ of the Fermi gas (see Appendix A7.2).

## 6.7. *Pair Potentials*

We next transform the band-structure energy back into $\mathbf{r}$ space. To do so, we insert the structure factor $S(\mathbf{q})$ into Eq. (6.6.6), to obtain

$$E_{bs} = \frac{1}{N^2} \sum_{\mathbf{q},i,j} F(q) \exp(i\mathbf{q} \cdot \{\mathbf{R}_i - \mathbf{R}_j\}) . \tag{6.7.1}$$

If we now define an indirect interaction

$$V_{\text{ind}}(R) = \frac{2\Omega}{(2\pi)^3} \int F(q) \exp(-i\mathbf{q} \cdot \mathbf{R}) d\mathbf{q} = \frac{\Omega}{\pi^2} \int_0^\infty q^2 F(q) \frac{\sin qR}{qR} dq \tag{6.7.2}$$

the second line exploiting the spherical symmetry of $F(q)$, then $E_{bs}$ becomes

$$NE_{bs} = \frac{1}{2}\sum_{i,j}{}' V_{\text{ind}}(\mathbf{R}_i - \mathbf{R}_j) + \frac{\vartheta}{(2\pi)^2}\int F(q)d\mathbf{q} \tag{6.7.3}$$

the prime of the sum meaning that the term $i = j$ is not to be included. The second term on the right-hand side on $NE_{bs}$, just as for the free-electron energy contribution, depends only on the total volume and does not therefore change when the ions are rearranged at constant volume. In this rearrangement, the electrostatic energy changes are due to the change in $\frac{1}{2}\sum' Z^2e^2/|\mathbf{R}_i - \mathbf{R}_j|$ and hence one can add $V_{\text{ind}}$ and $Z^2e^2/R$ to obtain the effective pair interaction

$$\Phi(R) = V_{\text{ind}}(R) + \frac{Z^2e^2}{R}. \tag{6.7.4}$$

We stress again that beyond the sum of these pair potentials $\Phi(R_{ij})$, there is to be added a volume-dependent (only) term in the total energy, This representation of the energy of simple sp metals is compared with two alternative representations of interatomic force fields in Appendix.

## 6.8. *Structural Stability of Transition Metals*

Alonso and March (1989) have discussed the trends in the crystal structure of transition metals and Table 6.1 is taken from their book. This shows the trend *hcp* → *bcc* → *hcp* → *fcc* with increasing period number. The magnetic 3d-metals Mn, Fe and Co provide exceptions, and are therefore in parentheses in Table 6.1. It is d-band effects that are responsible for the above structural trend.

We referred to the density of electronic states for free electrons, $N_o(E) \propto E^{1/2}$, in Eq. (6.1.2). The d-band densities of states for bcc, hcp and fcc

Table 6.1. Crystal structures of the transition metals.

| | $n(s+d)$ | | | | | | | | |
|---|---|---|---|---|---|---|---|---|---|
| Period | 3 | 4 | 5 | 6 | 7 | 8 | 9 | 10 | 11 |
| 3d | Sc | Ti | V | Cr | (Mn) | (Fe) | (Co) | Ni | Cu |
| 4d | Y | Zr | Nb | Mo | Tc | Ru | Rh | Pd | Ag |
| 5d | (La) | Hf | Ta | W | Re | Os | Ir | Pt | Au |
| Structure | hcp | hcp | bcc | bcc | hcp | hcp | fcc | fcc | fcc |

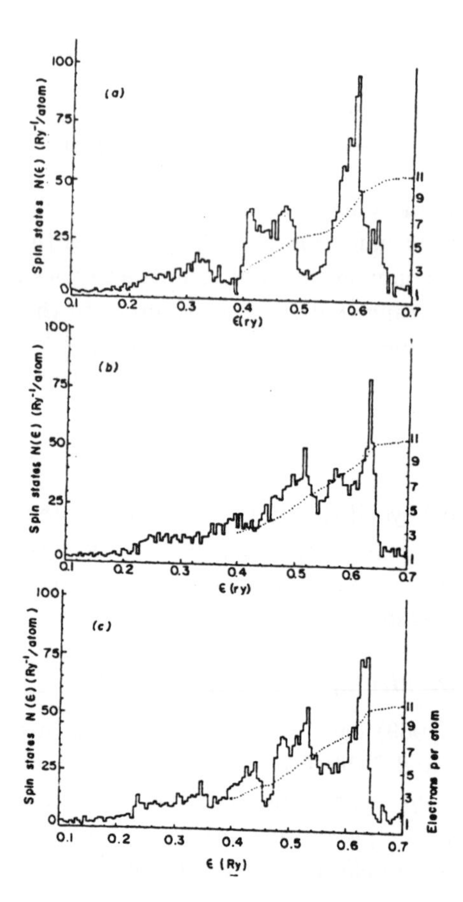


Fig. 6.5. Density of states for a model transition metal in the (a) bcc, (b) fcc and (c) hcp structures. The dotted curves represent the integrated density of states. Redrawn after Pettifor, (1970): see also Alonso and March (1989): Fig. 39.

structures (see Fig. 6.5, Fig. 39 of Alonso and March) depend, from energy band theory, on the crystal structures. In particular, the d-band of the bcc lattice is split into a bonding and an anti-bonding regime, separated by a pronounced minimum. The close-packed fcc and hcp densities of states have same general similarities, but are both quite distinct from the bcc case.

The relation between the shape of the density of states and crystal structure is already rather clear in the early work of Ducastelle and Cyrot-Lackmann (1970), who focussed on the relation between the local atomic environment and the moments of the density of status.

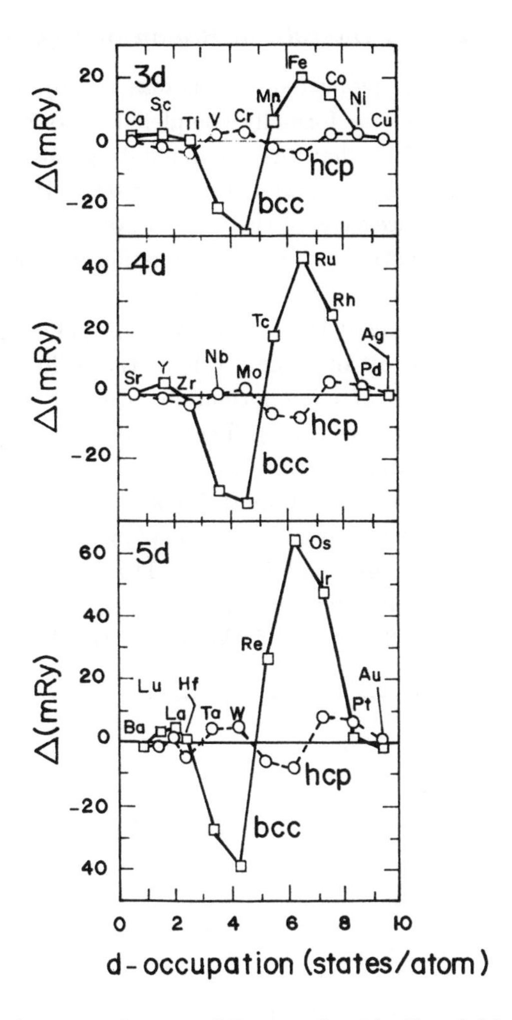


Fig. 6.6. Calculated structural energy differences for 3d, 4d and 5d transition metals at the experimental equilibrium volume as functions of d-occupation numbers. Energies are measured with respect to the fcc phase. Redrawn after Skriver, 1985.

A subsequent study of the structural energy differences in the 3d, 4d and 5d series is that of Skriver. His results are reproduced in Fig. 6.6 (Fig. 40 of Alonso and March), and display correctly, for the non-magnetic metals, the empirical structural trends (see also Alonso and March (1989) for a detailed discussion).

## 6.9. *Electron Density in Interstitial Region in Metals*

Density functional theory, pioneered by Thomas (1926), Fermi (1928) and Dirac (1930), whose work was formally completed by Hohenberg and Kohn (1964), is based on the fact that the ground-state gs electron density $n(r)$ determines the gs properties of a many-electron system. Plots of $n(\mathbf{r})$ provide evidently a coordinate-space view of the type of bonding occurring in metals which complements band-structure type information discussed earlier in this chapter.

More specifically (see also Alonso and March, 1989) the electron density, $n_b$ say, at the edge of an atomic cell in the solid is a key quantity for semi-empirical theories of alloy formation (see, for example, Miedema *et al.*, 1980; and other references there). The boundary density can be usefully identified with the interstitial electron density $n_{\text{out}}$ obtained in an electronic band-structure calculation of the 'muffin tin' type. The calculations of Moruzzi *et al.* (1978) reveal distinct trends as a function of atomic number $Z$, as shown in Fig. 6.7 (Fig. 49 of Alonso and March). The maxima resulting in the transition-metal series are, it turns out, due to the parabolic variation of atomic volume with $Z$ (see Fig. 6.8; Fig. 43 of Alonso and March, 1989).*

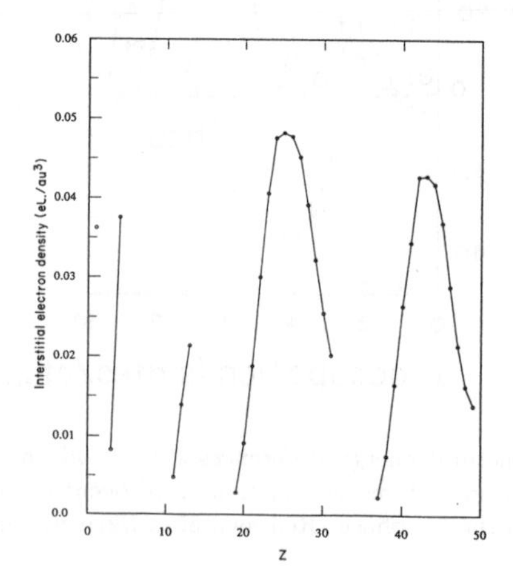


Fig. 6.7. Interstitial electron density $n_{\text{out}}$ versus atomic number $Z$. After Moruzzi *et al.* 1978: see Fig. 49 of Alonso and March (1989).

*The properties displayed in Fig. 6.8 for the 3d- and 4d-metals are excellent examples of calculations employing state-of-the art techniques for self-consistently solving the Schrödinger equation in crystals: are also the brief account in Section 6.11.

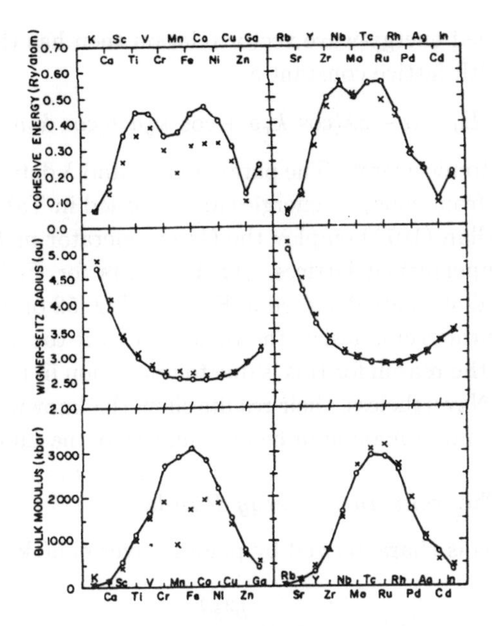


Fig. 6.8. Cohesive properties: top panel, cohesive energy; middle panel, Wigner-Seitz radius; bottom panel, bulk modulus. Crosses indicate experimental values. Redrawn after Moruzzi *et al.*, 1978.

It is worth noting here that calculated electron densities in metals with atomic number $Z <\sim 50$ are plotted in the book by Moruzzi *et al.* (1978). Localized approaches to construct electron density in periodic crystals have also, as with electronic energy band results, led to good agreement with X-ray experiments (Matthai *et al*, 1978; see also Jones and March, 1985).

## 6.10. *Trends in Vacancy Formation Energy with d-Shell Filling in Transition Metals*

Lannoo and Allan (1971) have discussed the trends of the vacancy formation energy $E_v$ as the d-shell is filled through the transition-metal series. These workers employ the tight-binding approach and furthermore they assume that the perturbation due to the vacancy extends only to near-neighbor atoms. Two further calculational aids are used: (i) the lattice is taken to be simple cubic and (ii) the assumption is made that d-orbital angularity need not be explicitly included.

The unperturbed energy-wave vector relation then has the form, with $\mathbf{k} = (k_x, k_y, k_z)$ and with lattice constant a:

$$E(\mathbf{k}) = E_0 - \alpha - 2\lambda(\cos\ k_x a + \cos\ k_y a + \cos\ k_z a) \tag{6.10.1}$$

$E_0$ denoting the atomic energy. The parameters $\alpha$ and $\lambda$ determined by Lannoo and Allan (1971) from cohesive energies are recorded in Table 6.2.

Lannoo and Allan (1971) employ the Green-operator method to relate the perturbed and unperturbed lattices. Their results for $E_v/20\lambda$ vs fractional occupation of the d-shell are depicted in Fig. 6.9 (Fig. 52 of Alonso and March, 1989). As these authors emphasize, the values of $E_v$ are too large quantitatively and they propose the reason for this is due to relaxation in the neighborhood of the vacancy site. Nevertheless, their results show the known general behaviour as the d-band fills up; a maximum being found near the middle of the series.

## 6.11. *Bloch's Theorem and Energy Bands*

The free Fermi gas is characterized by planes waves $\exp(i\mathbf{k}\cdot\mathbf{r})$, related to the eigenvalue $\epsilon_{\mathbf{k}}$ by

$$\epsilon_k = \frac{\hbar^2 k^2}{2m}\,. \tag{6.11.1}$$

Table 6.2. Parameters $\alpha$ and $\lambda$ in dispersion relation $E(\mathbf{k})$ of Eq. (6.10.1)

| | 1st series | 2nd series | 3rd series |
|---|---|---|---|
| $12\lambda$ (eV) | 5 | 6 | 7 |
| $\alpha$ (eV) | 0.25 | 0.30 | 0.35 |

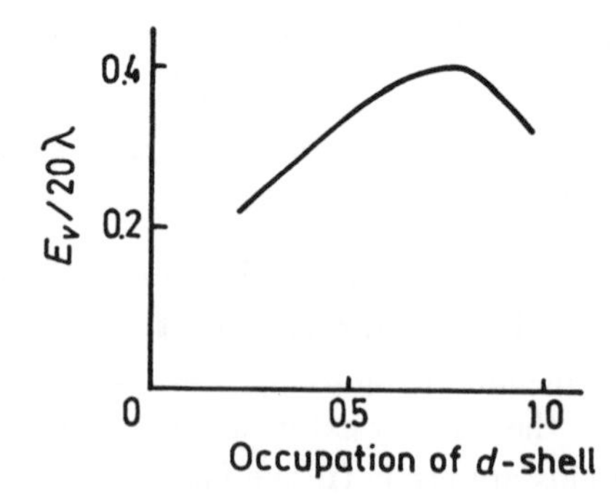


Fig. 6.9. Variation of vacancy-formation energy $E_v$ with occupation of d-shell (schematic). Redrawn after Lannoo and Allan, 1971.

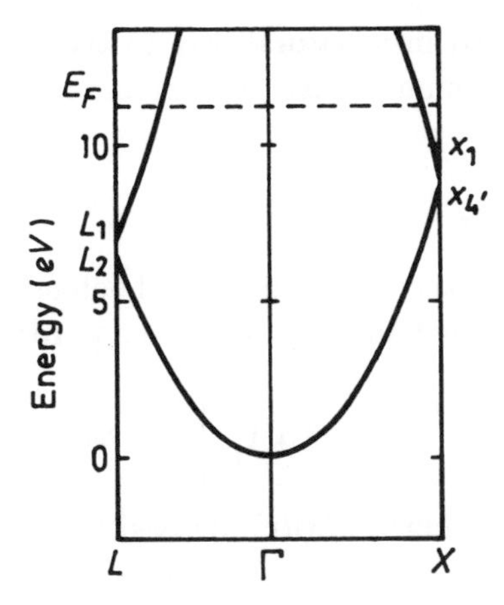


Fig. 6.10. Band structure of fcc aluminium. Redrawn after Pettifor, 1983.

For quite a few metals for which d or f bands are unimportant, the so-called nearly free electron (NFE) approximation is useful and such behaviour has been observed experimentally in studies of the Fermi surface. The introduction of a weak periodic potential introduces a gap in the free-electron band structure, a result for Al being shown in Fig. 6.10 (Fig. 20 of Alonso and March, 1989). At the Brillouin zone boundary $X$, the gap $\Delta\epsilon^X$ is related to the Fourier component of the crystal potential,* $V(200)$ by

$$\Delta\epsilon^X = 2|V(200)|\,. \tag{6.11.2}$$

More generally, when the crystal potential is 'switched on' to the free electron gas, the plane wave $\exp(i\mathbf{k}\cdot\mathbf{r})$ goes into the Bloch wave eigenfunction

$$\psi_{\mathbf{k}}(\mathbf{r}) = \exp(i\mathbf{k}\cdot\mathbf{r})u_{\mathbf{k}}(\mathbf{r}) \tag{6.11.3}$$

where $u_{\mathbf{k}}(\mathbf{r})$ is a periodic function with the period of the lattice. Using the reciprocal lattice vectors, $\{\mathbf{G}\}$ say, the periodic function $u_{\mathbf{k}}(\mathbf{r})$ can be expanded in the Fourier series

$$u_{\mathbf{k}}(\mathbf{r}) = \sum_{\mathbf{G}} v_{\mathbf{kG}}e^{i\mathbf{G}\cdot\mathbf{r}}\,. \tag{6.11.4}$$

To speed convergence of such a plane wave expansion Herring (1940) worked with plane waves that that had previously been orthogonalized to the core

*Compare the plane wave expansion (6.11.4) of another periodic function, where $U_kG$ are its Fourier components.

(OPW). This method led in a natural way to the concept of pseudopotentials (see Phillips and Kleinman, 1959; Cohen and Heine, 1970) which will be described below.

## 6.12. *Pseudopotentials*

In the pseudopotential method, the true crystal potential, $V(\mathbf{r})$ say, is replaced by a much weaker pseudopotential $V_{ps}(\mathbf{r})$ chosen to recover the original eigenvalues $\epsilon$, i.e.

$$\left[-\frac{\hbar^2}{2m}\nabla^2 + V_{ps}\right]\phi = \epsilon\phi\,.$$

The true crystal wave function $\psi(\mathbf{r})$, derived from the potential $V(\mathbf{r})$, can then be written as

$$\psi = \phi + \sum_c b_c\phi_c$$

where $\phi$ is the smooth pseudo wave function introduced above which $\phi_c$ are core states as in the OPW method mentioned above. The effect of this procedure is to keep the valence electron out of the core region, and can be expressed by writing

$$V_{ps} = V_c + V_R$$

where $V_R$ is the effective repulsive potential leading to exclusion of the valence electron from the core. This sum $V_c + V_R$ cancels to a large extent, resulting in a weak pseudopotential.

$V_R$ has, of course, to be obtained. An alternative to theoretical calculation is to fit the pseudopotential to experiment (see Cohen and Heine, 1970) — the so-called empirical pseudopotential method (EPM). Figure 6.11 shows the form of a typical pseudopotential (Fig. 21 from Alonso and March, 1989). The upper part of this figure shows the $\mathbf{r}$ space behaviour. Its Fourier transform (called $V(\mathbf{q})$) is expected to be small for large wave vectors $\mathbf{q}$, as depicted in the lower part of this figure.

Model potentials have also been constructed. One that is widely used is the so-called Ashcroft empty-core potential (Ashcroft, 1966). The assumption made is that of complete cancellation between the attractive Coulomb interaction and the repulsive contribution inside the core, and pure Coulombic form outside:

$$V^{\text{ion}}_{\text{empty-core}}(r) \left.\begin{array}{ll} = 0 & r < R_c \\ = -\dfrac{Ze^2}{r} & r > R_c \end{array}\right\}. \qquad (6.12.1)$$

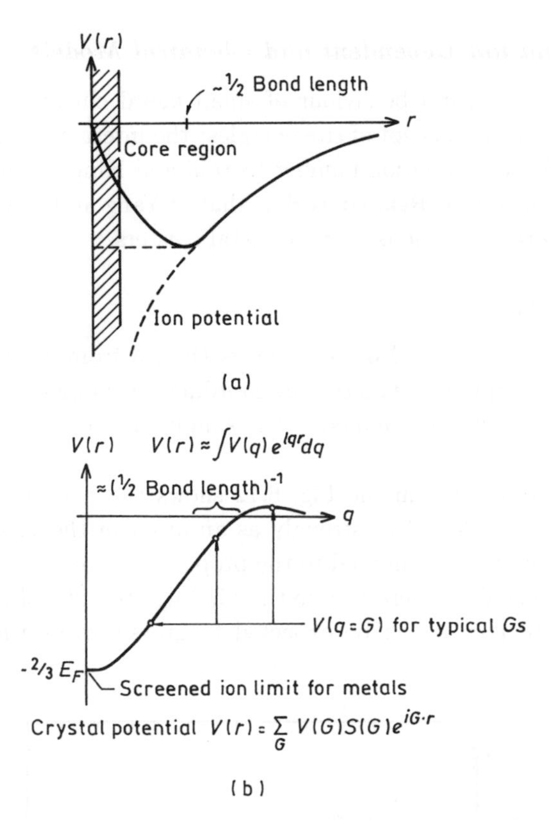


Fig. 6.11. (a) Schematic drawing of a typical pseudopotential. Redrawn after Cohen, 1970 (b) Schematic pseudopotential in reciprocal space. $\mathbf{G}$ is a reciprocal lattice vector, $S(\mathbf{G})$ is the structure factor and $V(\mathbf{G})$ is the pseudopotential form factor. Redrawn after Cohen, 1970.

The Fourier transform is readily obtained:

$$V^{\text{ion}}_{\text{empty-core}}(\mathbf{q}) = -\frac{4\pi Z e^2}{\Omega q^2} \cos\, qR_c\,. \tag{6.12.2}$$

The core radius $R_c$ can be fixed, for example, by fitting $q_0$, the position of the node in $V^{\text{ion}}_{\text{empty-core}}(q)$ to $q_0$ of empirical pseudopotentials, yielding $R_c = \pi q_0^{-1}/2$. The Fourier components $V_{ps}(\mathbf{q})$ of the lattice potential are found after screening the above ionic potential by the electron gas.

Finally, *ab initio* pseudopotentials have been derived (see, Yin and Cohen, 1982) but we shall not go into further details here.

## 6.13. *Coordination Dependent and Chemical Models*

To predict the structural behaviour of small metal clusters, there has been some emphasis of the concept of site energies: the interaction energy per atom with a particular coordination number (Strohl and King, 1989; Schoeb *et al.*, 1992; Yang *et al.*, 1993). Related work is that of Yang and DePristo (1994) on the factors determining the isomers of metal clusters.

### 6.13.1. *Metal Clusters*

In the spirit of the above, Fig. 6.12 shows (Fig. 3 from Yang and Depristo, 1994) the site energy as a function of coordination number for Pt clusters.* The point at coordination number 12 is simply the bulk cohesive energy of fcc Pt.

The long dashed line in the Fig. 6.12 shows the extreme limit in which an atom in dimer is found as strongly as an atom in the 12-coordinated fcc bulk crystal structure. Compared to the properties of real metal clusters, this line can be expected to overestimate the binding strength of low-coordinated atoms. In particular, it can be expected to grossly overestimate the dimer finding energy.

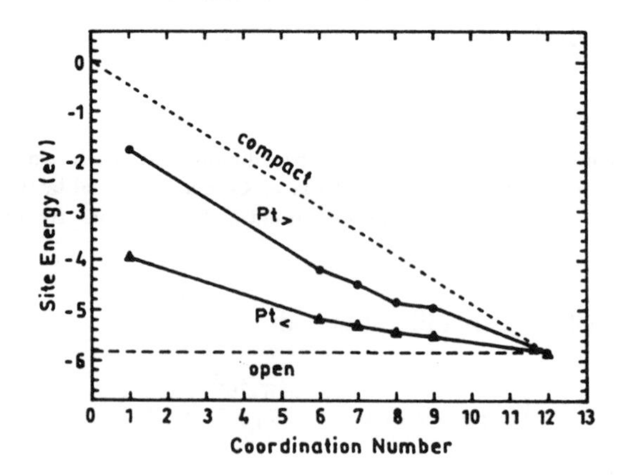


Fig. 6.12. Site energy (interaction energy per atom) as a function of coordination number as predicted by MD/MC-CEM calculations using the $Pt_>$ and $Pt_<$ electron densities. The two dashed lines encase the typical region of site energies in real systems.

---

*It is relevant here to note that S. Y. Lien and K.-Y. Chan (Mol. Phys. *86*, 939, 1995) have used the many-body potential of A. P. Sutton and J. Chen (Phil. Mas. Lett. *61*, 139, 1990) for Pt to treat adsorbed Pt on graphite.

The short dashed line in Fig. 6.12, following again the work of Yang and DePristo (1994) corresponds to another (never realized!) limiting situation in which the bond in the dimer is only as strong as one bond in the fcc structure.

More realistic curves in Fig. 6.12, as calculated from different forces fields by Yang and Depristo (1994) are labelled $Pt_>$ and $Pt_<$. As these workers stress, the structures of small metal clusters can be very different in spite of using similar interatomic force fields. Such potentials are based on the same basic assumptions and approximations and lead to the same cohesive energy and lattice constant for the bulk fcc crystal. Yang and Depristo (1994) show how they can differ in yielding low-coordinated site energies.

### 6.13.2. *Binary Metallic Alloys*

In subsequent work, related ideas have been developed by Zhu and Depristo (1995) for binary metallic alloys. In particular, they consider such alloys formed from Ni, Cu, Rh, Pd and Ag. The site energy (see above) for fixed coordination is now expressed as a quadratic function of the number of existing mixed metal bonds. The three parameters entering this functional form are (over) determined by the mixing energy as a function of composition for bulk bimetallic fcc systems. The model developed predicts accurately the microstructures of clusters of $Ni_{101}Cu_{100}$ and $Cu_{101}Pd_{100}$ which are prototypes for bimetallic clusters. For $Ni_{101}Pd_{100}$ however, the model has some limitations due to the atomic size mismatch of 10%, which distorts the cluster shape from a perfect lattice structure.

To conclude this section on clemical models, we will summarize the work of Zhang *et al.* (1994). This study is concerned with the fracture of transgranular cleavage of $Fe_3Al$ and the intergranular fracture of FeAl.

Zhang *et al.* make use of an empirical approach to bond energy to demonstrate that the fracture of transgranular cleavage of $Fe_3Al$ and the intergranular fracture of FeAl are due to their characteristic crystal structures.

### 6.13.3. *Structure and Bonding of $Fe_3Al$ and FeAl*

Fe-Al alloys are brittle. The Fe-Al alloys containing more than 17.9% Al were so brittle that their hardness could not be determined, since cracks appeared in the samples during testing. The brittle property of Fe-Al alloys is of interest to many materials scientists.

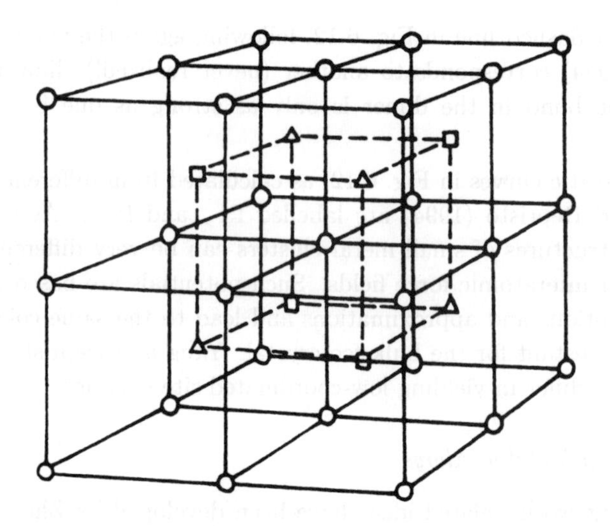

Fig. 6.13. Distribution of atoms in $Fe_3Al$ and FeAl. $Fe_3Al$: □, FeI; ○, FeII; △, Al; FeAl: ○, Fe; △ & □, Al.

We note that the space group of $Fe_3Al$ with $DO_3$ structure and fcc lattice is $O_h^5$ (Fm3m) and its structure formula can be written as $Fe^I\ Fe_2^{II}Al$. For $FeAl_*$ with $B_2$ structure and sc lattice, the space group is $O_h^1*$ (Pm3m). The configuration of atoms in $Fe_3Al$ and FeAl is depicted in Fig. 6.13 (Fig. 1 of Zhang *et al.*). It may be noted, regarding bonding in $Fe_3Al$ that the main bonds are $Fe^{II}$–Al, $Fe^I$–$Fe^{II}$, $Fe^I$–Al and $Fe^{II}$–$Fe^{II}$. The first two of those are in the $\langle 111 \rangle$ direction while the others are in the $\langle 100 \rangle$ direction. The two kinds of atoms in FeAl take the corner and bc sites respectively. Of the three types of bond in FeAl, the Fe-Al bonding is in the $\langle 111 \rangle$ direction, that between Fe and Fe and Al-Al being in the $\langle 100 \rangle$ direction.

Zhang *et al.* use an empirical approach to the bond energy and we shall not go into calculational detail here but rather give a qualitative description of their main findings.

The main slip system and the cleavage planes of a crystal with $DO_3$ structure are $\{110\}\langle 111 \rangle$ and $\{100\}$ respectively.

The slip of perfect crystals of $Fe_3Al$ and FeAl means that two parallel slip planes $\{110\}$ glide in the $\langle 111 \rangle$ direction. The energy needed for slip can be expressed through the energy which is necessary for one atom on the $\{110\}$ plane gliding in the $\langle 111 \rangle$ direction. The energy approximately equals the total energy of all bonds between the atom and the atoms on the near parallel slip plane. Similarly, the energy needed by the cleavage of the crystal along the $\{100\}$ plane can be characterized by the energy necessary to break all bonds between an atom and the atoms on the near parallel $\{100\}$ plane. Details are given by Zhang *et al.* (1994). Those workers calculate the energy which is necessary for an atom gliding on one of the $\{110\}$ plane in the $\langle 111 \rangle$ direction, or an atom separated from one of the $\{100\}$ plane. The two values for $Fe_3Al$ are $\sim$ 100 kJ/mol and 130 kJ/mol respectively. The corresponding values for FeAl are $\sim$ 95 and 120. From these results, Zhang *et al.* (1994) conclude that the FeAl crystal cleaves more readily than does $Fe_3Al$.

These workers, from such bond energy arguments, conclude that the fracture of transgranular cleavage of $Fe_3Al$ and the intergranular fracture of FeAl are consequences of the crystal structures from the energetics of the cleavage mechanism.

# Chapter 7

# Theory of Pair Potentials in Simple s-p Metals

We shall build up, in this chapter, from the simplest possible starting point, the theory of effective pair potentials between ions, mediated by the conduction electrons, in a simple s-p metal like Na or K.

## 7.1. *Thomas-Fermi Theory of Interaction Between Test Charges in Initially Uniform Electron Gas*

We consider first a positive test charge of magnitude $z_1 e$ introduced into an initially uniform degenerate electron gas of density $n_0$. The electrostatic potential, $V(r)$ say, generated by this introduction of the test charge into the overall neutral jellium model satisfies the linearized Thomas-Fermi equation

$$\nabla^2 V = q^2 V \tag{7.1.1}$$

which is the degenerate analogue of the classical linear Debye-Hückel theory of screening in an electrolyte.

In Eq. (7.1.1), $q^{-1}$ is the inverse Thomas-Fermi screening length $l_{TF}$. In simple physical terms, $l_{TF}$ is the product of a characteristic velocity and a characteristic time. These are not difficult to identify in a Fermi gas. Since its properties are dominated by the Fermi level, we must anticipate that the characteristic velocity is the Fermi velocity $v_F$. As to the characteristic time, Langmuir recognized that if the electron gas were disturbed, to create an imbalance of charge, the electrons would rush in to screen out the charge imbalance, overshoot, and oscillate with angular frequency given by

$$\omega_{\mathbf{plasma}} \equiv \omega_p = \left(\frac{4\pi n_0 e^2}{m}\right)^{1/2} \tag{7.1.2}$$

with $m$ the electron mass. The characteristic time then is $2\pi/\omega_p$, the period of these so-called plasma oscillations. Hence

$$l_{TF} \simeq v_F \frac{2\pi}{\omega_p} \tag{7.1.3}$$

But from Fermi gas theory, phase space theory yields

$$n_0 = \frac{8\pi}{3h^3} p_F^3 \tag{7.1.4}$$

and since the Fermi momentum $p_F = mv_F$ we can eliminate $n_0$ from $\omega_p$ in favor of $v_F$. The final result is then

$$q^2 = \frac{4k_F}{\pi a_0} \left( a_0 = \frac{\hbar^2}{me^2} \right) \tag{7.1.5}$$

where $k_F$ is the Fermi wave number, related to $p_F$ by $p_F = \hbar k_F$. $l_{TF} = q^{-1}$ turns out for a good metal to be $\sim 1$ Å, and the potential obtained by solving Eq. (7.1.1) for the screening of the test charge $z_1 e$ is simply

$$V = \frac{z_1 e}{r} \exp(-qr) \tag{7.1.6}$$

7.1.1. *Test Charge Interaction Energy: Basis of Electrostatic Model*

The screened potential $V$ created by a test charge $z_1 e$ will evidently interact with a second test charge $z_1 e$ at separation $R$. According to the electrostatic model (Lazarus) the interaction energy, $\Delta E(R)$ say, will be simply

$$\Delta E(R) = (z_2 e)\, V(R) = \frac{z_1 z_2 e^2}{R} \exp(-qR) \tag{7.1.7}$$

This result has been derived from the simplest density functional theory; namely the Thomas-Fermi method, by Alfred and March (1957; see Appendix A7.1)

7.1.2. *Wave Theory of Interaction Between Test Charges*

The linearized Thomas-Fermi (TF) method for the displaced charge $\rho(\mathbf{r}) - \rho_0$ by the potential energy $V(\mathbf{r})$, namely

$$(\rho(\mathbf{r}) - \rho_0) = \frac{q^2 V}{4\pi} \tag{7.1.8}$$

leads to the disappointing result that two test charges $z_1 e$ and $z_2 e$ in a Fermi gas repel one another at all distances $R$, with screened Coulomb form $\Phi(R) = (z_1 z_2 e^2/R)\exp(-qR)$, where $q^{-1}$ is the TF screening length: $\sim 1$ Å in a good metal like Cu.

Corless and March (1961) pointed out that if Eq. (7.1.8) (akin to 'geometrical optics') is replaced by a correct linearized wave theory, then such an interaction $\Phi(R)$ does have attractive regions. Below we first derive the correct $\mathbf{r}$ space equation transcending the semiclassical form (7.1.8), following the work of March and Murray (1960, 1961).

(a) Integral form of Schrödinger equation

If the self-consistent potential energy in which all electrons are taken to move is $V(\mathbf{r})$ as above, then the original plane waves $\Omega^{-1/2}\exp(i\mathbf{k},\mathbf{r})$, normalized in a piece of metal (uniform Fermi gas) of volume $\Omega$, are distorted into wave functions $\psi_{\mathbf{k}}(\mathbf{r})$ say, the wave vector $\mathbf{k}$ here labelling the unperturbed state from which $\psi_{\mathbf{k}}$ derives when the perturbed potential $V(\mathbf{r})$ is introduced. One must then solve

$$\nabla^2\psi_{\mathbf{k}} + \frac{2m}{\hbar^2}[\epsilon_{\mathbf{k}} - V(\mathbf{r})]\psi_{\mathbf{k}} = 0 \tag{7.1.9}$$

Let us proceed by analogy with Poisson's equation of electrostatics, namely

$$\nabla^2\phi = -4\pi\rho \tag{7.1.10}$$

A formal solution of Eq. (7.1.10) can be written down immediately as

$$\phi(\mathbf{r}) = \int \frac{d\mathbf{r}'\rho(\mathbf{r}')}{|\mathbf{r}-\mathbf{r}'|} \tag{7.1.11}$$

This can be regarded as arising from taking the right-hand side of Eq. (7.1.10), multiplied by a special solution of the left-hand side equated to zero (i.e. Laplace's equation) and integrated over $\mathbf{r}'$. Obviously $1/|\mathbf{r}-\mathbf{r}'|$ is such a solution, singular at $\mathbf{r} = \mathbf{r}'$. This is one of the simplest examples of a Green function, which leads directly to the solution (7.1.11).

Let us apply a slight generalization of the above example of solution of Poisson's equation to the Schrodinger Eq. (7.1.9) taking the term in $V(\mathbf{r})$ over to the right-hand side and writing $\epsilon_{\mathbf{k}}$ explicitly as the free electron energy $\hbar^2k^2/2m$. This then yields

$$\nabla^2\psi_{\mathbf{k}} + k^2\psi_{\mathbf{k}} = \frac{2m}{\hbar^2}V(\mathbf{r})\psi_{\mathbf{k}} \tag{7.1.12}$$

We can now solve Eq. (7.1.12) when the right-hand side is put to zero, and we find the (outgoing) spherical wave solution

$$G(\mathbf{r},\mathbf{r}') = \frac{\exp(ik|\mathbf{r}-\mathbf{r}'|)}{|\mathbf{r}-\mathbf{r}'|} \tag{7.1.13}$$

This obviously reduces correctly to $1/|\mathbf{r}-\mathbf{r}'|$ as $k \to 0$.

Then one has, for the full solution of Eq. (7.1.12)

$$\psi_{\mathbf{k}}(\mathbf{r}) = \Omega^{-1/2}\exp(i\mathbf{k}.\mathbf{r}) - \frac{m}{2\pi\hbar^2}\int d\mathbf{r}' G(\mathbf{r},\mathbf{r}')V(\mathbf{r}')\psi_{\mathbf{k}}(\mathbf{r}') \tag{7.1.14}$$

which is often referred to as the Lippmann-Schwinger integral form of the Schrödinger equation. Evidently the first term on the right-hand side of Eq. (7.1.14) is the (free electron) wave function before the potential $V(\mathbf{r})$ is introduced and the second term is constructed in precise analogy to Eq. (7.1.11).

(b) Solution for first-order change in electron density

Equation (7.1.8) evidently corresponds to an electron density change $\rho(\mathbf{r}) - \rho_o \propto q^2 V$ on introducing the potential $V$ into an originally uniform Fermi gas of density $\rho_o$. We wish now to calculate $\rho(\mathbf{r}) - \rho_o$ from the full wave theory, but still to first order in $V$. We can do this correctly by noting that since the potential $V(\mathbf{r}')$ is already present in the last term of Eq. (7.1.14), we can merely replace $\psi_{\mathbf{k}}(\mathbf{r}')$ in that term by the unperturbed wave function $\Omega^{-1/2}\exp(i\mathbf{k}.\mathbf{r})$. To form the electron density $\rho(\mathbf{r})$, we have to sum the wave function product $\psi^*_{\mathbf{k}}(\mathbf{r})\psi_{\mathbf{k}}(\mathbf{r})$ over all $\mathbf{k}$ out to the Fermi surface. Recalling that, to be consistent, we must only retain terms of first order in $V$, we find, with $k_F$ as usual denoting the Fermi wave number:

$$\begin{aligned}\sum_{|\mathbf{k}|<k_F} \psi^*_k(\mathbf{r})\psi_{\mathbf{k}}(\mathbf{r}) = &\sum_{|\mathbf{k}_F|<k_F} \Omega^{-1} - \sum_{|\mathbf{k}|<k_F} \Omega^{-1}\frac{m}{2\pi\hbar^2}\int d\mathbf{r}'V(\mathbf{r}')\\ &\times [G(\mathbf{r},\mathbf{r}')\exp(i\mathbf{k}\cdot\mathbf{r}'-\mathbf{r}) + G^*(\mathbf{r},\mathbf{r}')\\ &\times \exp(-i\mathbf{k}\cdot\mathbf{r}'-\mathbf{r})]\end{aligned} \tag{7.1.15}$$

Replacing the summation over $\mathbf{k}$ by an integration, using the fact that there are $\Omega/(2\pi)^3$ states per unit volume of $\mathbf{k}$ space and two spin directions, leads to the displaced electron density $\rho(\mathbf{r}) - \rho_o$ as

$$\begin{aligned}\rho(\mathbf{r}) - \rho_o = \frac{-2m}{(2\pi)^4\hbar^2}\int d\mathbf{r}'V(\mathbf{r}') \times \int_{|\mathbf{k}|<k_F} d\mathbf{k}[G(\mathbf{r},\mathbf{r}')\exp(i\mathbf{k}\cdot\mathbf{r}'-\mathbf{r})\\ + G^*(\mathbf{r},\mathbf{r}')\exp(-i\mathbf{k}\cdot\mathbf{r}'-\mathbf{r})]\end{aligned} \tag{7.1.16}$$

Since, from Eq. (7.1.13), the Green function $G$ depends only on the magnitude $k$ of the wave vector $\mathbf{k}$, it is straightforward to show that integrating over the angles of $\mathbf{k}$ simply replaces $\exp(\pm i\mathbf{k}\cdot\mathbf{r}' - \mathbf{r})$ by $\sin k|\mathbf{r}-\mathbf{r}'|/k|\mathbf{r}-\mathbf{r}'|$ (this latter function is just the $s(l=0)$ term in the expansion of a plane wave in spherical waves). One then obtains, combining $G$ and $G^*$, from Eq. (7.1.16)

$$\rho(\mathbf{r}) - \rho_o = \frac{-2m}{(2\pi)^4\hbar^2}\int d\mathbf{r}'V(\mathbf{r}') \times \int_0^{k_F} dk 4\pi k^2 \left[\frac{\sin k|\mathbf{r}-\mathbf{r}'|}{k|\mathbf{r}-\mathbf{r}'|}\cdot\frac{2\cos k|\mathbf{r}-\mathbf{r}'|}{|\mathbf{r}-\mathbf{r}'|}\right] \tag{7.1.17}$$

Performing the integration over $k$, the final result for the displaced charge can be written (March and Murray, 1960)

$$\rho(\mathbf{r}) - \rho_o = \frac{-mk_F^2}{2\pi^3\hbar^2}\int d\mathbf{r}'V(\mathbf{r}')\frac{j_1(2k_F|\mathbf{r}-\mathbf{r}'|)}{|\mathbf{r}-\mathbf{r}'|^2} \tag{7.1.18}$$

where

$$j_1(x) = x^{-2}[\sin x - x\cos x] \tag{7.1.19}$$

is the first-order spherical Bessel function.

To make contact with the semiclassical Eq. (7.1.8), we next observe that if $V$ varies sufficiently slowly in space then $V(\mathbf{r}')$ can be replaced approximately by $V(\mathbf{r})$ to yield

$$\rho(\mathbf{r}) - \rho_o = \frac{-mk_F^2}{2\pi^3\hbar^2}V(\mathbf{r})\int d\mathbf{r}'\frac{j_1(2k_F|\mathbf{r}-\mathbf{r}'|}{|\mathbf{r}-\mathbf{r}'|^2} \tag{7.1.20}$$

which then leads back to Eq. (7.1.8).

Combining the full first-order wave theory result (7.1.20) for the displaced charge with Poisson's equation, we find the self-consistent field equation.

$$\nabla^2 V = \frac{2me^2}{\hbar^2}\frac{k_F^2}{\pi^2}\int d\mathbf{r}'V(\mathbf{r}')\frac{j_1(2k_F|\mathbf{r}-\mathbf{r}'|)}{|\mathbf{r}-\mathbf{r}'|^2} \tag{7.1.21}$$

(c) Form of displaced charge round a localized potential $V(\mathbf{r})$

Without solving Eq. (7.1.21), it can readily be demonstrated that the displaced charge can have a very different character at large $\mathbf{r}$, depending on whether we use the wave theory result (7.1.20) or the semiclassical form (7.1.8). As an illustrative example, let us insert in Eq. (7.1.20) for $\rho(\mathbf{r}) - \rho_o$ the choice when

$V(\mathbf{r})$ is very short-range, idealized by $V(\mathbf{r}) = \lambda\delta(\mathbf{r})$, with $\delta(\mathbf{r})$ the Dirac delta function. Then Eq. (7.1.20) immediately yields

$$\rho(\mathbf{r}) - \rho_o = \frac{-mk_F^2}{2\pi^3\hbar^2}\frac{\lambda j_1(2k_F r)}{r^2} \tag{7.1.22}$$

But from the definition of $j_1(x)$ in Eq. (7.1.19) it is seen that at large $x$, $j_1(x) \sim -\cos x/x$ and hence from Eq. (7.1.22):

$$\rho(\mathbf{r}) - \rho_o \sim \frac{\cos(2k_F r)}{r^3} \tag{7.1.23}$$

This is then the new feature arising from the wave theory and if we use Poisson's equation with this asymptotic form of the displaced charge, we find that the potential $V(r)$ corresponding to $\rho(r) - \rho_o$ in Eq. (7.1 23) also behaves as $\cos(2k_F r)/r^3$ at sufficiently large $r$. That such oscillations exist in the long-range form of the displaced charge round a given fixed perturbation in a Fermi gas was first pointed out explicitly by Blandin, Daniel and Friedel (1959: see also March and Murray, 1960). However, such behaviour was certainly at least implicit in much earlier work by Bardeen.

Below, we shall apply Eq. (7.1.1), following Corless and March, to calculate the interaction energy $\Phi(R)$ of two test charges embedded in a Fermi gas. However, before doing that, it is important to demonstrate how the above theories, both semiclassical and wave theory, can be utilized to extract the dielectric function of a Fermi gas.

(d) Dielectric function

As already discussed, the screened potential energy $V(\mathbf{r})$ round a test charge ze in a Fermi gas follows from the semiclassical equation $\nabla^2 V = q^2 V$ as

$$V(r) = \frac{-ze^2}{r}\exp(-qr) \tag{7.1.24}$$

To introduce the dielectric function, we take the Fourier transform of $V(\mathbf{r})$ according to

$$\tilde{V}(\mathbf{k}) = \int d\mathbf{r}\exp(i\mathbf{k}\cdot\mathbf{r})V(\mathbf{r}) \tag{7.1.25}$$

when we find from Eq. (7.1.24) the result

$$\tilde{V}(k) = \frac{-4\pi ze^2}{k^2+q^2} \tag{7.1.26}$$

More interestingly, the wave theory can also be dealt with analytically in **k** space, unlike the situation in **r** space, where recourse had to be made to numerical methods (March and Murray, 1961). By Fourier transforming Eq. (7.1.21) (using a convolution property) we then find

$$\tilde{V}(k) = \frac{-4\pi z e^2}{k^2 + \frac{k_F}{\pi a_o} g\left(\frac{k}{2k_F}\right)} : a_o = \frac{\hbar^2}{me^2} \tag{7.1.27}$$

where

$$g(x) = 2 + \frac{(x^2-1)}{x} \ln\left|\frac{1-x}{1+x}\right| \tag{7.1.28}$$

This important result is derived in Appendix 7.2. It is to be noted that, in the long wavelength limit $k \to 0$, the function $g(x)$ tends to the value 4, and we find the same result

$$\tilde{V}(k=0) = \frac{-4\pi z e^2}{q^2} \tag{7.1.29}$$

from both semiclassical and wave theories, after using eq. (7.1.5).

Often, it proves valuable to express the above results in term of the wave-number dependent dielectric function $\epsilon(k)$ of the Fermi gas. This may be conveniently introduced for the present purposes by writing

$$\tilde{V}(k) = \frac{-4\pi z e^2}{k^2 \epsilon(k)} \tag{7.1.30}$$

It follows then from the screened Coulomb form that the semiclassical Thomas-Fermi (TF) dielectric function is given from Eq. (7.1.26) by

$$\epsilon_{TF}(k) = \frac{k^2 + q^2}{k^2} \tag{7.1.31}$$

while the wave theory yields

$$\epsilon(k) = \frac{k^2 + \frac{k_F}{\pi a_o} g\left(\frac{k}{2k_F}\right)}{k^2} \tag{7.1.32}$$

This latter expression appears first to have been given explicitly by Lindhard (1954), though it is certainly at least implicit in an earlier study by Bardeen (1937) on electron-phonon interaction.

## 7.2. *Density Functional Theory of Pair Potentials*

Let us state immediately the density functional theory result for the effective pair interaction energy $\Phi(R)$ derived using the superposition density for single-centre screened ions. The result takes the form

$$\Phi(R) = \Delta G(R) + \Delta U(R) \tag{7.2.1}$$

where $\Delta U(R)$ is the total potential-energy change which is determined solely by the total valence screening charge $Q(R)$. This quantity is plotted in Fig. 7.1 for Na (Perrot & March, 1990a) (Fig. 1 of their paper) and in Fig. 7.2 for Be (Fig. 4 of Perrot & March, 1990b).

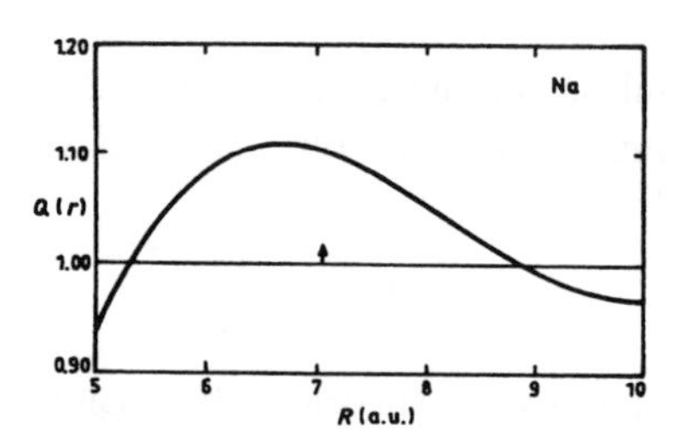


Fig. 7.1. Total valence screening charge $Q(R)$ taken from PM (1990) for liquid Na near freezing. $Q(R)$ determines the pair interaction $\phi(R)$ through Eq. (7.2.7) in linear response theory. Arrow denotes position of principal minimum in "exact" pair potential $\phi$.

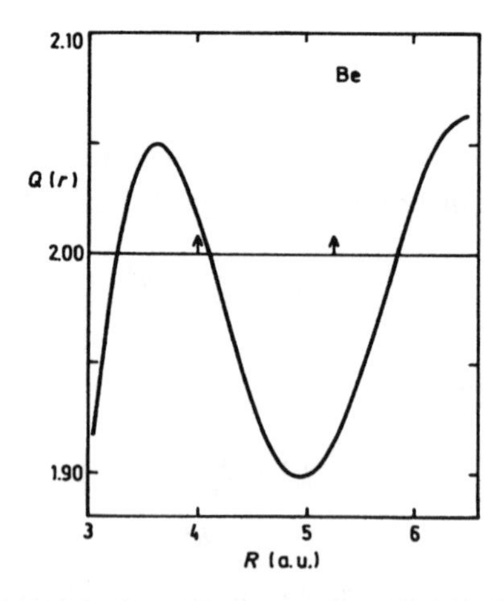


Fig. 7.2. Same as Fig. 7.1, but for liquid Be at a density equal to the solid density. Arrows denote first minimum and following maximum in "exact" pair potential $\phi(R)$. See Fig. 7.5 for the corresponding curve for $K^+$.

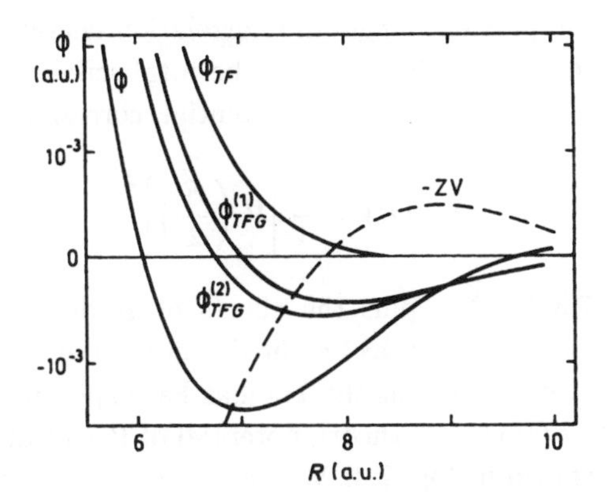


Fig. 7.3. This shows pair potentials calculated from local density (TF) theory plus density gradient corrections $T_2$ and $T_4$ to kinetic energy, for liquid Na near freezing. Various curves were obtained, using $Q(R)$ in Fig. 7.1, from different degrees of approximation as follows. $\phi_{\mathrm{TF}}$, calculated from Eq. (7.2.2); $\phi^{(1)}_{\mathrm{TFG}}$, from $T_2$ in Fig. 7.4 with $\lambda = \frac{1}{9}$ in which $T_2$ only is included; $\phi^{(2)}_{\mathrm{TFG}}$, from theory containing both $T_2$ and $T_4$; $\phi_{\mathrm{LR}} = -ZV; \phi(R)$ is the pair potential obtained by Perrot and March (1990a).

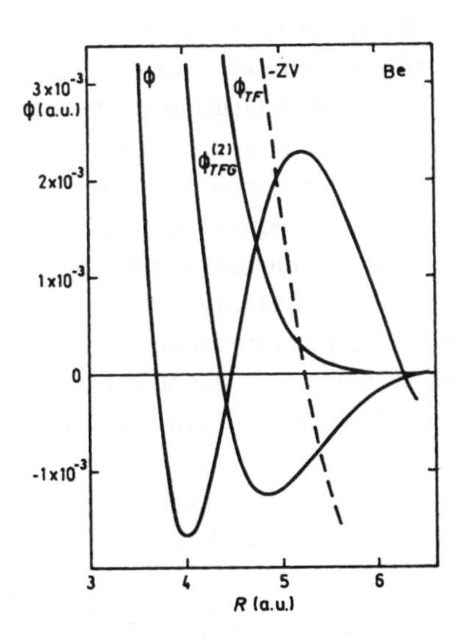


Fig. 7.4. Same as Fig. 7.3 but for Be. Density gradient correction $T_2$ to kinetic energy, referred to in Fig. 7.3, is given in terms of electron density $\rho(r)$ by $T_2 = (\lambda_8) \int (\nabla\rho)^2/\rho^{\mathrm{dr}}$. $T_4$ is by given Perrot and March (1990).

Thus, the kinetic plus exchange and correlation energy change $\Delta G(R)$ is responsible for differences in $\Phi(R)$, given the total screening charge $Q(R)$ for a particular metal. The Thomas-Fermi potential, corresponding to

$$\Phi_{TF}(R) = \frac{R}{4}\left[V\left(\frac{R}{2}\right)\right]^2 \tag{7.2.2}$$

is shown in Fig. 7.3 for Na, plus the effects of gradient correlations. The Perrot-March (1990a) potential $\phi(r)$ is the deepest curve in Fig. 7.3: obviously gradient corrections are improving the Thomas-Fermi potential, but are still far from the full density functional theory potential of Perrot and March. Similar results for Be are shown in Fig. 7.4. We outline next the derivation of $\phi(r)$.

### 7.2.1. *Density Functional Theory of Pair Potential*

The work of Alfred and March (1957) and the later study of Corless and March (1961) gave the semiclassical and the wave theory result of the interaction between test charges in a Fermi gas. These can be viewed as simple density functional theories: both however taking the two-centre density and potential as appropriate superposition of the corresponding one-centre case.

Perrot and March (1990a, b referred to below as PM) gave a generalization of these treatments which still represents the density, but now for realistic ions rather than test charges, as a superposition of one-centre quantities.

Below, we shall give an outline of the main steps in the derivation of PM (1990). The basic building block is now the screened ion in a Fermi gas. PM worked out the effective interionic potentials for Na and for Be, their main focus in these applications being liquid metals. For Na, a direct test of this interionic potential proves possible, by confrontation with results obtained by inverting the experimentally measured structure factor $S(k)$ (Johnson and March, 1963; Reatto, 1988). We shall see below that the main features show quite remarkable accord between the two approaches, though significant quantitative differences remain.

Following PM, let us start from the density functional theory equation for a single ion in an uniform electron gas of density $\bar{\rho}$. This can be written formally as

$$G'[\Delta + \bar{\rho}] - G'[\bar{\rho}] = -V(\mathbf{r}) \tag{7.2.3}$$

where $G'$ is a shorthand notation for the (functional) derivative $\delta G/\delta\rho(\mathbf{r})$, which in the TF method - the forerunner of modern DFT - is simply proportional to $\rho(\mathbf{r})^{2/3}$. As written above, the quantity $G = T + E_{xc}$ includes the sum of kinetic energy and exchange plus correlation effects. We assume to be concrete that $E_{xc}$ is taken in the local density approximation following PM. $V$ is the usual electrostatic potential due to the displaced charge $\Delta(\mathbf{r}) = \rho(\mathbf{r}) - \bar{\rho}$: namely

$$V = \frac{-Z}{r} + \frac{1}{r} * \Delta \tag{7.2.4}$$

where the notation $*$ is merely shorthand for the convolution product representing the electrostatic potential created by $\Delta(\mathbf{r})$.

### 7.2.2. *Solution of One-Centre Problem for Na Atom Embedded in a Cavity in Jellium*

To give a definite illustration coming from the (numerical) solution of Eq. (7.2.3), we take the example of a Na atom embedded in a jellium cavity at an average uniform electron density $\bar{\rho} = 0.0036046$ a.u. The radius $\mathrm{R_a}$ of the cavity is such that $(4/3)\pi R_a^3\bar{\rho} = z$, where $z$, the valence, is unity for

Table 7.1. Self-consistent density functional calculation of the density displaced by an Na atom embedded in a cavity in jellium at electron density $\bar{\rho} = 0.0036046$ a.u. The radius $R_a$ of the cavity is such that $\frac{4}{3}\pi R_a^3\bar{\rho} = Z^* = 1$.

| | xc Ichimaru (1982) | xc Hedin-Lundqvist (1969) |
|---|---|---|
| Eigenvalues ($Ry$) | | |
| $E_{1s}$ | −74.9628 | −74.9314 |
| $E_{2s}$ | −3.6336 | −3.6293 |
| $E_{2p}$ | −1.6281 | −1.6244 |
| Average of $r$ (a.u): $\langle r\rangle$ | | |
| 1s | 0.1439 | 0.1439 |
| 2s | 0.7916 | 0.7919 |
| 2p | 0.8090 | 0.8096 |
| Average of $r^2$ (a.u): $\langle r^2\rangle$ | | |
| 1s | 0.0279 | 0.0279 |
| 2s | 0.7600 | 0.7609 |
| 2p | 0.8578 | 0.8591 |
| Total energy of embedding ($Ry$) | | |
| $\Delta E$ | −322.8526 | −322.8237 |

Na. Then Table 7.1 collects the eigenvalues and some other properties of this one-centre density functional theory with two different choices for the exchange correlation energy $E_{xc}$. The first follows Ichimaru (1982) and the second uses also a local density form given by Hedin and Lundqvist (1969). These forms of $E_{xc}$, plus the charge $11e$ on a bare Na nucleus, and the average density $\bar{\rho}$ of uniform valence electrons represent the basic input used in constructing Table 7.1. The differences between the two forms of $E_{xc}$ is seen to be very small for all properties recorded in Table 7.1, Fig. 7.5 (Fig. 11 of March (1992) book) depicts the total valence screening charge $Q(r)$ of the valence electrons in Na: to graphical accuracy the difference between the two forms of $E_{xc}$ is not discernable. $Q(R)$ must tend to unity as $R \to \infty$ in monovalent metal.

### 7.2.3. *Pair Interaction Between Ions*

We turn now to treat the two-ion problem, but still within the 'superposition' approximation used to treat test charges in both semiclassical and wave theories. Then following PM, we take the two ions to be at positions $\mathbf{R_i}$ and $\mathbf{R_j}$, and write $\Delta_i = \Delta(\mathbf{r} - \mathbf{R_i})$ and $\Delta G(\Delta + \bar{\rho}) = \mathbf{G}[\Delta + \bar{\rho}] - G[\bar{\rho}]$. For the pair

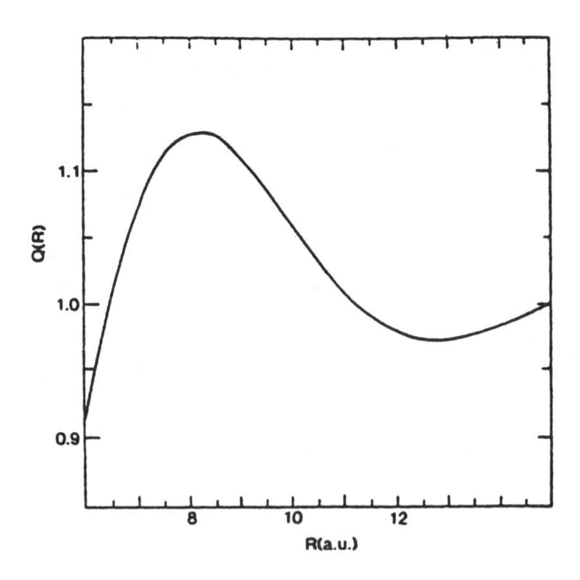


Fig. 7.5. Total valence screening charge $Q(R)$ inside a sphere of radius $R$ centred on an $K^+$ion. (After Johnson *et al*, 1994).*

*See also Chapter 8, Section 8.6.

interaction $\Phi$ between ions $i$ and $j$ PM write

$$\begin{aligned}\Phi = {} & \Delta G[\Delta_i + \Delta_j + \bar{\rho}] - \Delta G[\Delta_i + \bar{\rho}] \\ & - \Delta G[\Delta_j + \bar{\rho}] + \left(\frac{-z}{r_i}\right) \cdot \Delta_j + \left(\frac{-z}{r_j}\right) \cdot \Delta_i \\ & + \frac{1}{2}(\Delta_i + \Delta_j) \cdot \frac{1}{r} * (\Delta_i + \Delta_j) - \frac{1}{2}\Delta_i \cdot \frac{1}{r} * \Delta_i \\ & - \frac{1}{2}\Delta_j \cdot \frac{1}{r} * \Delta_j + \frac{z^2}{R}\end{aligned} \tag{7.2.5}$$

The * notation has already been defined: here the dot means the integral of the product of functions through the whole of space: technically the scalar product.

As PM stress, it is true that if one works with the exact density of the $Na_2$ 'molecule', written as $\Delta_i + \Delta_j + \delta$, rather than with just the superposition approximation $\Delta_i + \Delta_j$, then the final result for $\Phi(R)$ to be given below is correct to $O(\delta^2)$, due to the stationary properties of the energy functional.

To develop $\Phi(R)$ further, we note from Eq. (7.2.4) for $V$ that

$$\begin{aligned}zV(R) & = \frac{-z^2}{R} + \int \frac{z}{|\mathbf{r}' - \mathbf{R}|}\Delta(\mathbf{r}')d\mathbf{r}' \\ & = \frac{-z^2}{R} + \int \frac{z}{\mathbf{r}''}\Delta(\mathbf{r}'' + \mathbf{R})d\mathbf{r}'' \\ & = \frac{-z^2}{R} + \left(\frac{z}{r_i}\right) * \Delta_j\end{aligned} \tag{7.2.6}$$

Utilizing this Eq. (7.2.6) in Eq. (7.2.5), one readily finds

$$\begin{aligned}\Phi(R) = {} & \Delta G[\Delta_i + \Delta_j + \bar{\rho}] - \Delta G[\Delta_i + \bar{\rho}] \\ & - \Delta G[\Delta_j + \bar{\rho}] + \left(\frac{-z}{r_j} \cdot \Delta_i\right) \\ & + \frac{1}{2}\Delta_i \cdot \frac{1}{r} * \Delta_j + \frac{1}{2}\Delta_j \cdot \frac{1}{r} * \Delta_i \\ & - zV(R)\end{aligned} \tag{7.2.7}$$

In PM, some particular approximations to the result Eq. (7.2.7) were examined. However, let us go next to the full result of PM for $\Phi(R)$ in Na metal. Their pair potential is the upper curve at large r in Fig. 7.6. The other curve

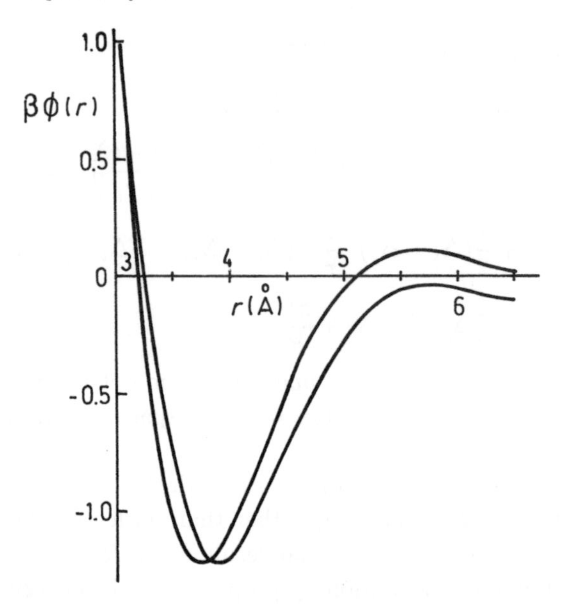


Fig. 7.6. Pair potentials for liquid Na at $T = 100°\text{C}$ and density of 0.929 g cm$^{-3}$. Electron theory: upper curve for large $r$. Inversion of liquid structure factor: lower curve for larger $r$.

Table 7.2. Characteristics of diffraction potential compared with electron theory form.

| Positions of turning points and nodes (Å) | Principal minimum | First maximum | Second minimum | First node |
|---|---|---|---|---|
| Diffraction $\phi(r)$ | 3.9 | 5.76 | 7.44 | 3.3 |
| Electron theory $\phi(r)$ | 3.73 | 5.67 | 7.37 | 3.20 |

will also be considered in some detail below. Note that in Fig. 7.6 the pair potential (best for liquid Na just above its freezing point because of the choice of PM for $\bar{\rho}$) $\varphi(r)$ is expressed in units of $k_BT \equiv \beta^{-1}$ where $T = 100°\text{C}$ (the mass density equivalent of $\bar{\rho}$ being 0.929 g cm$^{-3}$) characteristics of this electron theory pair potential $\varphi(r)$ have been collected in Table 7.2 (Table II of Perrot and March, 1990a). We turn immediately to make contact between this DFT calculation of the pair potential in liquid Na metal just above its freezing point and that which can be extracted from the experimentally measured structure factor $S(k)$, where $S(k)-1$ is the Fourier transform of $g(r)-1$, $g(r)$ being the ionic pair correlation function of the liquid metal.

### 7.2.4. *Inversion of Measured Liquid Structure Factor S(k) for Na Near Melting to Extract a 'Diffraction' Pair Potential*

Johnson and March (1963) proposed a route by means of which a pair potential $\phi(r)$ could be extracted from diffraction measurement of the liquid structure factor $S(k)$. However, it must be stressed that, for successful application of their proposal, it is essential to have diffraction experiments (neutrons preferably but otherwise X-ray) over a very wide range of $k$. The small angle scattering region turns out to be particularly important: one must measure $S(k)$ by diffraction at sufficiently small $k$ so that safe extrapolation into the value of $S(o)$ given by fluctuation theory, namely

$$S(o) = \rho_i k_B T K_T \tag{7.2.8}$$

where $\rho_i$ is the ionic number density and $K_T$ the isothermal compressibility, is ensured.

Johnson and March (1963) focussed all attention on the so-called force equation. The ions in, for example, liquid Na just above its freezing point can be treated classically and we write therefore the Boltzmann form for $g(r_{12})$ in terms of the so-called potential of mean force $U(r_{12})$:

$$g(r_{12}) = \exp(-\beta U_{12})\,; \quad \beta = (k_B T)^{-1}\,. \tag{7.2.9}$$

From this result, one can write the total force acting on atom 1 as the negative of the gradient of $U(r_{12})$ with respect to $\mathbf{r_1}$. This can evidently be divided into two contributions: (i) That due to the direct pair force $\frac{-\partial\phi(r_{12})}{\partial\mathbf{r_1}}$ between atoms 1 and 2 at separation $r_{12}$ and (ii) That due to the remaining atoms.

To deal with the contribution (ii) above, let us introduce the so-called three-atom correlation function $n_3(\mathbf{r_1}\,\mathbf{r_2}\,\mathbf{r_3})$, defined such that the probability that volume elements $d\mathbf{r_1}$, $d\mathbf{r_2}$ and $d\mathbf{r_3}$ around $\mathbf{r_1}$, $\mathbf{r_2}$ and $\mathbf{r_3}$ are occupied by atoms is $n_3(\mathbf{r_1}\,\mathbf{r_2}\,\mathbf{r_3})$. Then remembering that the probability of finding atom in the volume element $d\mathbf{r_3}$, when atoms 1 and 2 are certainly in volume elements $d\mathbf{r_1}$ and $d\mathbf{r_2}$ around $\mathbf{r_1}$ and $\mathbf{r_2}$ is

$$\frac{n_3(\mathbf{r_1}\,\mathbf{r_2}\,\mathbf{r_3})d\mathbf{r_3}}{\rho_o^2 g(r_{12})}$$

with $\rho_o$ the atomic number density, we may write the desired force equation as

$$\frac{-\partial U(r_{12})}{\partial\mathbf{r_1}} = \frac{-\partial\phi(r_{12})}{\partial\mathbf{r_1}} - \int \frac{n_3(\mathbf{r_1}\,\mathbf{r_2}\,\mathbf{r_3})}{\rho_i^2 g(r_{12})}\frac{\partial\phi(r_{13})}{\partial\mathbf{r_1}}d\mathbf{r_3} \tag{7.2.10}$$

This equation, as it stands is an exact consequence of classical statistical mechanics, given of course the assumption of pair forces. But to use it directly to derive $\phi(r_{12})$ from the measured liquid structure factor $S(k)$, from which $g(r)$ can be found by suitable Fourier transform, one must have knowledge concerning $n_3$. This three-atom function is therefore at the heart of the liquid structure problem.

### 7.2.5. *Approximate Analytic Structural Theories*

Kirkwood (1935) argued for the simplest possible approximation that $n_3$ could be 'decoupled' as the product of pair terms, i.e.

$$n_3(\mathbf{r_1}\,\mathbf{r_2}\,\mathbf{r_3}) \cong \rho_o^3\, g(r_{12})\, g(r_{23})\, g(r_{31}) \tag{7.2.11}$$

Inserting this into the force equation one is led to the so-called Born-Green (1946) theory of liquid structure. Rushbrooke (1960) demonstrated that one could then integrate the force equation to find

$$\frac{U(r)}{k_BT} = \frac{\phi(r)}{k_BT} - \rho \int E(\mathbf{r}-\mathbf{r}')\, h(\mathbf{r}')\, d\mathbf{r}' \tag{7.2.12}$$

where $h(r) = g(r) - 1$ is the total correlation function, while $E$ is defined by

$$E(r) = \int_r^\infty \frac{\phi'(t)}{k_BT} g(t)dt \tag{7.2.13}$$

Since $g(t) \to 1$ for sufficiently large $r$, we see from this equation that $E(r) \to \frac{-\phi(r)}{k_BT}$ for sufficiently large $r$. But this can be shown (far from the liquid-gas critical point) to be the behaviour of the Ornstern-Zernike direct correlation function introduced in the footnote.* If the definition of $c(r)$ is used, and $E(r)$ in Eq. (7.2.13). is replaced by $c(r)$, then a second liquid structure theory - the so-called hyperenetted chain (HNC) theory because of its diagrammatic derivation - emerges. This yields

$$\frac{\phi(r) - U(r)}{k_BT} = h(r) - c(r)\,(\text{HNC equation}) \tag{7.2.14}$$

This equation, it turns out (nor the Born-Green approximation) is adequate for quantitative work on Na near the melting point. It is customary to add

*The Fourier transform $\tilde{c}(k)$ of $c(r)$ is defined from the liquid structure factor $S(R)$ introduced above as $\tilde{c}(k) = [s(k) - 1]/s(k)$.

a so-called bridge function $B(r)$ to the right-hand side of Eq. 7.2.14. The further assumption that $B(r)$ is insensitive to the detail of $\phi(r)$, coupled with computer simulation, has been successfully used by Reatto (1988) and co-workers to extract the 'diffraction' potential shown as the second curve in Fig. 7.6. Table 7.2 compares the main features of the diffraction potential, following PM with the electron theory results. The agreement is remarkable. Even the difference between the two curves at large r can be discussed from first-principles theory (Blazej and March, 1993, see also Blazej, Flores and March, 1995), but we will not go into further details here.

### 7.2.6. *Pair Potentials for Iridium and Rhodium**

As Ivanov *et al.* (1994) emphasize, iridium and rhodium show unusual mechanical properties for fcc metals. In particular, they exhibit brittle failure as single crystals, this occurring after a long stage of plastic deformation (Douglass, Crier and Jaffee, 1961; Hieber, Mordike and Haessen, 1967; Ried and Routbort, 1972; Gandhi and Ashby, 1979).

As stressed elsewhere in the present Volume (see especially Chap. 8) the possibility of describing both perfect lattice properties, and defects, both point and extended, of transition metals by pair interactions is highly questionable. However, as shown by Greenberg *et al.* (1990), in the specific case of Ir, such a pair interaction $\Phi(r)$ can be set up, using perturbation theory with a local pseudopotential, the contribution of three-body interactions to the lattice properties turning out to be small corrections. A number of forms of $\Phi(r)$ were put forward by Greenberg *et al.* (1990) but because of a lack of experimental data on the phonon spectra a final choice did not prove possible between these various interactions.

In the work of Ivanov *et al.* (1994), experimental results using the technique of inelastic neutron scattering are presented for the phonon dispersion relations in Ir. These authors then select and study interatomic potentials $\Phi(r)$ for Ir and also for its analogue Rh. As discussed below, these potentials were then utilized to simulate point defects, stacking faults and some dislocations.

For Ir, Greenbery *et al.* (1990) constructed $\Phi(r)$ for Ir of the form (Ivanov *et al*, 1994)

$$\Phi(r) = \frac{Z^2}{r} - \frac{\Omega_o}{\pi^2 r} \int_0^\infty q \sin(qr)\, F(q) dq \qquad (7.2.15)$$

*Though this chapter is dominantly about potentials in s-p metals, it will be useful in relation to mechanical properties to discuss these more complex metals to conclude.

where

$$F(q) = \frac{\Omega_o q^2}{8\pi} V_{ps}^2(q) \left[\frac{1}{\epsilon(q)} - 1\right] \tag{7.2.16}$$

Here $Z$ is the effective ionic charge, $\Omega_o$ the atomic volume, $V_{ps}(q)$ the Fourier transform of the local pseudopotential and $\epsilon(q)$ is the dielectric function.

For $\epsilon(q)$, Greenberg *et al.* (1990) employed the Geldart-Taylor approximation with the correlation according to Ceperley. They used the Animalu-Heine form for $V_{ps}(q)$: (see Ivanov *et al.*, 1994):

$$V_{ps}(q) = \frac{-4\pi Z}{q^2\Omega_o} \left\{\cos(qr_o) - u\left[\frac{\sin(qr_o)}{qr_o} - \cos(qr_o)\right]\right\}$$

$$\times \exp\left[-\xi\left(\frac{q}{2k_{fo}}\right)^4\right] \tag{7.2.17}$$

$k_{fo}$ being the Fermi wave number at zero temperature $T$ and zero pressure $P$. Equation (7.2.17) for the pseudopotential carries three parameters, $r_o$, u and $\xi$. These were fitted to (i) the equilibrium condition $P(\Omega_{\text{exp}}) = 0$, $\Omega_{\text{exp}}$ being the experimental value of the volume per atom, (ii) the Debye temperature $\theta_{\text{exp}} = 425\text{K}$ and (iii) a reasonable value of the hard-sphere packing parameter $\eta : 0.45 < \eta < 0.53$:

$$\eta = \frac{\pi d^3}{6\Omega_o}, \quad \int_0^\infty dr \left\{1 - \exp\left[\frac{-\Phi(r)}{T}\right]\right\} = d \tag{7.2.18}$$

at the melting temperature $T = T_m$. In the study of Greenberg *et al.* (1990). $Z$ was treated as a fourth free parameter, and they concluded that $\Phi(r)$ above gave an adequate description of physical properties provided $3.5 \leq Z \leq 4.5$. Ivanov *et al.* (1994) appeal to the first-principles energy band calculations of Dacorogna *et al.* (1982) to fix $Z$ as 4.5.

Much less experimental data is available for Rh. However, $Z$ was taken as 3.86 from Dacorogna *et al.* (1982) while $\xi$ was fixed at the same value as for Ir. The other two parameters were then fixed from $P(\Omega_{\text{exp}}) = 0$ and $B(\Omega_{\text{exp}}) = B_{\text{exp}}$, with $B$ the bulk modulus and $B_{\text{exp}} = 26.9 \times 10^{11}$ dyne cm$^{-2}$. The pseudopotential parameters for both metals are collected in Table 7.3, taken from Ivanov *et al.* (1994).

The phonon dispersion curves using this $\Phi(r)$ for Ir are shown in Fig. 7.7 (Fig. 2 from Ivanov *et al.*, 1994), compared with the experimental points. The agreement is better than semi-quantitative. The potential is plotted in Fig. 7.8 (Fig. 4 of Ivanov *et al.*, 1994), together with that for Rh.

Table 7.3. $\Omega_o, r_o$ and $V_o = uZ/r_o$ in atomic units (after Ivanov *et al.*, 1994).

| Metal | $\Omega_0$ | $Z$ | $\xi$ | $r_o$ | $u$ | $-V_o$ |
|---|---|---|---|---|---|---|
| Ir | 95.52 | 4.5 | 0.30 | 2.700 | −1.243 | 2.070 |
| Rh | 92.58 | 3.86 | 0.30 | 2.637 | −1.247 | 1.823 |

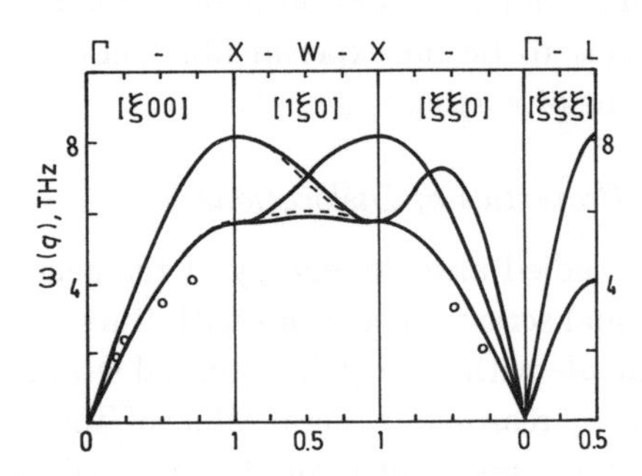


Fig. 7.7. Phonon dispersion curves in Ir compared with experimental points (0), after Ivanov *et al.* (1994). Broken line — nearest-neighbour only. Results along symmetry directions are shown.

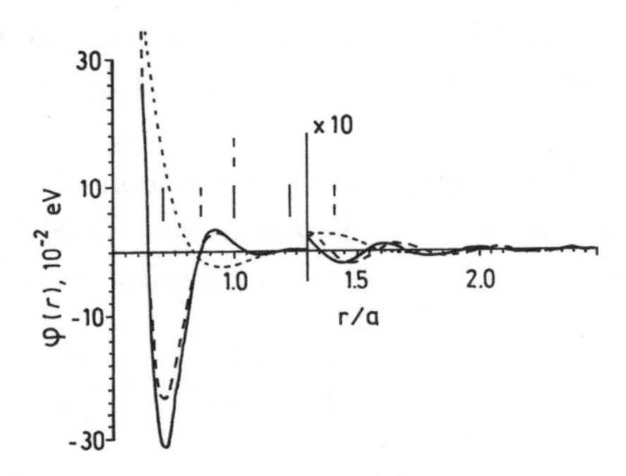


Fig. 7.8. Pair potentials for Ir (continuous curve) and Rh (dashed curve) taken from Ivanov *et al.* (1994).

Table 7.4. Point defect energies in eV for vacancy(v) and interstitial(I) from pair potentials ($f \equiv$ formation, $m \equiv$ migration).

| Metal | $E_v^f$ | $E_v^m$ | $E_I^f$ | $E_I^m$ |
|---|---|---|---|---|
| Ir | 1.64 | 3.05 | 6.20 | 0.51 |
| Rh | 1.12 | 2.15 | 4.44 | 0.39 |

Point defect energies obtained by Ivanov *et al.* (1994) with the above potentials are collected in Table 7.4. If the monovacancy approximation is assumed for self-diffusion then the activation energy* $E_v^f + E_v^m = 4.69$ eV for Ir is in excellent agreement with the experimental value of 4.55 eV given by Arkhipov *et al.* (1986), $f$ denoting formation and in $m$ migration.

Stacking fault energies were also calculated from $\Phi(r)$, in the $\{111\}$ plane, and values of 20 mJm$^{-2}$ for Ir and $\sim$ 90 mJm$^{-2}$ for Rh were obtained. Unfortunately there seems to be no experimental data to compare with these predictions, at the time of wirting.

### 7.2.7. *Structure and Properties of Dislocations*

Ivanov *et al.* (1994) studied also the energy of the generalized stacking fault and splitting of edge and screw dislocations with Burgers vector $\mathbf{b} = 1/2\langle 110\rangle$. Computations were made with their pair potential $\Phi(r)$ for Rh since this force law yields, as above, a reasonably high value of the stacking fault energy in the $\{111\}$ plane and consequently an effective splitting of dislocations in simulation.

Simulation of screw and edge dislocations by Ivanov *et al.* (1994) demonstrate that these split effectively. The splitting width was determined in terms of the Peierls-Nabarro model. The above width was 14a and 8a for edge and screw dislocations respectively, and these values agree with estimates from elasticity theory.

The distribution of the edge and screw components of the displacements on the slip plane for the screw dislocation was carried out also by Ivanov *et al.* (1994) in terms of elasticity theory by means of the formula

$$\frac{1}{2}[10\bar{1}] = \left(\frac{1}{4}[10\bar{1}] + \frac{x}{12}[12\bar{1}]\right) + \left(\frac{1}{4}[10\bar{1}] + \frac{x}{12}[\bar{1}\bar{2}1]\right) \tag{7.2.19}$$

There exist a splitting, determined by the minimum sum of the energy of elastic interaction between partial dislocations and by the energy of the generalized stacking fault. For the Rh pair potential, the most favorable splitting of the edge and screw dislocations according to Ivanov *et al.* (1994) is at $x = 0.95$. Refined force fields beyond the pair potential approximation will be discussed in the following chapter.

---

*See, e.g., *Atomic Transport in Solids* by A. R. Allnatt and A. B. Lidiard (Cambridge Univ. Press UK, 1993).

# Chapter 8

# Transcending Pair Potentials: Glue Models of Interatomic Forces

## 8.1. *Introduction*

As some background to the present chapter, we note the following points:

(1) In the past, much effort was put into crystallographic characterization of extended crystal defects: in particular dislocations and grain boundaries.

(2) Properties and behaviour of interfaces, and to a large extent dislocations, *cannot* be determined solely by such geometrical characteristics. For dislocations it is the atomic structure in their cores which is a dominant factor.

(3) Investigation of such core structure requires atomic modelling.

(4) One needs a method to evaluate the *total* energy E of a system of many particles as a function of their positions.

## 8.2. *Embedded Atom and Related Approaches*

In the previous chapter, the theory of pair potentials in simple metals like Na and K has been developed. But in many metals, one can now transcend this description. Below, an account will be given of so-called glue models of interatomic force fields.*

---

*The description 'glue models' is intended to embrace a variety of methods with different physical origins. These include the following: (i) M. J. Stott and E. Zaremba (Phys. Rev. *B22*, 1564, 1980), (ii) M. S. Daw and M. I. Baskes (Phys. Rev. Lett. *50*, 1285, 1983), (iii) M. W. Finnis and J. E. Sinclair (Phil. Mag. *A50*, 45, 1984), (iv) K. W. Jacobsen, J. K. Notskov and M. J. Puska (Phys. Rev. *B35*, 7423, 1987), (v) F. Ercolessi, M. Parrinello and E. Tosatti (Phil. Mag. *A58*, 213, 1988) and (vi) D. G. Pettifor (Phys. Rev. Lett. *63*, 2480, 1989).

We begin by summarizing the work on Al by Robertson *et al.* (1993). As these workers point out, all the simple glue models to be found in the literature may be written as follows:

$$E = \sum_i E^i \tag{8.2.1}$$

$$E^i = \sum_{j \neq i} \phi(\mathbf{r_{ij}}) + U(n^i) \tag{8.2.2}$$

$$n^i = \sum_{j \neq i} \rho(\mathbf{r_{ij}})\,. \tag{8.2.3}$$

As Robertson *et al.* (1993) stress, the various glue models appropriate to nontransition metals then differ in the forms chosen for $\phi$, $U$ and $\rho$, as well as in the choice of the experimental data used to parametrize the functions.

### 8.2.1. *First-Principles Energy Calculations for Al Structures*

Robertson *et al.* (1993) have constructed a data base of some 170 first-principles total energy calculations of Al structures. These correspond to coordination numbers ranging from zero to 12 and nearest-neighbour distance from 2.0 to 5.7 Å. Their calculations elucidate the nature of metallic bonding.

Using this database, Robertson *et al.* have tested a large range of glue models set out above. They find that none yields a RMS error less than 0.1 eV per atom. These workers argue that this minimum error is due to bonding effects which are outside the glue model framework. Given this, they conclude that this sets a limit to the ultimate reliability of any glue scheme.

### 8.2.2. *Coordination Number Dependence of Energy/Atom for Al Structures*

Robertson *et al.* (1993) test the above glue formalism by focusing only on structures for which (i) the near-neighbour distance is fixed at some chosen value, $r_o$ say, and (ii) the constituent atoms have equivalent environments so that the energy of any structure can be partitioned between the atoms in accord with Eq. (8.2.1).

As a first step, they then neglect neighbours beyond the nearest. The energy per atom, $E_a$ say, for this set of structures should then be given by

$$E_a = c\phi(r_o) + U(c\rho(r_o)) \tag{8.2.4}$$

where $c$ denotes the coordination number.

Table 8.1. The energies and the coordination numbers for the eighteen structures of the original database. (After Robertson *et al*, 1993).

| Structure | Coordination number | Energy per atom (eV) |
|---|---|---|
| Atom | 0 | −54.95 |
| Dimer | 1 | −55.66 |
| Line | 2 | −56.28 |
| Graphite | 3 | −56.95 |
| Girder | 4 | −57.04 |
| Square layer | 4 | −57.29 |
| Diamond | 4 | −57.42 |
| Square slab | 5 | −57.64 |
| Close packed layer | 6 | −57.49 |
| Simple cubic | 6 | −57.91 |
| fcc 110 slab | 6 | −57.54 |
| Close packed slab | 7 | −57.89 |
| fcc 100 slab | 8 | −57.85 |
| Vacancy lattice | 8 | −58.10 |
| Simple hexagonal | 8 | −58.12 |
| bcc | 8 | −58.24 |
| fcc 111 slab | 9 | −57.97 |
| fcc | 12 | −58.31 |

Robertson *et al.* calculate $E_a$ for $r_o = 2.85$ Å, which is near to the equilibrium separation in fcc Al, for the eighteen structures recorded in Table 8.1 (Table 1 of Robertson *et al.*, 1993).

### 8.2.3. *Methodology*

The energies recorded there were calculated using density functional theory. A local pseudopotential for Al was employed (Cheng *et al.*, 1987), together with a plane wave basis with a cut-off energy of 190 eV. The exchange and correlation contribution to the one-body potential was formed from the results of Ceperley and Alder (1980) as fitted by Perdew and Zunger (1981).

The data in Table 8.1 was fitted by Robertson *et al.* to a function of the form

$$E_a = E_o + A\sqrt{c} + Bc\,. \tag{8.2.5}$$

The choice of the function $U$ in Eq. (8.2.4) as proportional to $\sqrt{c}$ was originally proposed by Finnis and Sinclair (1984). Robertson *et al.* (1993) note that the fit (8.2.5) has a RMS deviation of 0.20 eV, with

$$A = -1.41 \text{ eV}, \; B = 0.09 \text{ eV}, \; E_o = -54.74 \text{ eV}. \tag{8.2.6}$$

## 8.3. *Embedded Atom Method: Analytic Model for fcc Metals*

Johnson (1988) has studied the implications of the embedded-atom method* by using a simple nearest-neighbour analytic model for the fcc lattice. The model has as input information: (i) the atomic volume (ii) the cohesive energy (iii) the bulk modulus (iv) the average shear modulus and (v) the vacancy formation energy. Since the embedded-atom method developed from density-functional theory, microscopic information is also required in addition to the inputs (i)–(v) above. In Johnson's model, this latter information is the slope at the nearest-neighbour distance of the spherically averaged free-atom electron density calculated with Hartree-Fock theory. This is fitted by an exponentially decreasing function. Additionally a universal equation relates crystal energy and lattice constant (Rose *et al.*, 1984). Johnson's model also employs an exponential repulsion between nearest-neighbour atoms. Also, in his model, the anisotropy ratio of the cubic shear moduli is constrained to be 2.

As in the previous section, the energy $E$ per atom, with only nearest-neighbour interactions included can be written as

$$E(r_0) = F(\rho(r_0)) + 6\phi(r_0) \tag{8.3.1}$$

with

$$\rho(r_0) = 12 f(r_0). \tag{8.3.2}$$

The difference in the various models then enters through the choice of $\phi(r)$ and $f(r)$.

Johnson's choice for his analytic model corresponds to:

$$f(r) = f_e \exp\left\{-\gamma\left(\frac{r}{r_{oe}} - 1\right)\right\}, \quad r \leq r_c \tag{8.3.3}$$

where $r_c$ is a cut-off parameter, while $r_{oe}$ is the equilibrium near-neighbour distance. The pair potential $\phi(r)$ in Eq. (8.3.1) is taken as a Born-Mayer

---

*See especially M. S. Daw and M. I. Baskes (Phys. Rev. *B29*, 6443, 1984).

repulsion:

$$\phi(r) = \phi_e \exp\left\{-\gamma\left(\frac{r}{r_{oe}} - 1\right)\right\}, \quad r \leq r_c\,. \tag{8.3.4}$$

Finally the embedding function $F(\rho)$ in Eq. (8.3.1) is chosen by Johnson (1988) as

$$F(\rho) = -E_c\left[1 - \frac{\alpha}{\beta}\ln\left(\frac{\rho}{\rho_e}\right)\right]\left(\frac{\rho}{\rho_e}\right)^{\alpha(\beta)} - \Phi_e\left(\frac{\rho}{\rho_e}\right)^{\frac{\gamma}{\beta}} \tag{8.3.5}$$

where $\rho_e = 12f_e$ and $\Phi_e = 6\phi_e$, while $E_c$ is the experimental cohesive energy.

Johnson emphasizes that, since only $\rho/\rho_e$ enters Eq. (8.3.5), the parameter $f_e$ cancels out of the model. Furthermore, for the elastic constants for any defect calculation in which the relaxations are small, the precise choice of the cut-off $r_c$ plays no role as long as $r_c$ is well within the gap between first and second neighbours, as is true in Johnson's work.

### 8.3.1. *Average Shear Constant*

Finally, the average shear constant $G_e$ is calculated in terms of the parameters $\gamma$, $\beta$ and $\phi_e$ entering the model as

$$G_e = \frac{2\gamma(\gamma - \beta)}{5\Omega_e}\phi_e\,. \tag{8.3.6}$$

The magnitude of $\beta$ is determined from atomic wave function calculations. The cohesive energy, compressibility and lattice parameter (thus also the equilibrium atomic volume $\Omega_e$ entering Eq. (8.3.6)) are given exactly, while, as mentioned above, the shear anisotropy is fixed at a value of 2.

### 8.3.2. *Analytic Results for Vacancy and Divacancy Energetics*

Johnson points out that in close-packed fcc metals the energies associated with vacancy formation, divacancy binding and formation of planar surfaces are dominated by contributions before atomic relaxation is allowed. Such unrelaxed energies can be calculated analytically within his model and we summarize his findings below:

$$E^{\text{unrelaxed}}_{\text{vac. form.}} = \frac{15\Omega_e G_e}{\beta\gamma}\left[1 - \frac{1}{24}\frac{\gamma}{\beta} - \frac{1}{864}\frac{\gamma}{\beta}\left(2 - \frac{\gamma}{\beta}\right)\right]$$
$$+ \text{smaller term proportional to } \frac{\Omega_e B_e}{\beta^2}\,. \tag{8.3.7}$$

where $B_e$ is the equilibrium bulk modulus. The term displayed explicitly in Eq. (8.3.7), for the parameter values adopted by Johnson ($\gamma \approx 8, \beta \approx 6$) is about 1.2 eV while the term involving $B_e$ contributes only $\approx 0.1$ eV, for the example of Cu as the prototype metal considered by Johnson (1988).

The divacancy binding energy $E_{2v}^{B,\text{unrelaxed}}$ is also given approximately by Johnson (1988, his Eq. (34)) as

$$E_{2v}^{B,\text{unrelaxed}} = \frac{5}{2}\frac{\Omega_e G_e}{\beta\gamma}\left[1 + \frac{1}{8}\frac{\gamma}{\beta} + \frac{11}{864}\frac{\gamma}{\beta}\left(2 - \frac{\gamma}{\beta}\right)\right]$$
$$+ \text{ term proportional to } \frac{\Omega_e B_e}{\beta^2}. \tag{8.3.8}$$

Similarly, Johnson calculates unrelaxed surface energies for the 3 fcc low-index planes, but we refer the reader to the original paper for his formulae for these.

However, for his prototypical metal Cu he finds for 111, 100 and 110 low-index planes the unrelaxed values 991, 1194 and 1292 erg/cm$^2$ respectively. The sequence from lowest to highest surface energy is $\{111\}$ to $\{100\}$ to $\{110\}$, as expected for the fcc lattice.

The surprising result which emerges from Johnson's analytic model based on the embedded-atom ideas is that the shear modulus, rather than the bulk modulus or the cohesive energy, is the dominant parameter determining the vacancy formation energy. However, the unrelaxed divacancy binding energy, and the unrelaxed surface energies even more so, contain contributions which depend on the embedding function at electron densities significantly smaller than the equilibrium value. As Johnson stresses, the curvature thus plays a more important role and the shear modulus is not as dominant in these cases.

Finally, we have noted that, following Johnson's numerical estimates, the shear modulus term is also larger in the divacancy binding energy than the bulk modulus (though now the factor is only $\approx 4$ between the two pieces). Roughly therefore the vacancy formation energy is $\approx 15\ \Omega_e G_e/\beta\gamma$ while the divacancy formation energy (somewhat less precisely) is $\frac{5}{2}\ \Omega_e G_e/\beta\gamma$, i.e. the ratio: divacancy binding energy to vacancy formation energy $\approx 1/6$.

### 8.3.3. *Hexagonal-Close-Packed Metals*

The embedded atom method has subsequently been extended to the hexagonal-close-packed (hcp) metal Zr (Goldstein and Jónsson, 1995). These workers incorporate the non-ideal *c:a* ratio and the elastic responses in their fitting procedure. As in the work of Johnson (1988) reported above, simple functional forms are assumed for the pair interaction, atomic electron density and embedding function. Experimental data used to parametrize the functions are (i) cohesive energy (ii) equilibrium lattice constants (iii) single crystal elastic constants and (iv) vacancy formation energy. The equation of state set up by Rose *et al.* (1984) is employed to reproduce the pressure dependence of the cohesive energy, taking account of the anisotropic elastic response of the crystal.

Their interatomic force field has been applied to calculate stacking fault and self-interstitial formation energies.

## 8.4. *Inequality Relating Vacancy Formation Energy in a Hot Crystal (Near Melting) to Rigidity*

Following the conclusion of Johnson (1988), discussed in the previous section, that at $T = 0$ the vacancy formation energy $E_v$ correlates with the shear modulus in a simple embedded atom model, March (1989) has studied a related problem, but now for $E_v$ near the melting temperature $T_m$. His work is related to the considerations on $E_v$ in Appendices 6.1 and 7.4.

In common with Johnson (1988), the work of March (1989) neglected ionic relaxation around the vacant site, which is permissible in close-packed crystals (for simple bcc metals such as Na and K, see Appendix 5.2 for a treatment of such relaxation effects round a vacant site). Since one is now dealing with the hot crystal near melting ($T = T_m$), appeal is also made to the liquid pair distribution function $g(r)$ at $T_m$. This is useful if melting leaves the local coordination of the hot, highly anharmonic, solid largely intact (it would exclude, e.g. Si and Ge where covalently bonded semiconductors become liquid metals on melting, with therefore very different force laws in the two phases). In this 'liquid-like' language March (1987) has shown that

$$E_v = \Omega \frac{\partial E}{\partial V}\bigg|_{T_m} + \text{term involving } \frac{\partial g(r)}{\partial \rho} \tag{8.4.1}$$

Rashid and March (1989) have shown in various cases that the term in Eq. (8.4.1) involving the dependence on ionic number density $\rho$ of the liquid pair function $g(r)$ is small compared with the term involving 'departures from Joule's Law', i.e. the dependence of the total internal energy $E$ on the total volume $V$ of the liquid (compare Appendix 7.4).

Thermodynamics can now be used to yield in the liquid

$$\left(\frac{\partial E}{\partial V}\right)_T \simeq T\left(\frac{\partial p}{\partial T}\right)_V , \tag{8.4.2}$$

neglecting the pressure $p$ as small compared with $\partial E/\partial V$. Secondly, one invokes the specific heat difference through

$$C_p - C_v = -T\left(\frac{\partial p}{\partial T}\right)_V^2 \left(\frac{\partial V}{\partial p}\right)_T \tag{8.4.3}$$

Using the fluctuation theory result for the long wavelength limit $k \to 0$ of the liquid structure factor $S(k)$ (essentially the Fourier transform of $g(r)$, and putting this together with Eqs. (8.4.1–8.4.3) yields, with $\gamma = C_p/C_v$,

$$\frac{E_v}{k_B T_m} = \left[\frac{(\gamma-1)C_V/k_B}{S(0)}\right]^{\frac{1}{2}}\Bigg|_{T_m} \tag{8.4.4}$$

March (1989) now introduces elastic moduli in the liquid phase. With application of a high-frequency stress, a liquid responds as a solid and among other effects will exhibit rigidity. March (1989) has used the arguments of Schofield (1966) who relates the rigidity $G$ to the liquid structure factor and an (assumed) pair potential. Schofield establishes an inequality which is used by March (1989) to show that

$$\frac{\gamma}{S(0)|_{T_m}} \leq \frac{5}{3}\frac{G}{\rho k_B T}\Bigg|_{T_m} . \tag{8.4.5}$$

Using this inequality in conjunction with Eq. (8.4.4) leads to

$$\frac{E_V}{k_B T}\Bigg|_{T_m} \leq \left[\frac{5C_V}{3k_B}\left(\frac{\gamma-1}{\gamma}\right)\frac{G\Omega}{k_B T}\right]_{T_m} \tag{8.4.6}$$

where $\Omega$ is the atomic volume.

Equation (8.4.6) was developed a little further by March (1989) by following Johnson (1988) and adopting scaling of $G\Omega$ with $k_B T_m$. Then, one can eliminate $k_B T_m$ from the inequality (8.4.6) in favour of a 'constant', J say, equal to $G\Omega/k_B T_m$, to yield

$$E_v \leq \left[\frac{5}{3}\frac{C_v}{k_B}\left(\frac{\gamma-1}{J\gamma}\right)\right]^{1/2} G\Omega\Bigg|_{T_m} , \tag{8.4.7}$$

This inequality is the main result of this section.

The above treatment of $E_v$ at the melting temperature $T_m$ leads to the weaker link than Johnson's result (now $T = 0$) between $E_v$ and rigidity $G$ via the inequality (8.4.7). Nevertheless, it is to be stressed that there is no conflict between Johnson's finding from a simplfied version a 'glue' model and the above result (8.4.7), which, it must be reiterated, has only been established at the melting temperature $T_m$.

## 8.5. *Screw Dislocation Core Structures for Niobium and Molybdenum*

Vitek (1995) has reported screw dislocation core structures which were calculated by means of Finnis-Sinclair type potentials for bcc niobium and molybdenum. The calculated core structures given by Vitek (1995) are displayed in Figs. 8.1 and 8.2 (Figs. 3 and 4 from Vitek, 1995). To aid understanding of these figures, and again following Vitek (1995), the orientations of all the $\{110\}$ and $\{112\}$ planes belonging to the $[111]$ zone are shown in Fig. 8.3 (Fig. 5 of Vitek, 1995).

Figures 8.1(a) and 8.2(a) show the screw components while Figs. 8.1(b) and 8.2(b) display the edge components. In the first case the length of the arrows used is normalized by $|\frac{a}{6}[111]|$ and when the magnitude of the displacement is equal to this value the length of the corresponding atom is equal to the separation of the neighbouring atoms (Vitek, 1995) in the second case the vectors have been magnified by a factor of ten. The main feature of the cores,

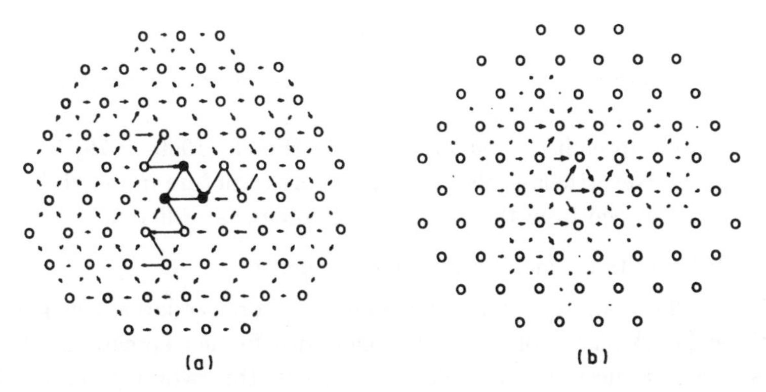


Fig. 8.1. Core structure of the 1/2[111] screw dislocation in molybdenum. (a) Screw component. (b) Edge component.

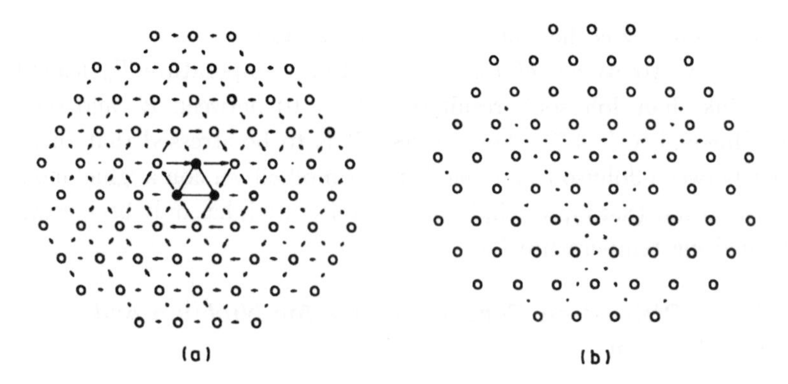


Fig. 8.2. Core structure of the 1/2[111] screw dislocation in niobium. (a) Screw component. (b) Edge component.

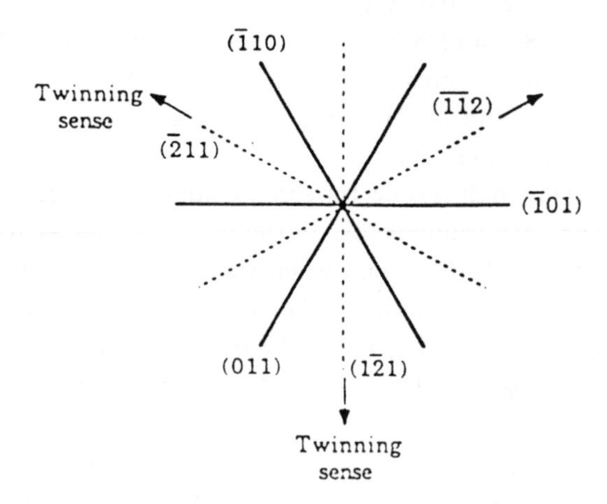


Fig. 8.3. Orientation of the [110] and [112] planes belonging to the [111] zone.

the non-planar spreading, first proposed by Hirsch (1960, 1979) is the same in both cases. Nevertheless, the calculated core structure in molybdenum and niobium can be seen from the figures to be somewhat different.

(i) Detailed results for molybdenum core structure

As Vitek (1995) points out, for molybdenum the core is relatively widely spread into three {110} planes of the [111] zone, with further spreading into {112} planes. This induces in these planes shears in the twinning sense. This is revealed in Fig. 8.2(a) (Vitek, 1995) by the fact that the arrows on these sides of the {110} planes where the corresponding {112} planes are sheared in the twinning sense, are large than on the other sides. Further, as is evidenced in Fig. 8.2(b), the displacements inside the core also possess significant edge components. Vitek (1975, see also 1995) gives symmetry arguments that the dislocation core may exist in two related variants and therefore this core configuration is termed a degenerate structure.

(ii) Results for niobium core structure

In contrast to molybdenum above, the niobium core structure turns out to be invariant with respect to all the symmetry operation of the lattice. This different core structure is thus termed non-degenerate. This time the core spreads symmetrically into the three {110} planes of the [111] zone, but is much narrower than for molybdenum. Vitek (1994, 1995) observes, from the evidence in Fig. 8.2(b) that there are now no significant edge components of the core displacements. Furthermore, twinning-antitwinning asymmetry is not visible in the core spreading. Nevertheless, studies of the effect of external stresses on this core by Vitek *et al.* (1995) exposed this asymmetry in the orientation dependence of the crss.

To summarize the above, following Vitek (1994, 1995), the results of his work reported above show that both the degenerate and non-degenerate states are feasible for the core structure of the bcc screw dislocations and indicate that the former is appropriate for molybdenum and the latter for niobium. Vitek also emphasizes that while the twinning-antitwinning asymmetry appears in both cases, the edge component of the core displacements is only in evidence in the degenerate case. Due to this component, the non-glide shear stresses, perpendicular to the Burgers vector, can have a pronounced effect on the behaviour of the dislocation core and induce significant departures from the Schmid law (Vitek *et al.*, 1992; 1995). It is relevant in the present context to note that the experimentally observed orientation dependence of the crss is substantially less for niobium than it is for molybdenum (Bowen *et al.*, 1967; Vesely, 1968; Duesbery and Foxall, 1969).

## 8.6. *Quantum-Chemical Model of Cold Metallic Lattice Energies as Function of Coordination Number c*

Following these examples of "glue" models, we turn in this section to a different route, via quantum-chemical modelling, to represent the energy of crystal lattices as a function of near-neighbour distance $r_0$ and coordination number $c$. The treatment below follows the study of March, Tosi and Klein, and presents results of March and Rubio (1997) on metallic $K$ lattices with different coordination number (see the related Fig. 7.5 for the screening charge around an ion $K^+$ in a Fermi gas of density equal to that of the conduction electrons in metallic $K$).

The basic equation of the quantum-chemical approach is for the cohesive energy $E_c$

$$E_c = \frac{1}{2} c R(r_0) - f(c) g(r_0) . \tag{8.6.1}$$

March *et al.* (1995) note that $R(r_0)$ and $g(r_0)$ are now characterized by the appropriate curves of the quantum-chemical dimer. This is $K_2$ in the example worked out by March and Rubio. One needs 'potential energy curves' for $K_2$ for the ground singlet ($\Sigma_g$) and triplet $^3\Sigma_u$ excited states. $R(r_0)$ is the triplet $^3\Sigma_u$ potential curve in Eq. (8.5.1.) while $g(r_0)$ is the 'exchange' part, which is half of the difference between the triplet $^3\Sigma_u$ and the singlet $^1\Sigma_\nu$ potential-energy curves.

As noted by March *et al.* (1995), Malrieu *et al.* (1984) have given Eq. (8.5.1) for fcc ($c = 12$), bcc ($c = 8$) and simple cubic (sc)($c = 6$) respectively. These results correspond to $f(c)$ defined in Eq. (8.6.1) as 4,4 and 3. The work of March and Rubio (1997) leads them, by study of the diamond lattice of $K$ atoms, and of chains to $f(4) \simeq 2.9$, $f(2) \simeq 1.2$. Thus (compare Eq. (8.2.5) for Al at one chosen $r_0$) $f(c)$ varies more slowly than linearly with $c$.

Without going into full details, March and Rubio (1997) have compared the model (8.6.1) using specific 'potential energy curves' for $K_2$ from the chemical literature, with density functional calculations on metallic $K$ lattices of different coordinate number, the latter approximates the exchange-correlation potential $V_{xc}cr$ in the Slater-Kohn-Sham one-electron Schrodinger equation by local density formula (i.e. homogeneous electron liquid, with the constant density $\rho_0$ there replaced by the local electron density $\rho(\mathbf{r})$ in the periodic metal crystals). The agreement between the quantum-chemical model and the density functional calculations is impressive over a range of near-neighbour distances $r_0$ for the lattices of different coordination numbers $c$ examined by March and Rubio (1997).

## 8.7. *Further Work on Dislocations and Grain Boundaries*

### 8.7.1. *Body-Centred Cubic Metals*

As Vitek (1994, 1995) notes, some of the first atomistic studies of dislocations were carried out on bcc metals. The reason was that their plastic behaviour shows striking contrast with that of fcc metals (Kubin, 1982; Christian, 1983; Duesbery, 1989).

In the fcc metals, the slip systems are always ⟨110⟩ {111}, the critical resolved shear stress (crss) is small and essentially independent of orientation and temperature ($\sim 10^{-5}$ G at 4.2K) obey the Schmid law already discussed in Chap. 2. As Vitek (1994, 1995) points out, it is not always recognized that the above law contains two distinct assertions. The first (compare Chap. 2) is that plastic flow begins when the resolved shear stress on a possible slip system reaches a constant, critical value. The second assertion is that this critical stress is not affected by any other component of the applied stress tensor (see Vitek, 1994, 1995).

For the bcc metals, slip occurs in the close-packed ⟨111⟩ directions but the slip surfaces, not always planar, vary with orientation and sense of the applied stress. This violates the Schmid law and suggests an intrinsic anisotropic resistance to the slip in these bcc materials (Vitek, 1994, 1995), The crss is now strongly dependent on temperature, with values as large as $10^{-2}$ G at 4.2K and depends on the orientation of the applied stress. Since prominent deformation features are the same in alkali and in transition metals (Duesbery, 1989; see also Depersio and Escaig, 1977; Sakia *et al.*, 1979), Vitek (1994, 1995) stresses that this points to insensitivity to detailed bonding and force law in the case of the bcc structure.

Early atomistic studies of Vitek *et al.* (1970), Vitek and Bowen; see also Vitek (1975) confirmed the proposal of Hirsch (1960, 1979) that the major aspects of the deformation characteristics of bcc metals can be explained by the non-planar spreading of the cores of 1/2⟨111⟩ screw dislocations into several planes of the ⟨111⟩ zone.

However, in spite of the generalities referred to above in bcc metals, the deformation behaviour of all transition bcc metals is not the same.

It is of interest to reiterate some examples of grain boundaries carried out with Finnis-Sinclair potentials, already discussed in Chapter 3.

### 8.7.2. *Grain Boundaries in Metals and Alloys*

(1) $\sum = 3$ tilt boundary with a ⟨112⟩ rotation axis in Cu

Atomistic studies by Schmidt *et al.* (1995) have revealed that a particular $\sum = 3$ tilt boundary with *a* ⟨112⟩ rotation axis stabilized Cu in the bcc configuration. Cu has earlier been observed in the bcc form either as an epitaxial layer formed on a substate of bcc iron by Celinski *et al.* (1991) or as small precipitates in a bcc Fe matrix (Jenking *et al.*, 1991). The study of Schmidt *et al.* (1995) demonstrates that the bcc structure can also occur under purely material constraints in grain boundaries. The boundary under discussion is an asymmetrical tilt configuration with inclination of the boundary plane 84° with respect to the usual $(111)/(11\bar{1})$ coherent twin. Laub *et al.* (1994), by thermal studies referred to in Chapter 3, obtained indications that the boundary energy of $\sum 3\langle 112\rangle$ tilt boundaries has a minimum at this inclination of the boundary plane.

The atomistic calculations, using glue model potentials for Cu (Ackland *et al.*, 1987) demonstrate that a layer of bcc structure forms in this boundary and $\{110\}_{bcc}$ planes connect the $\{111\}_{fcc}$ planes of each grain (see also Vitek, 1995). Vitek stresses that the structure of the phase boundaries between bcc and fcc is a near coincidence type, involving a compromise in lattice strains. It is best suited to this specific boundary inclination and accounts for its lower energy relative to neighbouring inclinations. However, a necessary prerequisite is that the energy difference between bcc and fcc Cu is especially small (0.023 eV/atom: see Vitek, 1995). This was, in fact, demonstrated by the *ab initio* calculations of Paxton *et al.* (1990) and of Kraft *et al.* (1993). The structure discussed above has been confirmed experimentally by HREM (Schmidt *et al.*, 1995): see also Chapter 3.

(2) Grain boundaries in copper-bismuth alloys

As discussed in Chapter 3, the copper-bismuth system is well suited to study segregation and embrittlement phenomena.

Some of the experimentally observed facts are briefly summarized in Chapter 3: a phenomenon discussed there is the segregation induced faceting (Ference and Balluffi, 1988). As Vitek, (1995) stresses, this phenomenon is associated with the formation of a new two-dimensional phase. This motivated a combined atomistic theoretical and HREM investigation, which not only revealed the existence of this phase but also determined its detailed structure (Luzzi, 1991; and Yan *et al.*, 1993: see also Chapter 3).

Finnis-Sinclair type many-body potentials were used. The Cu-Bi interaction was fitted to (i) the lattice parameter (ii) bulk modulus of the theoretical $Cu_3Bi$ compound in the $Ll_2$ structure and (iii) the enthalpy of mixing for the Cu-Bi liquid solution at 1200K. The parameters for the $Ll_2Cu_3Bi$ structure were found from an *ab initio* LMTO calculation (Yan *et al.*, 1993; Vitek, 1995: see Chapter 3).

This form of many-body potentials was used to calculate the structure of $(111)/(11\bar{1})$ facets containing Bi. Vitek (1995) describes the structure of the

$(111)/(11\bar{1})$ facets containing Bi; first extracted from HREM observations (see Appendix 2.1) and then confirmed as the stable structure by molecular statics calculations (see Chapter 3).

As discussed in Chapter 3, this study demonstrates that a two-dimensional ordered phase may form at the $(111)/(11\bar{1})$ twin boundaries in Cu-Bi provided a sufficient amount of Bi is available at this boundary. Vitek stresses that these results provide concrete evidence that empirical many-body potentials are very useful even in a relatively complex Cu-Bi system, provided sufficient relevant input is employed to construct them. He emphasizes in the present example the importance of *ab initio* electronic structure calculations, which sample atomic configurations not attainable in the laboratory. Such input allows the size of a Bi atom when surrounded by Cu atoms to be correctly incorporated (see Chapter 3).

(3) Grain boundary in NiAl with Finnis-Sinclair potentials

The structural unit of the $\sum = 5$, (310) [001] grain boundary in NiAl is shown in Fig. 8.4. Atomistic calculations were performed by Fonda *et al.* (1995) using N-body empirical potentials constructed following the Finnis-Sinclair (1984) approach (compare Ackland *et al.*, 1987). The total energy $E$ of the system of $N$ atoms is expressed as in Eq. (3.7.3) of Chapter 3.

We reiterate that the Ni-Ni potential was adjusted to fit (i) to experimental lattice parameter (ii) cohesive energy (iii) elastic constants and (iv) vacancy formation energy of pure Ni (Ackland *et al.*, 1987). The procedure for constructing the Al-Al potential was analogous to that for Ni-Ni (Vitek *et al.*, 1991). For separations less than the first nearest-neighbour distance in the Al fcc structure, the Al-Al repulsive interaction was increased in order to make the Al antisite defect on Ni site energetically unfavorable relative to the formation

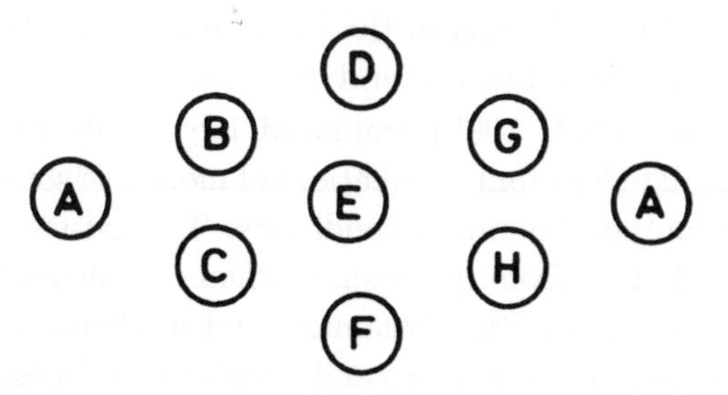


Fig. 8.4. Structural unit of the $\Sigma = 5$, (310) [001] grain boundary in NiAl. Sites A–D and $F$ are Ni atoms and sites $E$, $G$ and $H$ are Al atoms in the stoichiometric boundary.

of double vacancies on Ni sites. Fonda *et al.* (1995) base this requirement on the evidence that Al enrichment in NiAl is associated with the formation of constitutional vacancies on Ni sites (Bradley and Taylor, 1937; Wasilewski, 1968).

The Ni-Al interactions were constructed as follows (Fonda *et al.*, 1995). $\phi_{NiAl}$ was written as the geometric mean of the $\phi$ potentials for pure Ni and pure Al; this is consistent with its interpretation in terms of hopping integrals (Ackland *et al.*, 1988). The direct interaction $V_{NiAl}$, finally, was constructed to reproduce (i) the known lattice constant (ii) cohesive energy and (iii) elastic constants of the NiAl $B_2$ compound. At atomic separations smaller than the first nearest-neighbour distance, $V_{NiAl}$ was adjusted by Fonda *et al.* to obey equation of state requirements. With these potentials, the NiAl $B_2$ structure is energetically favoured over the $Ll_0$ structure and is also stable up to 2000K. In addition, these potentials give the energy of the [110] antiphase boundary to within the experimentally known range (Miracle, 1993: see also Chapter 3).

## 8.8. *Friction, Mechanical Properties and Interatomic Interactions**

Bhushan and coworkers (1990, 1994) measured the identation hardness of surface films with an identation depth as small as 1 nm for Si(111) and used AFM measurements to show that ion implantation of silicon surfaces increases their hardness and thus their wear resistance. They studied friction mechanisms on an atomic scale, a well characterized freshly cleaved surface of highly oriented pyrolytic graphite (HOPG), the atomic-scale friction force of which exihited the same periodicity as the corresponding topography, but the peaks in friction and those in topography were displaced relative to each other. A Fourier expansion of the interatomic potential was used to calculate the conservative interatomic forces between atoms of the AFM (see Appendix 2.5) tip and those of the graphite surface. Maxima in the interatomic force do not occur at the same location in the normal and lateral directions.

Landman *et al.* (1993, 1995) employed large scale molecular-dynamics simulation to calculate Newtonian equations of motion of some thousand atoms subjected to appropriate boundary conditions. The nature of the many-body interactions were derived from quantum-mechanical calculations. They determined the structural, energetic, dynamical and mechanical properties of the system on nanometric distance scales and femtosecond timescales. They found

*See also A. M. Stoneham, M. M. D. Ramos and A. P. Sutton, Phil. Mag. *A67*, 797 (1993).

that for a clean gold substrate contacted by a nickel tapered and faceted tip, and for a nickel surface and a gold tip, an instability occurs as the tip approaches the sample to distance of about 4 Å. At this point, a jump to contact occurs with gold atoms being displaced by about 2 Å in about 1 picosecond and with adhesive bonding between the two materials driven and accompanied by atomic-scale wetting of nickel by gold atoms. They argued that the latter is the result of differences in their surface energies, just as it is for the case of surface wetting by a liquid film. Retraction of the tip from the surface after contact causes inelastic deformation of the sample formation of a connective neck of atomic dimensions and eventual rupture. Similar simulations between the surfaces of crystalline ionic solids ($CaF_2$) and between semiconductor surfaces (silicon) showed similar jumps to contact and force hysteresis, but less easily deformed plastically and rather tending to brittle failure. Some of these results have been used to calculate the crss of sheared interfaces and its temperature dependence: these predictions can be tested by experiments.

## 8.9. *Empirical Potentials vs Density Functional Calculations for Mechanical Properties*

In concluding this chapter, we refer to the article by Heine (1994), who discusses in general terms the *ab initio* simulation of complex processes in solids. He discusses the question 'Will we ever be able to trust empirical potentials' and points out that it can be countered by the challenge 'Can we ever trust *ab initio* LDA (local density approximation, density functional) calculations.' To the first question, he writes in the above article that in the area of metals, the various types of glue models considered in this present chapter include metallic many-atom bonding and are vastly better than the models that preceded them. But LDA is also being transcended and density gradient corrections appear to improve results to the point where such calculations have become one focus for theoretical chemistry. Heine takes the example of a grain boundary between two grains of given orientation, and notes that are four overall degrees of freedom aside from relaxing individual atomic fractions near the boundary. And that just sets the lowest energy configuration of one type of boundary, before one considers any sliding or impurities. 'How to use efficiently the new generation of machines with parallel architecture' will become an important issue as one approach to extending understanding of mechanical properties of metals.

# Chapter 9

# Positron Annihilation: Experiment and Theory

## 9.1. *Background*

The extreme sensitivity of positrons to crystal imperfections in metals and alloys (vacancies, impurities and their aggregates, various types of dislocations, stacking faults, grain boundaries as well as surfaces) makes the positron annihilation method capable of yielding unique information on the concentration, configuration and internal structure of lattice defects in solids. (Hautojärvi and Corbel, 1995; Nieminen and Manninen, 1979; Nieminen, 1995). For example, Schaefer *et al.* (1992) and Badura *et al.* (1995) reported their studies of high-temperature thermal vacancy formation in the intermetallic aluminides $Fe_3Al$ and $Ni_3Al$ by means of positron lifetime spectroscopy. The intermetallic ordered alloys are considered to be of importance as future structural materials due to their favourable high temperature mechanical properties. Deng *et al.* (1995) studied the effects of Nb and Mg on microdefects in NiAl alloys. The experimental results show that there are large-open-volume defects occurring on the grain boundaries in NiAl alloy and the weakness of the grain boundary is due to less valence electrons participating in metallic bonds there. The densities of valence electrons in bulk and grain boundary increase with the addition of Nb atoms, but decrease with the addition of Mg atoms.

Shen *et al.* (1986) calculated the positron annihilation effects of a simple dislocation core model and showed that the effects at pure dislocation lines are large enough to be measurable even though much weaker than at vacancies. Shi *et al.* (1990) combined positron lifetime measurements at 100K with TEM observations on the dislocation structure in deformed zinc single crystals, and showed that the lifetime of positrons trapped and annihilated in basal dislocation lines increases by about 20% compared with that in the bulk. Positrons would be trapped in deeper traps (jogs) via the dislocation line when the jog concentration along it is sufficiently high. Shirai *et al.* (1992) studied the systematic change in positron lifetime with the magnitude of Burgers vectors of dislocations in Al, Cu, Au and Cu-8%Ge alloy. Their results showed that positron annihilation parameters observed predominantly come from dislocation lines themselves and also that positron lifetime spectroscopy can be used to differentiate between different types of dislocations.

Another area where positrons can be employed is in studies of the electronic structure of metals and alloys. A comprehensive study of spin-dependent $\rho(\mathbf{p})$* in Fe is reported by Genould *et al.* (1988). Singru (1995) reviewed the study of the angular correlation of positron annihilation radiation in transition metals in relation to their Fermi surface shape and electron momentum distribution.

Several texts (West, 1973: Hautajärvi, 1979; Brandt and Dupasquier, 1983; Dupasquier and Mills, 1995) providing an introduction to the subject of positron annihilation in solids exist in the literature. In addition, the proceedings of ICPA-1 to ICTP-10, the successive International Conferences on Positron Annihilation, provide a valuable collection of research and review papers describing the studies of metals using positrons. Readers are recommended to consult these sources for full details of this field. In the following sections of this chapter, we can give only a brief introduction to experiment and theory,† while the work on defects and mechanical properties will be reported in somewhat more detail.

## 9.2. *Interaction of Positrons with Vacancies in Metals*

The positron distribution in metals is closely related to potential energy as follows. The potential energy $U(r)$ comes from the Coulomb attraction of the electrons and the Coulomb repulsion on the ion cores. Consequently, near the ion core regions, its wave function amplitude is small but in the interstitial space between the ions, it increases rapidly to become a maximum. Figure 9.1 schematizes the potential energy of a positron in a direction of the crystal through both the minima of the potential energy and the lattice sites. In this figure, the energy of the positron ground state is denoted by $\varepsilon_0$ (see Seeger, 1973; 1976; also see Lung and March, 1986).

---

* $\rho(\mathbf{p})$ is used for the momentum density

† See also J. A. Alonso, J. Jiang, C. W. Lung, J. Y. Wong and N. H. March, An. Fis. (Spain) *93*, 136 (1997).

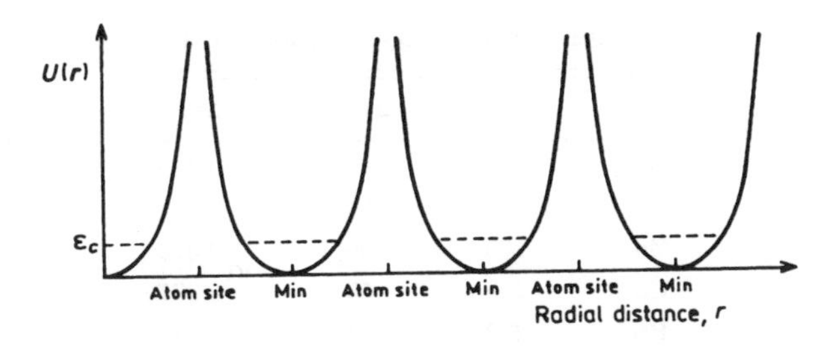


Fig. 9.1. Potential energy of a positron, $U(\mathbf{r})$, in a crystal direction going through the atom sites and the positions of minimum potential energy. The energy of the positron ground state in the potential is denoted by $\varepsilon_C$.

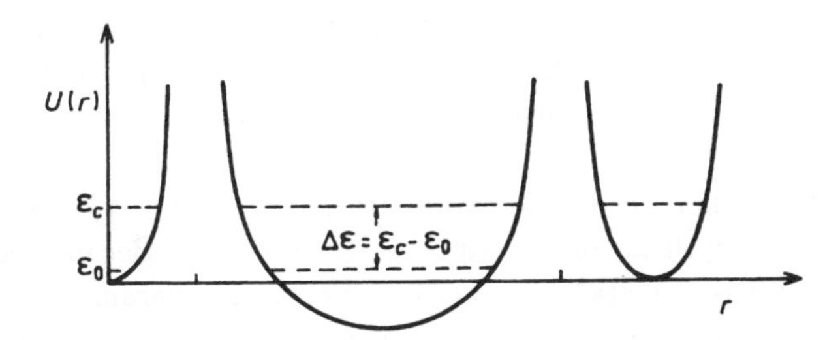


Fig. 9.2. Potential energy of a positron near a vacancy. The difference between the positron ground state in the perfect crystal and the energy of the bound state is denoted by $\Delta\varepsilon$.

If a vacancy is introduced into a perfect crystal, i.e., an ion in the perfect crystal is removed to the surface and the gas of the conduction electrons allowed to fill the additional lattice cell at the surface, the conduction (and valence) electrons at and near the vacancy redistribute in such a way that the effective negative charge created by the removal of the ion core is completely screened. After this process, a repulsive (positive) self-consistent electron potential at the vacancy is formed. Then, positrons feel an attractive potential. A bound state for positrons may be formed, and thus the positron wave function may become localized at the vacancy. The annihilation characteristics of a positron binding to a vacancy are different from those of the perfect crystal. Since the electron density at a vacancy is reduced compared to the perfect lattice, the lifetime of a trapped positron,* $\tau_t$, is longer than that of a free positron $\tau_f$

*Early independent proposals of the trapping model are by B. Bergersen and M. J. Stott (Solid St.Commun. 7, 1203, 1969) and D. C. Connors and R. N. West (Phys. Lett. *30A*, 24, 1969).

Table 9.1. $\Delta\tau/\tau_f$ values of various metals.

| metal | $\tau_f$ (ps) | $\Delta\tau$ (ps) | $\Delta\tau/\tau_f$ |
|---|---|---|---|
| Al | 172 | 65 | 0.378 |
| In | 206 | 64.5 | 0.313 |
| Zn | 175 | 56 | 0.320 |
| Cd | 200 | 55.5 | 0.2775 |
| Pb | 220 | 54 | 0.245 |
| Tl | 210 | 20 | 0.095 |
| Mg | 235 | 20 | 0.085 |

(the so-called long-slit angular-correlation curves are narrower than for a free positron). Figure 9.2 depicts the potential energy of a positron near a vacancy.

$\varepsilon_0$ is the energy of the positron in the bound state, $\Delta\varepsilon = \varepsilon_c - \varepsilon_0$ being the binding energy.

The relative difference of the two lifetimes ($\Delta\tau = \tau_t - \tau_f$) can sometimes characterize the depth $\Delta\varepsilon$ of the bound-state energy $\varepsilon_0$. Because of the difference of $\tau_f$ in different metals, we may use $\Delta\tau/\tau_f$ as a parameter representing the binding strength of a positron to a vacancy. Table 9.1 shows $\Delta\tau/\tau_f$ values of various metals taken from Seeger (1973). In general, this rule seems to be working well, though Tl and Mg need further study.

The relationship of $\tau_t$ and $\Delta\varepsilon$ may be discussed in another way (see Seeger, 1973; also Lung and March, 1986). Trapping effects can only be observed if the positron does not escape from its trap before it annihilates. From the analogy to chemical rate theory, Seeger (1973) assumed the escape rate from a trap of depth $\Delta\varepsilon$ to be of the order of magnitude

$$\nu_t = \frac{k_B T}{h} \exp\left(-\frac{\Delta\varepsilon}{k_B T}\right) \tag{9.2.1}$$

where $h$ is Planck's constant, $k_B$ Boltzmann's constant, and $T$ the absolute temperature. Thus the criterion for observing trapping effects is qualitatively

$$\tau_t \nu_t < 1\,. \tag{9.2.2}$$

Then, from (9.2.1) and (9.2.2)

$$\Delta\varepsilon > k_B T \ln\left(\frac{k_B T \tau_t}{h}\right)\,. \tag{9.2.3}$$

Table 9.2. $\Delta\varepsilon$ values for various metals calculated using pseudopotential theory.

| Metal | Li | Na | K | Mg | Zn | Cd | Al | In | Sn | Pb |
|---|---|---|---|---|---|---|---|---|---|---|
| $\Delta\varepsilon$ (eV) | 0.1 | 0.1 | 0.1 | 0.9 | 1.1 | 0.8 | 2.0 | 1.4 | 1.4 | 1.2 |

Table 9.3. Positron-annihilation parameters in various defect models (Al) (Shen *et al.*, 1986).

| Model | $\tau_t$ (ps) | $\tau_f$ (ps) | $\Delta\varepsilon$ (eV) | $\Delta\tau/\Delta\varepsilon$ (ps/eV) |
|---|---|---|---|---|
| Vacancy | 251 | 168 | 2.2 | 37.7 |
| Cylindrical hole | 229 | 168 | 2.1 | 29 |
| Arponen *et al.* (1973) | 229 | 168 | 2.8 | 21.8 |
| Peierls-Nabarro model | 183 | 168 | 1.1 | 13.6 |

With $\tau_t = 2\times10^{-10}$ s, the right-hand side of (9.2.3) gives 0.15 eV for $T = 250$K (typical of alkali metals) and 0.72 eV for $T = 1000$K (typical of noble metals). For monovalent metals this means that the noble metals are more likely to exhibit trapping at vacancies than the alkali metals. A larger valency is also favourable for positron trapping (Table 9.2), since the depth of the potential well increases with increasing screening charge in the Mott-Fumi model. At the time of writing, trapping effects have been observed in Cu, Ag, Au, Mg, Zn, Cd, Al, In, Tl and Pb.

Shen *et al.* (1986) have calculated the positron trapping and annihilation parameters with two simplified defect-core models representing dislocations. The calculation with the Peierls-Nabarro model indicates that a relatively small but measurable lifetime change is associated with a relatively large value of the positron-binding energy $\Delta\varepsilon$. As the extended character of the defect model increases, the $\Delta\tau/\Delta\varepsilon$ value decreases monotonically (Table 9.3). This means that for different types of defects, one cannot conclude whether $\Delta\varepsilon_1$ is larger than $\Delta\varepsilon_2$ or not if one only knows that $\tau_{t1}$ is larger than $\tau_{t2}$, even if the inequality (9.2.3) holds.

## 9.3. *Trapping Model*

We should consider situations in which some transitions between the various possible states occur in times comparable with that for annihilation. A complete analysis requires, in principle, an understanding or a microscopic model of the mechanisms by which such transitions occur. Complex and comprehensive analysis has been set out by Brandt and Paulin (1972), and Seeger (1973, 1976). A simple rate equation approach involving time-independent transition rates has been presented by West (1973, 1979).

The following is the calculation by Seeger (1973, 1976), in which the specific trapping rate, $\mu_j(t)$ is time dependent for discussions on the positron diffusion mechanism and the enthalpy of formation of vacancies.

Let $n_f$ and $\tau_f$ denote the concentration and lifetime of free positrons. We consider positrons in different types of traps and denote the concentration of traps of type $j$ by $c_j$, the specific trapping rate by $\mu_j(t)$, the concentration of positrons localized at these traps by $n_j$, the positron lifetime in the trapped state by $\tau_j$ and the rate of escape of positrons from the trap by $\gamma_j$ (per unit concentration). Making the plausible assumption that the initial distribution (at time $t = 0$) of positron-trap pairs is uniform, the solution of a set of $m+1$ rate equations is obtained. These can be used to calculate the mean lifetime $\bar{\tau}$ of positrons.

$$\frac{dn_f(t)}{dt} = -\left(\frac{1}{\tau_f} + \sum_{j=1}^{m} \mu_j(t)\, c_j\right) n_f(t) + \sum_{j=1}^{m} \gamma_j\, n_j(t)$$

$$\frac{dn_j(t)}{dt} = -\frac{n_j(t)}{\tau_j} + \mu_j(t)\, c_j\, n_f(t) - \gamma_j\, n_j\,(t) \tag{9.3.1}$$

$$(j = 1, 2, \ldots, m)\,.$$

If $\gamma_j = 0$, Eq. (9.3.1) may be integrated for given $(\mu_j(t) \doteq \mu_j)$

$$n_f(t) = n_f(0)\exp(-t/\tau_0)$$

$$n_j(t) = \tau_0\tau_j \frac{\mu_j c_j}{\tau_0 - \tau_j} n_f(0)\exp(-t/\tau_0) + \left[n_j(0) - \tau_0\tau_j\frac{\mu_j c_j}{\tau_0 - \tau_j} n_f(0)\right]\exp(-t/\tau_j)$$

where

$$\frac{1}{\tau_0} = \frac{1}{\tau_f} + \sum_{j=1}^{m} \mu_j c_j\,. \tag{9.3.2}$$

The mean lifetime is

$$\bar{\tau} = -\frac{[n_f(0) + \int_0^\infty \sum_{j=1}^m n_j(t)dt]}{n_f(0) + \sum_{j=1}^m n_j(0)}\,. \tag{9.3.3}$$

Suppose that no trapped positrons exist at the initial time, i.e.

$$n_j(0) = 0\,, \quad j = 1, 2, 3, \ldots, m\,, \quad n_f(0) = 1\,.$$

Equation (9.3.3) may be simplified to become

$$\bar{\tau} = \tau_f \cdot \frac{1 + \sum_{j=1}^{m} \tau_j \mu_j c_j}{1 + \tau_f \sum_{j=1}^{m} \mu_j c_j}\,. \tag{9.3.4}$$

Equation (9.3.4) is equivalent to Eq. (3.12) in West's paper (1979) and Hautojärvi and Corbel (1995), in which $\lambda_1 = \tau_f^{-1}$, $K_{1j} = \mu_j c_j$. If $m = 1$, that means one type of trapping only

$$\bar{\tau} = \tau_f \cdot \frac{1 + \tau_1 \mu_{1v} C_{1v}}{1 + \tau_f \mu_{1v} C_{1v}} \tag{9.3.5}$$

or in familar form

$$\frac{\bar{\tau} - \tau_f}{\tau_t - \bar{\tau}} = \tau_f \mu_{1v} C_{1v}\,, \quad \text{or} \quad \frac{\bar{\tau} - \tau_{\mathrm{f}}}{\tau_{\mathrm{t}} - \tau_{\mathrm{f}}} = \frac{\mu_{1\mathrm{v}} \mathrm{C}_{1\mathrm{v}}}{\tau_{\mathrm{f}}^{-1} + \mathrm{C}_{1\mathrm{v}} \mu_{1\mathrm{v}}}\,. \tag{9.3.6}$$

The process of positron capture at defects is basically a quantum-mechanical one, especially in the case of weak trapping and the positron near the traps not locally depleted.

However, there is a possibility that the mobility of the untrapped positron sets a limit to the trapping rate, shifting the process from the transition-limited regime discussed above to the diffusion-limited one. In this case, the diffusion theory is applied and the process would be temperature dependent.

Manninen and Nieminen (1981) considered the thermodynamics of detrapping for vacancies. Models for positron trapping at dislocations without detrapping (Hautojärvi, 1979; Xiong, 1986) and with detrapping (Xiong and Lung, 1988; Hautojärvi and Cobel, 1995) were proposed for analyses of dislocation processes in materials. Different initial conditions are assumed to predict positron annihilation properties at grain boundaries of Zn-23wt% Al (Yan *et al.*, 1991).

### 9.3.1. *Positron-Diffusion Mechanism and Enthalpy of Formation of Monovacancies*

If the traps are entirely characterized by the radii $r_j$ of spherical attractive potential wells then (Seeger, 1973, 1976; or see Lung and March, 1986)

$$\mu_j(t) = \frac{4\pi\gamma_j D}{\Omega}[1 + r_j(\pi D t)^{-1/2}] \tag{9.3.7}$$

where $\Omega$ is the atomic volume and $D$ the positron-diffusion constant. If $t$ is sufficiently large then*

$$\mu \approx \frac{4\pi D r_{1v}}{\Omega} . \tag{9.3.8}$$

In the case where diffusion and transition compete, the full time-dependent diffusion equation must be solved (Hautojärvi and Corbel, 1995). For more details, see Dupasquier (1995), for the monovacancy case. Equation (9.3.5) can then be written as

$$\bar{\tau} = \tau_f \frac{1 + 4\pi r_{1v} C_{1v} \tau_{1v} D/\Omega}{1 + 4\pi r_{1v} C_{1v} \tau_f D/\Omega} . \tag{9.3.9}$$

The concentration of vacancies in thermal equilibrium at absolute temperature $T$ is given by

$$C_{1v} = \exp\left(-\frac{G_{1v}^F}{k_B T}\right) = \exp\left(\frac{S_{1v}^F}{k_B}\right) \exp\left(-\frac{H_{1v}^F}{k_B T}\right) . \tag{9.3.10}$$

From (9.3.10) and (9.3.9), we find

$$H_{1v}^F = k_B T \left[\frac{S_{1v}^F}{k_B} \ln\left(\frac{4\pi D}{\Omega} r_{1v} \tau_f\right) + \ln\left(\frac{\tau_{1v} - \bar{\tau}}{\bar{\tau} - \tau_f}\right)\right] . \tag{9.3.11}$$

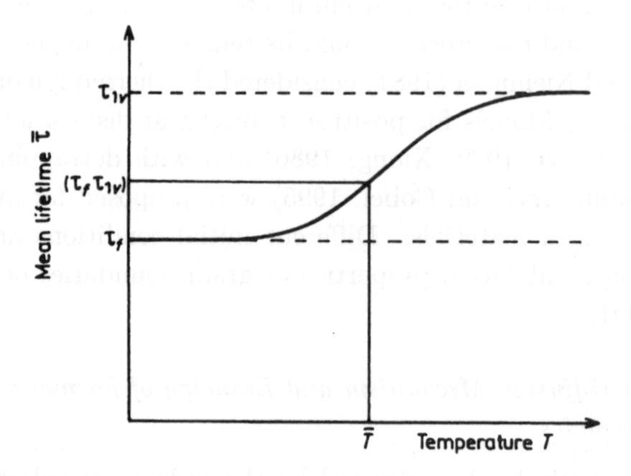


Fig. 9.3. Temperature dependence of mean positron lifetime $\bar{\tau}$ due to trapping by monovacancies in thermal equilibrium.

*In Eq. (9.3.8) $r_{1v}$ is the radius of the spherical attractive potential well of the monovacancy.

Equation 9.3.9 shows that the mean lifetime $\bar{\tau}$ increases with increasing vacancy concentration $C_{1v}$ or with increasing temperature from the value $\tau_f$ characteristic of a defect-free crystal to the value $\tau_{1v}$ characteristic of a situation where the vacancy concentration is so high that all positrons are trapped in a time short compared with the lifetime. This leads to a temperature dependence as shown in Fig. 9.3. We have assumed that $\tau_f$ and $\tau_{1v}$ are temperature independent. In reality, both quantities may be weakly temperature-dependent because of thermal-expansion effects; $r_{1v}$ and $\Omega$ are also weakly temperature-dependent for the same reason.

The diffusion coefficient $D$ of the positrons appears in (9.3.11). We can introduce a positron mobility $\mu$ through the Nernst-Einstein relationship.

$$D = k_B T \mu / e \,. \tag{9.3.12}$$

Owing to experimental difficulties, there exist few direct measurements of positron mobility in condensed matter. Sometimes the results may be influenced by the effect of the electric field on the kinetic mobility prior to thermalization. Bergersen *et al.* (1974) have calculated the various contributions in a number of simple metals. The main contribution of scattering is from acoustic-phonon scattering. The contribution due to impurity scattering of conduction electrons is very small (see Nieminen and Manninen, 1979). However, there are good reasons to expect that typical diffusion constants are of the order $D_t \approx 0.1 - 1.0$ cm$^2$/s at around room temperature. The average diffusion length before annihilation can be obtained by Fick's law $R^2 \approx 6D\tau$ where $\tau$ is the lifetime of positrons and $D$ is the diffusion coefficient. In general, during the lifetime $\tau \geq 10^{-10}$ s, the thermalized positron diffuses over a volume of about (1000 Å)$^3$ (see Paulin, 1983). However, this is the calculated result under the condition that high purity and defect-free materals have been assumed. Actually, there are many defects and impurity atoms in industrial materials. The relation $R^2(t) \sim t$ would not hold in these systems; the diffusion length of thermalized positrons in this case would be different from that in high-purity defect-free materials.

Lung (1995) discussed the positron diffusion process in fractals medium. The common feature of these models is that transport can be anomalous, the mean square displacement scales with time as $R_f^2(t) \sim t^{2/d_w}$, where $d_w > 2$ is the fractal dimension of the random walk. Lung reported that $R_f(t)$ would be much shorter than $R$, the average diffusion length in high-purity and defect-free materials. This will be discussed in detail in Sec. 9.5.

### 9.3.2. *Trapping Rate*

A quantum-mechanical approach to the trapping rate of positrons by vacancies was first given by Hodges (see Nieminen and Manninen, 1979) using the Golden Rule transition rate

$$\sigma = \frac{2\pi}{\hbar} \sum_{i,f,t} P_i |M_{i,f,t}|^2 \delta(\varepsilon_i - \varepsilon_f - \varepsilon_t) \tag{9.3.13}$$

where $P_i$ is the occupation probability of an initial positron state $|i>$, $\varepsilon$ denotes energy, and the subscripts $i$, $f$, and $t$ denote the initial and final states of the medium and the trapped positron states respectively. The matrix element $M_{i,f,t}$ involves the overlap between the initial and final states as well as factors describing the specific energy-absorption mechanism. Usually, a binding energy of the positron to the trap of the order of 1 eV will be liberated and absorbed by elementary excitations of the solid. These are either electron-hole excitations or lattice vibrations. In metals, the former should provide the dominant mechanism, and (9.3.13) with a Boltzmann-type distribution $P_j$ of initial positron states then leads to an essentially temperature-independent trapping rate for traps with small binding energy. For instance, the trapping rate of positrons by dislocations may be discussed in terms of the Golden Rule (Mckee *et al.*, 1974).

$$\sigma = v^{-2/3} \frac{2\pi}{\hbar} \sum_{k_z, \mathbf{k}, \mathbf{k}'} |\langle 0, \mathbf{k}| v_+ | k_z, \mathbf{k}' \rangle|^2 f_{\mathbf{k}} (1 - f_{\mathbf{k}'}) \times \delta\left( \frac{\hbar^2 k'^2}{2m} - \frac{\hbar^2 k^2}{2m} - \varepsilon_D + \frac{\hbar^2 k_z^2}{2m} \right) \tag{9.3.14}$$

with

$$\langle 0, \mathbf{k}| v_+ | k_z, \mathbf{k}' \rangle = \frac{1}{v} \int d^3 r \int d^3 r' \Psi_0^*(r) \Psi_{k_z}(r) v_+(r - r') \exp[i(\mathbf{k} - \mathbf{k}') \cdot \mathbf{r}']$$

where $\varepsilon_D$ is the positron binding energy of the deepest state due to a dislocation line. $\Psi_0$ is the lowest positron state propagating throughout the sample, $\Psi_{kz}$ is the trapped positron state propagating with momentum $\hbar k_z$ along the dislocation line, and $v$ is the sample volume, $v_+(\mathbf{r})$ is the positron-electron interaction potential, which is assumed to be screened but static. The sum over $k_z$ must be restricted to the trapped states, that is $\hbar^2 k_z^2 / 2m < \varepsilon_D$.

## 9.4. *Positron-Annihilation Characteristics*

A first-principles description of the positron-annihilation process constitutes a difficult many-particle problem even for a perfect metal. We are not going to describe how the annihilation characteristics for an electron-gas are calculated, but rather our aim will be, by suitable approximations, to discuss how the electron-gas data can be used in estimating the annihilation characteristics in real metals and in defects.

In a homogeneous electron gas of density $n$ the positron annihilation rate can be written as (Nieminen *et al.*, 1979)

$$\lambda(n) = \lambda_0(n)\gamma(n) = \pi r_0^2 cn\gamma(n) \tag{9.4.1}$$

where $\lambda_0$ is the Sommerfeld free-electron formula, $r_0$ the classical electron radius and $c$ the velocity of light. $\gamma$ is the density-dependent enhancement factor due to the strong electron-positron correlation, which increases the electron density at the site of the positron.

If the electrons and the positron are treated as independent particles then the many-body wavefunction is a Slater determinant and the momentum distribution of the annihilation quanta is

$$\Gamma_0(p) = \frac{\pi r_0^2 c}{(2\pi)^3} \sum_j \left| \int d^3 r e^{-i\mathbf{p}\cdot\mathbf{r}} \Psi_j(\mathbf{r}) \Psi_+(\mathbf{r}) \right|^2 \tag{9.4.2}$$

where $\Psi_+$ is the ground-state positron wavefunction* and the sum goes over the occupied electron states $\Psi_j$. The conventional long-slit angular-correlation apparatus measures only one component, $p_z$, of the momentum distribution, so that the angular-correlation curve is

$$I(p_z) = \int dp_x \int dp_y \Gamma_0(p) \tag{9.4.3}$$

which reduces, in the isotropic case, to

$$I(p_z) = 2\pi \int_{p_z}^{\infty} dp \Gamma_0(p) = \frac{r_0^2 c}{4\pi}(p_F^2 - p_z^2)\theta(p_F - |p_z|)\,. \tag{9.4.4}$$

This is an inverted parabola, whose width is proportional to the Fermi momentum $p_F$. This describes well the observed angular correlation curves of simple

*For early work, see the calculation by B. Donovan and N. H. March (Phys. Rev. *110*, 582 (1958)) for Cu metal.

metals if one neglects the small broad Gaussian part due to the core electrons and higher momentum components of the valence electrons.

In defect solids, at the vicinity of a lattice defect the electron-density distribution deviates from that of the perfect lattice-for a positron localized at the defect, the overlap with the core electron is diminished as ions are missing from the defect. This leads to narrowing in the angular-correlation curve. On the other hand, the conduction electron density is also depleted at the defect region, with a concomitant narrowing in momentum distribution. Furthermore, the localization increases the positron momentum — this is reflected as a slightly pronounced tail of large momenta.

Two approximations have been proposed with the intention of estimating the momentum distribution directly from the electron density. One is called a local approximation — it replaces (9.4.4) by

$$I(p_z) = \frac{r_0^2 c}{4\pi} \int d^3r |\Psi_+(\mathbf{r})|^2 \lambda[n(\mathbf{r})][p_F(\mathbf{r})^2 - p_z^2]\theta(p_F^2(\mathbf{r}) - |p_z|^2) \qquad (9.4.5)$$

which means simply that at each point $\mathbf{r}$ the positron annihilation is as in a uniform electron gas and produce a free-electron parabola whose width is determined by the local density $n(\mathbf{r})$. Another approximation has been proposed by Arponen *et al.* (1973) which partly takes into account the non-local character of the momentum distribution and is called mixed-density approximation. The partial annihilation rate at total momentum $p$ is

$$\Gamma(\mathbf{p}) = \int d^3r \int d^3r' \exp[i\mathbf{p}\cdot(\mathbf{r}-\mathbf{r}')\Psi_+^*(\mathbf{r})\Psi_+(\mathbf{r}')$$
$$\times g(p_F(\mathbf{R})|\mathbf{r}-\mathbf{r}'|)\{\lambda[n(\mathbf{r})]\lambda[n(\mathbf{r}')]\}^{1/2} \qquad (9.4.6)$$

where $\mathbf{R} = \frac{1}{2}(\mathbf{r}+\mathbf{r}')$ and $g$ is a function related to the electron-electron pair-correlation function

$$g(z) = \frac{3}{z^3}(\sin z - z\cos z)\,. \qquad (9.4.7)$$

This approximation also takes into account the momentum associated with the localized positron state, this being ignored in the local model. The one-dimensional angular-correlation curves for positrons annihilating in vacancies in aluminium have been calculated by using the two approximation methods discussed above. The mixed-density approximation is in a fairly good agreement with the experimental results, whereas the local approximation clearly overestimates the narrowing of the curve from the free-electron parabola,

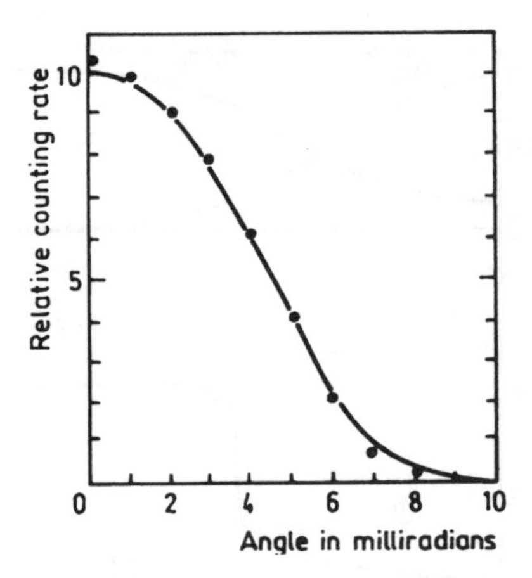


Fig. 9.4. (a) Angular correlation curves for valency electrons in an aluminium vacancy. (Kusmiss and Stewart, 1967)

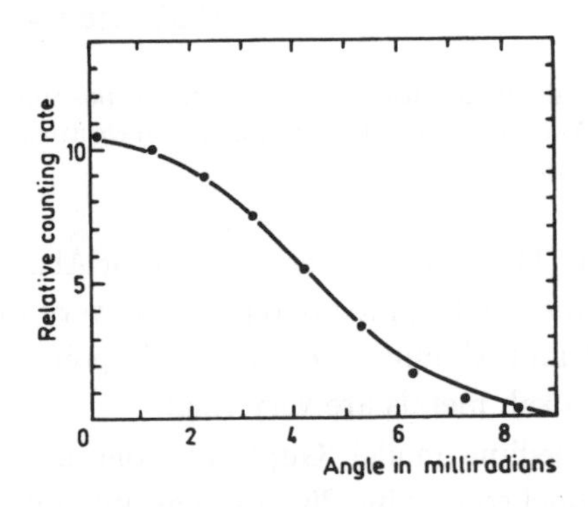


Fig. 9.4. (b) Angular correlation curve for positrons annihilating in dislocations of aluminium. (Arponen *et al.* 1973)

(Fig. 9.4). However, calculations by Shen *et al.* (1985) showed that the calculated angular-correlation curves are not sensitive to the choice of theoretical model of calculation.

### 9.4.1. *Vacancies and Vacancy Clusters*

Hodges (1970) used the pseudoatom picture while Manninen *et al.* (1975) used the density-functional method to study positron binding to vacancies. In Fig. 9.5, the trapping potential and the positron pseudo-wavefunction are

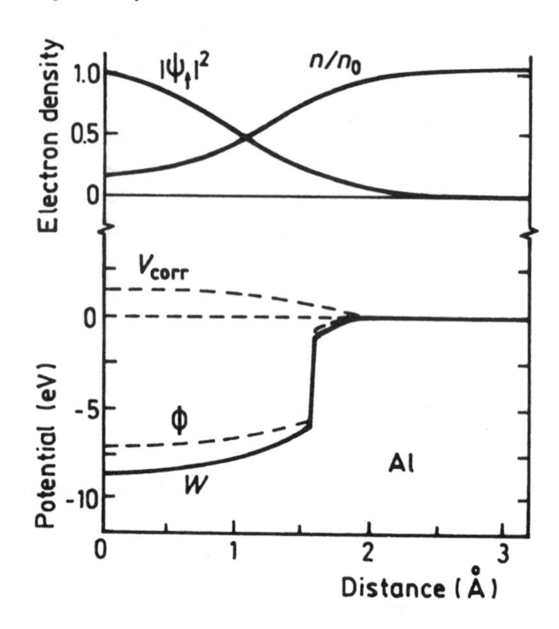


Fig. 9.5. Electron and positron densities, $n$ and $|\psi_+|^2$, and positron trapping potential $V$ in an aluminium vacancy. $V_{\rm corr}$ and $\phi$ are the correlation and electrostatic parts of the trapping potential. (Manninen *et al.* 1975)

shown for aluminium. The binding-energy values for Al are sensitive to calculational details. However, the trapping potentials are very shallow for the alkalis; both Hodges (1970) and Manninen *et al.* (1975) pointed out that positron-binding energies for alkali metals are very small.

To transcend the jellium model, Gupta and Siegel (1977; 1980) employed a supercell lattice model containing 26 Al atoms and a vacancy in the calculation of positron-annihilation characteristics in a crystal by the APW method. Angular-correlation curves in Al were found to be in good agreement with experiment, but the binding energy calculated by this method for Al seems somewhat high. Also, detailed electron-positron correlations have been neglected; these play a role for calculation of the lifetime and binding energy. Furthermore, the vacancy concentration of the supercell model is higher than that of any solid at the melting point.

The main results of the specific trapping rate, $\mu_{1V}$, calculated by Nieminen and Laakkonen (1979) can be summarised as follows ($\mu_{Iv} = K/c_{Iv}$):

(1) $\mu_{1V} = (10^{14} - 10^{15})\ \mathrm{s}^{-1}$ for a monovacancy.

(2) $\mu_{1V}$ increases with the positron binding energy. More electrons are excited above the Fermi surface at higher $E_b$.
(3) $\mu_{NV} \approx N_{\mu_{IV}}$ for small ($N < 10$) spherical clusters of $N$ vacancies.
(4) $\mu_{IV}$ is independent of temperature. The thermal energy and momentum of the free positron are negligible compared to the total energy and momentum transferred in the process.
(5) $\mu$ for vacancy clusters is temperature dependent when the size is comparable to $\lambda_{th}^{+}$; the thermal de Brozlie wavelength of the positron.

There are uncertainties in the experimental value for the trapping coefficient $\mu$. Subsequently, Kluin *et al.* (1991) have correlated positron lifetime, dilatometry, and lattice parameter experiments at high temperatures and obtained 'absolute' values for $\mu_{1V}$ which are slightly lower than the estimates above.

In metals, calculations have been made assuming that the positron does not drastically distort the defect one is examining. In the case of shallow positron traps, where the 'self-trapping' effects due to the positron can be important (see Nieminen, 1995), neglecting the positron-induced relaxations tends to underestimate the positron localization. The best way to calculate the positron-induced atomic and electronic relaxation is to include the positron in the Car-Parrinello scheme* via the two-component density-functional theory. An accurate plane-wave representation of the positron wave function can be obtained, potentials and forces calculated accurately, and the ionic relaxation pattern calculated self-consistently (Gilgien *et al.*, 1994; see also Nieminen, 1995).

The annihilation characteristics of trapped positrons can be used as fingerprints to identify defects. The experimental annihilation characteristics of monovacancies can be directly measured and compared to theoretical calculations. However, it is safe to compare with other experimental methods, say electric resistance measurements etc.

As for vacancy clusters, the annihilation parameters are very sensitive to their radii. Small spherical cavities can be treated as vacancy clusters. The calculated lifetime for 3D unrelaxed vacancy clusters in Al and Fe are shown in Fig. 9.6 as a function of the void radius. Nieminen (1979) pointed out that, since the number of vacancies is small, the trapping rate of the void is approximately proportional to the number of vacancies. When the void size increases, the lifetime rapidly approaches its saturation value, 500 ps, which corresponds

*See, for example, the review by P. A. Madden.

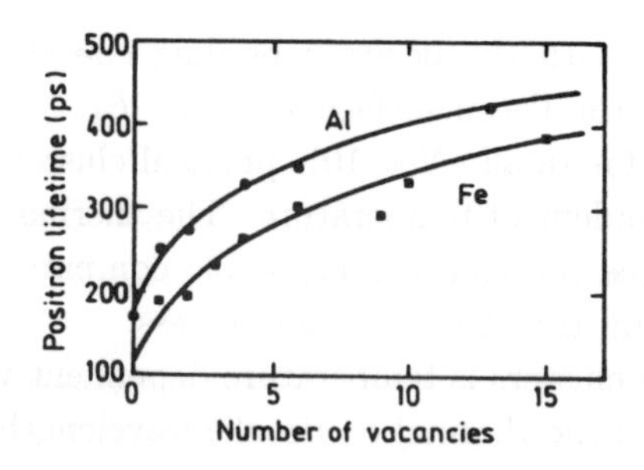


Fig. 9.6. Positron lifetime at three-dimensional vacancy clusters in Al and Fe. The points represent the results of theoretical calculations. The solid lines are only to guide the eye. From Puska and Nieminen, (1983).

to the zero-density limit of the annihilation rate. The deviations from the smooth curve reflect the discrete structures of the clusters.

Calculations by Corbel *et al.* (1986) show that the positron lifetime strongly depends on the geometry of a cluster. In a strongly relaxed vacancy cluster the lifetime may become even smaller than in a monovacancy. For a relaxed 2D cluster (loop) the lifetime is about the same as in a monovacancy. For a given lifetime, Fig. 9.6 gives always a lower limit for the number of monovacancies in the cluster. The model overestimates the lifetimes in di- and trivacancies, which in reality are far from spherical. It is also well known that larger clusters should be considered as essentially macroscopic surfaces.

The sensitivity of annihilation parameters to the size of vacancy clusters has been used to study the dependence of the apparent microvoid size on the annealing process in electron irradiated Mo (Eldrup *et al.*, 1976), the vacancy clustering induced by impurities in cold worked Ni (Dlubek *et al.*, 1979), and the annealing process in electron — and neutron — irradiated iron and Fe-C alloys (Hautojärvi *et al.*, 1980).

It is well known that the vacancy concentration in thermal equilibrium is given by (9.3.10). The formation enthalpy can be determined from the Arrhenius plot of the positron trapping rate $K = \mu_{1V} C_{1V}$ in (9.3.6). The advantage of the positron method is the high sensitivity to the vacancies at relatively low temperatures ($\sim 0.5\ T_m$) and concentration ($\sim 10^{-6}$ at$^{-1}$). Moreover, the technique can be applied also to dilute and concentrated alloys. Since several assumptions are needed in analysing experimental results on the temperature dependence of the positron parameters in the pretrapping and saturation trapping regions, the accuracy is limited to 5%. The role of divacancies near the melting point is open (Siegel, 1982; Schaefer, 1982), the vacancy migration energy enthalpies measured at high temperatures have often given contradictory results to the direct observations of vacancy migration in irradiated samples at low temperatures.

In general, small amounts of interstitial impurities have no effect on positron annihilation, since the positron is repelled from the region of high ion density. The substitutional impurities may attract or repel the positron, depending as a first estimate on the impurity size — the smaller the impurity, the more free space is left for the positron. Some impurities such as H (Jena *et al.*, 1981), He (Snead and Goland, 1975), Li (Kubica *et al.*, 1975), and C (Hautojärvi *et al.*, 1980) have also been investigated.

Hautojärvi *et al.* (1985) demonstrated experimentally that hydrogen impurities can be bound to vacancies, this being observed in Ta at 70K after low-temperature alpha-particle irradiation. The vacancy-hydrogen-complex formation shifts the vacancy migration to higher temperatures. Vacancy-hydrogen complexes still retain the capability to trap positrons. The calculation by Hautojärvi *et al.* (1985) are in quite good accord with the experimental observations on positron and hydrogen interactions with vacancies. From the result for samples with hydrogen impurities they estimated the upper limit on the vacancy-hydrogen binding energy to be 1.0–1.2 eV.

Theoretical calculations in conjunction with experiment can lead to estimates of the position of the impurity atom from the centre of the vacancy. For example, Hansen *et al.* (1984) showed that a N atom cannot be located near the centre of the vacancy in Mo. Hautojärvi *et al.* (1985) pointed out that the hydrogen trapped in a vacancy is partly delocalized over the entire vacancy, having maximum densities outside the centre close to the adjacent octahedral sites. Positrons can also trap at vacancies decorated by one or two hydrogen atoms, and the calculated annihilation characteristics are in agreement with positron-lifetime measurements.

Table 9.4 shows the calculated positron lifetime and binding energies in Al vacancies with one to six H atoms near the $O_h$ sites. The equilibrium site is assumed not to depend on the number of H atoms in the vacancy, and when more H atoms are considered they are assumed to surround the vacancy symmetrically. From Table 9.4 it can be seen that the lifetime difference between a clean vacancy and one decorated by a single H atom is quite small (hardly observable) when assuming a realistic position. The previously predicted lifetime (for one central H atom) of 204 ps is close to the lifetime in a vacancy with six H atoms (Hansen *et al.*, 1984).

Table 9.4. Calculated lifetime $\tau_{V-H}$ and binding energies $\Delta\varepsilon$ for positrons in H-decorated and Al vacancies (Hansen *et al.*, 1984).

| Number of atoms | $\tau_{V-H}$ (ps) | $\Delta\varepsilon$ (eV) |
|---|---|---|
| 0 | 251 | 2.2 |
| 1 | 244 | 2.1 |
| 2 | 231 | 1.9 |
| 4 | 212 | 1.5 |
| 6 | 193 | 0.9 |

The positron lifetime at the vacancy containing a neutral hydrogen in aluminium has been estimated as 223 ps by Sankar and Iyakutti (1985). However, Jena *et al.* (1981) have reported the lifetime of the positron in a vacancy-hydrogen (as proton) complex to be 188 ps. To our knowledge at the time of writing, there are no experimental data available for comparison.

The thermally activated positron detrapping of defects is also a subject of some interest. Manninen and Nieminen (1981) showed that the ratio of detrapping to trapping of vacancies is

$$\frac{\nu}{\sigma} = \Omega_0 \exp\left(-\frac{S_V}{k_B}\right)\left(\frac{m^* k_B T}{2\pi\hbar^2}\right)^{3/2} \exp\left(\frac{E_V - E_b}{k_B T}\right) \tag{9.4.8}$$

where $\Omega_0$ is the volume of a unit cell, $S_V$ and $E_V$ are the vacancy-formation entropy and energy respectively. $E_b$ is the positron-binding energy due to the vacancy. The temperature dependence and fluctuations of $E_b$ are ignored in (9.4.8). Actually, these are two opposing factors involved. On the one hand, trapping effects are weakened by thermally activated positron detrapping due to lattice vibrations, and, on the other hand, the nonlinear expansion of the crystal lattice would enhance the trapping effects. It is difficult at present to estimate these two effects accurately. Smedskjaer *et al.* (1985) measured the Doppler-broadening parameters of nominally 99.8wt% pure niobium in the temperature range 300–2580K. Their results showed no measurable effect due to the presence of oxygen in niobium. The results were discussed in terms of the two-state trapping model, paying attention to the possibility of thermally activated positron detrapping from the vacancy.

Meyberg *et al.* (1985) considered the influence of positron diffusion on trapping and detrapping reactions. They based their work on the differential equations

$$\frac{\partial \rho_f(r,t)}{\partial t} = D_+ \nabla^2 \rho_f(r,t) + [f(t) - \tau_f^{-1}]\rho_f(r,t)$$
$$\frac{d\rho_t(t)}{dt} = \left(\frac{\alpha}{v_c}\right) \rho_f(r_0,t) - [\beta + \tau_t^{-1}]\rho_t(t) \tag{9.4.9}$$

for the densities $\rho_f(r,t)$ and $\rho_t(t)$ of the probabilities of finding at time $t$ a free positron at a distance $r(> r_0)$ from a trap and a positron in a trap (i.e. at $r < r_0$). The first terms on the right-hand side of (9.4.9) describe the diffusion of free positrons (diffusivity $D^+$) and the transition rate to the trapped state (frequency $\alpha/v_c$). The rate at which $\rho_f(r,t)$ changes owing to reactions involving positrons or traps not belonging to the free- positron-trap pair considered occurs with the time-dependent frequency $f(t)$, whereas $\rho$ is the frequency of positron escape from the trapped state. Positron annihilation is accounted for by the terms containing the lifetimes $\tau_f$ or $\tau_t$ of free or trapped positrons. Equation (9.4.9) are coupled via the inner boundary condition at $r = r_0$

$$\frac{4\pi D^+ r_0^2}{v}\left[\frac{\partial \rho_f(r,t)}{\partial r}\right]_{r=r_0} = \frac{\alpha}{v}\rho_f(r_0,t) - \beta\rho_t(t)\,. \tag{9.4.10}$$

The crystal volume $v$ is assumed to be large. It turns out that in the limit $t \to \infty$ (except for very small $\tau_f$) the space-averaged concentrations $n_f$, $n_t$ and $c_t$ of free positrons, trapped positrons and traps obey a mass-action law

$$\frac{n_t}{n_f c_t} = K \tag{9.4.11}$$

where $K$ is independent of time.

### 9.4.2. *Dislocations*

Positron trapping at dislocations is not so easy to demonstrate unambiguously because of the difficulty of isolating these defects from other defects, say vacancies, which are created simultaneously during plastic deformation. Theoretically, the major problem is to find a realistic model for the atomic configuration around the dislocation core. Knowing this the conduction electron density and the positron state can be solved using the methods which have been successfully used for studies of vacancies. Usually, one assumes a simple dislocation model and calculates its annihilation parameters for comparison with experiments. The main problem is: does the dislocation line itself act as a trapping site for positrons? Arponen *et al.* (1973) proposed a model representing a cylindrical hole (dislocation core) associated with the elastic displacement field (outside) to explain the experimental results of Hautojärvi (1972) on the positron annihilation effects of deformed Al. From this model, a binding energy of about 2.8 eV was obtained for the interaction between the dislocation line and the positron. This is contrary to the previous calculation of Martin and Paetsch (1972), based on pairwise interactions between the positron and metal ions, whose positions were obtained from linear elasticity theory. This approach led to a very small value of $< 0.1$ eV in Al. According to them, the annihilation characteristics of such a state would be indistinguishable from those of a bulk state. A comparison with the estimate of Arponen *et al.*, shows that the calculated binding energy is very sensitive to the description of the core region. However, their description is too simple and far from the realistic structure of a dislocation core. This leads to a very large value of binding energy, $\sim 2.8$ eV, in Al (see Table 9.3). Smedskjaer *et al.* (1980) reported that the observed changes in the positron annihilation in parameters due to the presence of dislocations in deformed metals originate from annihilation in point-like defects (e.g. jogs) associated with the dislocation line. Park *et al.* (1985) reported their results of dislocation density measurements by transmission electron microscopy, etch pits and positron trapping effects measurement in iron single crystals. They concluded that positrons are trapped both in edge and screw dislocation lines. The lifetime of the former is 165 ps and of the later is 142 ps. This reflects the fact that these authors believe the dislocation line itself is acting as a trapping site for positrons though deformations limited in most cases to less than 10% in that work are not enough to hinder the creation of jogs. At the same time, Shen *et al.* (1985) used the method by Arponen *et al.*, but made the dislocation core model more realistic, to calculate the binding energy. They assumed the ion density to be the same as the Peierls-Nabarro (P-N) model which is more realistic than the cylindrical hole model that Arponen *et al.* (1973) used, although it is still not an accurate representation of a dislocation core in a real crystalline solid.

In the P-N model, the resulting displacement $u_x$ in the x-direction can be solved in a very simple, exact form

$$u_x = -\frac{b}{2\pi}\tan^{-1}\left(\frac{x}{\xi}\right) \tag{9.4.12}$$

where $\xi = a/2(1-\nu)$ is known as the half-width of a dislocation. To simplify it to the cylindrically symmetrical case, one obtains

$$u_r = -\frac{b}{2\pi}\tan^{-1}\left(\frac{r}{\xi}\right) \tag{9.4.13}$$

$$u_\theta = 0\,.$$

Then, the volume expansion ratio would be

$$\begin{aligned} f(r) &= \operatorname{div}\mathbf{u} \\ &= -\frac{b}{2\pi r}\tan^{-1}\left(\frac{r}{\xi}\right) - \frac{b}{2\pi\xi}\frac{1}{1+(\frac{r}{\xi})^2}\,. \end{aligned} \tag{9.4.14}$$

In Eq. (9.4.14), $f(r) \sim \frac{1}{r}$, as $r \gg \xi$. It is inconsistent with the linear elasticity theory of dislocations. From this model

$$\begin{aligned} \rho_i(r) &= \rho_0[1 + f(r)] \\ &= \rho_0\left[1 - \frac{b}{2\pi r}\tan^{-1}\left(\frac{r}{\xi}\right) - \frac{b}{2\pi\xi}\frac{1}{1+(\frac{r}{\xi})^2}\right]\,. \end{aligned} \tag{9.4.15}$$

This led to $E_b = 1.1$ eV and $\tau_d = 183$ ps.* One then reaches the following conclusions: (1) The lifetime for a positron trapped in a pure dislocation line is 15 ps larger than that for a positron in the perfect lattice. However, it is measurable in experiments. (2) The positron dislocation binding energy $E_b$ is not small but one order of magnitude larger than the results calculated by Martin and Paetsch (1972) without consideration of the dislocation core structure. The calculated binding energy is much lower than that of Arponen *et al.* (1973). It seems more reasonable though this model still exaggerates the displacement $u_y$ in the y-direction due to the use of $u_r$ instead of $u_x$ in (9.4.13) and then overestimates the binding energy a little.

Shi *et al.* (1990) observed the pyramidal and parallel basal dislocations induced from the deformation with compressive loading normal to the (0001) plane and to the $(10\bar{1}2)$ plane of the Zn single-crystal specimen by TEM and measured the positron lifetime data for the samples at 100K and 300K. Because the loading direction is normal to the base plane for sample Zn-A, pyramidal dislocations are induced by second-order slip systems $\langle 11\bar{2}3\rangle\{11\bar{2}2\}$, As the forest dislocations become tangled with each other and penetrate the base plane to form jogs at their intersections easily (Friedel, 1964), the jogs along

* $E_b$ is the positron binding energy, and $\tau_d$ is the lifetime of positron at the dislocation.

Table 9.5. Results of a free analysis with two components for T = 100K (Shi *et al.*, 1990).*

| Sample | $\tau_1$ (ps) | $I_1$ (%) | $\tau_2$ (ps) | $I_2$ (%) | $\bar{\tau}$ (ps) |
|---|---|---|---|---|---|
| Zn-A | $100 \pm 4$ | 36 | $227 \pm 3$ | 64 | 181 |
| Zn-B | $96 \pm 4$ | 22 | $117 \pm 2$ | 78 | 160 |

* $I_1$, and $I_2$ are relative intensities (see Section A9.1).

Table 9.6. Results of a free analysis with two components for T = 300K (Shi *et al.*, 1990). See also Eq. (A9.1.1).

| Sample | $\tau_1$ (ps) | $I_1$ (%) | $\tau_2$ (ps) | $I_2$ (%) | $\bar{\tau}$ (ps) |
|---|---|---|---|---|---|
| Zn-A | $129 \pm 3$ | 50 | $210 \pm 5$ | 50 | 170 |
| Zn-B | 160 | 100 | – | – | – |

the dislocations were quite dense as were seen by TEM. In a sample of Zn-B, because the locating axis is normal to the $(10\bar{1}2)$ plane which is at an angle of 43° with the (0001) plane, according to the Schmid rule, the greatest shear stress is almost in the $\langle 11\bar{2}0\rangle$ direction, and only the basal slip systems is operating (Lavrent'yer and Salita, 1979). It can be estimated from the image that the dislocation density is about $10^{10}$ cm$^{-2}$, and the jog concentration along them is less than $10^{-3}$ nm$^{-1}$. Due to the mobility of monovacancies in Zn (Schumacher, 1970), the existence of monovacancies could be ruled out (Hidalgo *et al.*, 1986) and only dislocation-type defects are considered for positron trapping in the deformed Zn samples. The results of a free analysis with two components for the lifetime spectra are listed in Table 9.5. The longer lifetimes in the table are labelled as $\tau_{2A}$ for sample Zn-A and $\tau_{2B}$ for sample Zn-B. The difference between them is about 50 ps, and $\tau_{2A}$ is close to the lifetime value $\tau_V$ of 240 ps for a positron annihilated in a vacancy. The lifetime $\tau_{2B}$ for sample Zn-B would be explained by the model for describing a positron-dislocation interaction because of the jog concentration in it being less than $10^{-3}$ nm$^{-1}$. This value, $\tau_d = 177 \pm 2$ ps is nearly 30 ps more than $\tau_b$,† in the bulk at 100K. The positron lifetime results of a free analysis with two components are listed in Table 9.6 for the samples measured at 300K. The mean lifetime of a positron annihilating in sample Zn-B is only 4 ps more than in well annealed Zn, even

† $\tau_b$ is the positron lifetime in the pure bulk metal.

taking into account the change in $\tau_b$ with temperature. No trapping component was found. For sample Zn-A, the value of $\tau_{2A}$ is almost independent of the sample temperature, but the component intensity decreases evidently. Shi *et al.* (1990) suggested that the decrease in intensity of longer-lifetime component as the temperature is raised is induced by the detrapping of the thermally activated positron from dislocation lines.

In summary, when the jog concentration along a dislocation is sufficiently high in the sample, the positron-dislocation interaction can be described by the generalised model (Smedskjaer *et al.*, 1980), in which the dislocation acts as a stepping stone to even deeper traps (jogs) at T = 100K. If the jog concentration is less than $10^{-3}$ nm$^{-1}$, most of the positrons trapped along the dislocation line will be annihilated in it with a lifetime $\tau_d$ of 177 ± 2 ps. It is shown that the dislocation parameters are different from those in the bulk. It is easy to imagine that in a certain range of jog concentration, positrons would trap both along dislocation line and jogs with appropriate distribution and that in a certain range of temperature, the positrons trapped along the dislocation line will become unmeasurable gradually. Different experiments are evidently needed.

Shirai *et al.* (1992) measured the positron lifetime at 110K for the lattice, vacancies and dislocations in pure Al, Cu, Au and Cu-8at%Ge alloy. Positron lifetimes obtained for defects introduced by deformation at room temperature are 215 ± 2 ps in Al, 159 ± 4 ps in Cu, 166 ± 1 ps in Au and 168 ± 1 ps in Cu-8at%Ge: the stress level of which is just above the yield point of each specimen. They reported that these lifetimes can be reasonably considered to come mainly from dislocations, since vacancies, which may be formed by deformation, can easily migrate and disappear at room temperature. It is well established that dislocations in materials which have lower stacking-fault energy have more tendency to dissociate into partial dislocations. The equilibrium width $d$ of the extended edge dislocation, stacking-fault energies and Burgers vectors were known in experiments. It was found that the positron lifetime of dislocations entirely depends on, and is roughly proportional to, the magnitude of the Burgers vectors of dislocations. They comment that these results strongly suggest that positrons are predominantly trapped along dislocation cores, though no discussions on the density of jogs and their influence were reported.

Xiong (1986), Xiong and Lung (1988) proposed a model for positron trapping and detrapping at dislocations and jogs, in which a critical temperature

$T_c$ was defined. When $T > T_c$, positrons are trapped at jogs directly from the free state, and when $T < T_c$, positrons can be trapped at dislocation lines. If the density of jogs is large enough, dislocation lines play the role of stepping stones and positrons would move along the line and be trapped into deeper traps (jogs) finally. If the density of jogs is not large enough, positrons would be trapped at both dislocation lines and jogs with an appropriate ratio for the distribution. If the density of jogs is small enough, positrons can be mainly trapped at dislocation lines as in the case of the sample Zn-B at T = 100K.

### 9.4.3. *Grain Boundaries and Interfaces*

Since grain boundaries are regions of low atomic density, they would be expected to serve as trapping sites for positrons. Until now, the structure of grain boundaries in metals and alloys has not been revealed clearly. However, some information on grain boundaries can be obtained by the positron annihilation technique. Mckee *et al.* (1980) studied the positron lifetime and S-parameter for Doppler broadening of positron annihilation radiation as a function of mean grain size in a Zn-Al alloy and they obtained clear evidence of trapping at grain boundaries. Hidalgo and de Diego (1982) proposed a model for positron trapping at grain boundaries. They suggested a linear relationship between any linear annihilation parameter and the inverse mean grain size, but this relationship is dependent on the condition that $L > 2L_+$ where $L$ is the mean grain size, and $L_+$ is the mean positron diffusion length.

Yan *et al.* (1991) studied the positron annihilation at grain boundaries in Zn-22wt%Al alloy. They found that the relationship between the mean positron lifetime and the inverse grain size is linear only when the grain size is larger than 0.5 $\mu$m, approximately. Figure 9.7(a) shows that the curve deviates from the straight line when $L < 0.5$ $\mu$m.

Since, the disordered regions (grain boundaries) are not distributed homogeneously, the model for prediction of experimental results must include the positron motion from the point at the end of non-thermal trajectory to the trapping region. The diffusion trapping model (DTM) is a more complete version of the well-known simple trapping model (STM). The latter can be obtained as a limiting case of the former. Seeger (1992) treated the exact solution of a simple trapping model for the trapping of positrons in grain boundaries. Explicit expressions are given for the mean positron lifetime as well as for the decomposition of the lifetime spectrum into lifetime components in terms of grain size, positron diffusivity, rate coefficients describing the positron

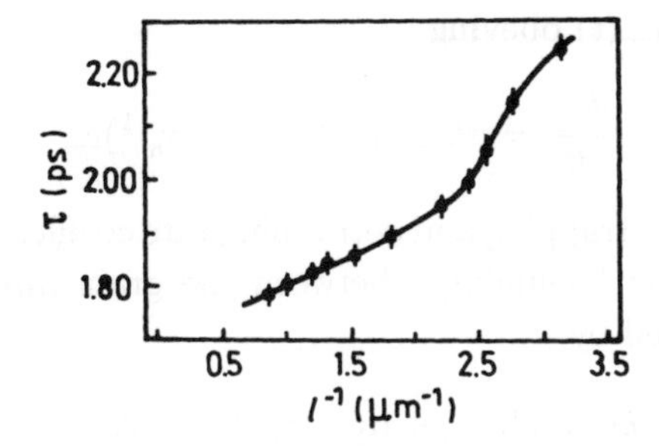


Fig. 9.7. (a) Mean positron lifetime $\tau$ versus inverse mean grain size $L^{-1}$.

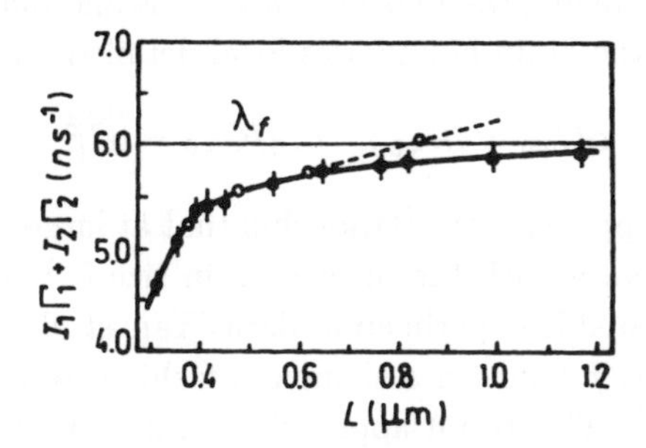


Fig. 9.7. (b) The relationship between $I_1\Gamma_1+I_2\Gamma_2$ and mean grain size $L$: •, our experimental results; ○, experimental results of Mckee *et al.* (1980). The horizontal line is the level for $\lambda_f$ (6 ns$^{-1}$).

trapping and detrapping at the grain boundaries, and positron trapping lifetimes in the bulk and in the grain boundaries. The model accounts qualitatively for the measurements on fine-grained ZnAl alloys.

According to the model (Seeger, 1992), the behaviour of the positrons is characterized by their bulk diffusivity $D_+$ their bulk ("free") lifetime $\tau_f$ and their lifetime $\tau_b(>\tau_f)$ in the grain boundaries. At time $t=0$ a homogeneous initial $e^+$ concentration $C=C_0$ has been implanted in the sample. The temporal and spatial evolution of the $e^+$ concentration, $C_+(r,t)$, is governed by

$$\frac{\partial C_+}{\partial t}=D_+\nabla^2 C_+-\frac{C_+}{\tau_f}. \tag{9.4.16}$$

The grain structure is modelled by assuming "spherical grains" of radius $r_0$ which are surrounded by grain-boundary layers of unspecified thickness. The $e^+$ trapped in the grain boundaries are described in terms of a planar concentration $c_+ = c_+(t)$ obeying

$$\frac{dc_+}{dt} = \alpha C_+(r_0, t) - (\beta + \tau_b^{-1})c_+ \tag{9.4.17}$$

where $\alpha$ and $\beta$ denote trapping and detrapping rate coefficients. The continuity of the $e^+$ flux at the "boundary" between the grain interior and the grain boundary is expressed by

$$D\nabla C_+|_{r=r_0} + \alpha C_+(r_0, t) = \beta c_+(t)\,. \tag{9.4.18}$$

Under the above-mentioned conditions the solution of (9.4.16) possesses spherical symmetry. Introducing the Laplace transforms and making the reasonable assumption that at the grain boundaries local detailed balancing applies:

$$\alpha C_{+0} = \beta c_{+0}\,; \quad c_{+0}/C_{+0} = \alpha/\beta\,. \tag{9.4.19}$$

Seeger obtained an approximate relationship that at large grain sizes the mean lifetime $\bar{\tau}$ varies linearly with the inverse grain size $l^{-1} \propto r_0^{-1}$, in agreement with most of the available experimental data (Yan *et al.*, 1991). As the grain size becomes smaller, $\bar{\tau}$ tends in a sigmoidal fashion towards a limiting value $\tau_b[1 - \beta\tau_f/(1 + \beta\tau_b)]$. This result approaching a limiting value is qualitatively consistent with the trend of the curve on fine-grained Zn-Al alloys, Fig. 9.6 (Yan *et al.*, 1991).

Dupasquier and Somoza (1995) pointed out that as a limiting case of the diffusion trapping model (DTM), the well-known simple trapping model (STM) is appropriate to describe positron trapping when the average distance from trap to trap is much smaller than the positron diffusion length $L_+$. This means that the material with grain sizes of the order of ten nanometers embedded in a disordered matrix is the limiting case of the DTM model. In materials with grain sizes of the order 1 $\mu$m, the motion of thermal positrons should be adequately described by combining the positron transport and trapping (DTM) in a set of different equations. The intermediate case is much more complex. The expressions for $\bar{\tau}$ contain $e^+$ diffusivities in a rather complicated form (Seeger, 1992). In general, materials are not of high-purity and defect free. The mean square displacement does not obey Fick's law $R^2 \sim t$, but scales with time as $R^2(t) \sim t^{2/d_w}$, where $d_w > 2$ is the fractal dimension of the random walk (Lung, 1995). The relation between positron diffusion constant $D_+$ and the characteristic diffusion length $L_+$, $L_+ = \sqrt{D_+\tau_{\text{bulk}}}$ is not well defined. Whether the grain dimension (the average distance from trap to trap) is much smaller or much larger than the positron diffusion length $L_+$ is quite difficult to judge.

Moreover, Dupasquier and Somoza (1995) pointed out that the information obtained by PAS measurements essentially concerns the volume of open spaces inside the intergrain regions. Since this is a disordered region, a continuous volume distribution is to be expected. With an ideal set-up with infinite resolving power, this continuous volume distribution should be reflected in a continuous distribution of lifetimes. However, to the best of our knowledge, there has been no attempt at analysing a positron lifetime spectrum in the grain boundary region or a nanocrystalline solid in terms of a continuous spectrum of lifetimes. Usually, one uses the traditional model with a finite (small) number of lifetimes.

In order to interpret their experimental results, Yan *et al.* (1991) proposed a simple model with the following assumptions:

(i) the grain boundary is a kind of three-dimensional composite defect which occupies a volume fraction $\alpha$,
(ii) all positrons thermalized are distributed uniformly in the alloy and thereby a fraction $\alpha$ of the positrons are already trapped at grain boundaries when they are thermalized,
(iii) there are no other kinds of defect and
(iv) the detrapping rate can be neglected.

Denoting by $n_f(t)$ and $n_t(t)$ the occupation probabilities of free positrons and positrons trapped at the grain boundaries, at time $t$, the rate equations can be written as

$$\begin{aligned} \frac{dn_f}{dt} &= -\lambda_f n_f(t) - k n_f(t) \\ \frac{dn_t}{dt} &= -\lambda_t n_t(t) + k n_f(t) \end{aligned} \tag{9.4.20}$$

where $\lambda_f$ and $\lambda_t$ are the positron annihilation rates in the bulk and at grain boundaries, respectively, defined by $\lambda_f = 1/\tau_f$ and $\lambda_t = 1/\tau_t$. $k$ is the positron-trapping rate for the grain boundaries.

Let $t = 0$ correspond to the time when positrons are completely thermalized; so that the initial conditions are

$$\begin{aligned} n_f(0) &= 1 - \alpha \\ n_t(0) &= \alpha \, . \end{aligned} \tag{9.4.21}$$

Table 9.7. Mean grain sizes, $L$, and thickness of grain boundaries, $w$, for the Zn-Al alloy samples (Yan *et al.*, 1991).

| Sample | Mean grain size $L(\mu m)$ | Thickness of grain boundaries $w(\mu m)$ |
|---|---|---|
| 0 | $0.319 \pm 0.003$ | $0.071 \pm 0.010$ |
| 1 | $0.361 \pm 0.003$ | $0.047 \pm 0.007$ |
| 2 | $0.387 \pm 0.003$ | $0.033 \pm 0.006$ |
| 3 | $0.411 \pm 0.004$ | $0.030 \pm 0.006$ |
| 4 | $0.452 \pm 0.004$ | $0.034 \pm 0.007$ |
| 5 | $0.548 \pm 0.005$ | $0.022 \pm 0.006$ |
| 6 | $0.653 \pm 0.007$ | $0.017 \pm 0.007$ |
| 7 | $0.757 \pm 0.007$ | $0.014 \pm 0.008$ |
| 8 | $0.824 \pm 0.010$ | $0.013 \pm 0.009$ |
| 9 | $0.990 \pm 0.010$ | $0.010 \pm 0.008$ |
| 10 | $1.155 \pm 0.010$ | $0.007 \pm 0.013$ |

From the solutions of Eqs. (9.4.20) and (9.4.21), the following formulae can be obtained:

$$I_1\Gamma_1 + I_2\Gamma_2 = \lambda_f - (\lambda_f - \lambda_t)\alpha \tag{9.4.22}$$

$$\tau = \tau_f(1 + k\tau_t)/(1 + k\tau_f) + [(\tau_t - \tau_f)/(1 + k\tau_f)]\alpha\,. \tag{9.4.23}$$

In the simple trapping model (STM), the value of $I_1\Gamma_1 + I_2\Gamma_2$ ($\Gamma_1 = 1/\tau_1, \Gamma_2 = 1/\tau_2$) should be equal to a constant $\lambda_f$ ($= 1/\tau_f$). The experimental values of $I_1\Gamma_1 + I_2\Gamma_2$ for the various samples have been calculated using the data analysis and the relationship between $I_1\Gamma_1 + I_2\Gamma_2$ and the grain size $L$ is shown in Fig. 9.7(b). The experimental results do not agree with the STM. The results of Mckee *et al.* (1980) are also plotted in Fig. 9.7(b) for comparison.

Making use of Eq. (9.4.22), the results in Fig. 9.7(b) can be interpreted very well qualitatively. The smaller the mean grain boundaries in the alloy, the greater is the difference between the value of $I_1\Gamma_1 + I_2\Gamma_2$ and the constant $\lambda_f$. The results of Mckee *et al.* (1980) in Fig. 9.7(b) also support this point. However, Eq. (9.4.22) exhibits a limiting value $\lambda_f$, as $\alpha = 0$. This is consistent with the experimental results (Yan *et al.*, 1991). Using Eq. (9.4.23), $\alpha$ and then the order of thickness of the grain boundaries, $w$, were estimated by Yan *et al.* (1991). All the values of $w$ for the various samples are listed in Table 9.7.

The thickness of the grain boundaries $w$ estimated by positron measurements seems much larger than by Auger spectroscopy (Balluffi, 1977; also see Fen *et al.*, 1987). The thickness $w$ is only of the order of a couple of lattice parameters. However, $w$ estimated by positron measurements is consistent with the data reported by Arharov (1955, 1958), (also see Fen *et al.*, 1987) which is about 100–1000 Å. Just as Seeger (1992) pointed out that the thickness of grain-boundary layers is unspecified, it seems that $w$ includes the disordered region ($\sim$ order of several lattice parameters) and a transition region of lattices with defects or impurities to the interior of a grain. This transition region is large and its thickness can be estimated by the phenomena of solute segregation or interior adsorption at grain boundaries and their nearby distorted regions in the grain (Arharov, 1955, 1958; McLean, 1957). The thickness of this transition region depends on the method of measurement. The sensitivity of positron annihilation at vacancies is of the order $10^{-7}$ atomic concentration. However, the sensitivity of Auger spectroscopy for elements is only $10^{-2} - 10^{-3}$ atomic concentration. It would appear that techniques with higher sensitivity would feel thicker transition region than that with lower sensitivity. For instance, Mishra (1980) investigated the grain boundary of Mn-Zn ferrite and analysed the chemical contents at the grain boundary with Scanning Electron Microcopy (SEM)/Energy Dispersive X-ray Spectrum (EDX), and found that Ca and Si segregate in a region of 2000 Å thick (see also Cui, 1990). Of course, there are other factors leading to grain boundary segregation. Here, as a sensitive technique for vacancy concentration, it would be reasonable to believe that the transition region measured by positron annihilation is of $10^3$ Å thick in order of magnitude.

In their experimental results, Yan *et al.* (1991) found that the thickness of the transition region, $w$ of the grain boundaries increases with decreasing grain size. Perhaps, the larger the curvature of the grain, the larger the number of defects near the grain boundary in the grain is needed for relaxing the stress due to the mismatch at the grain boundary.

Slow positron beam apparatus has also been used for studies on interfaces. It has been shown that it is possible to obtain the depth of the oxide layer and some information on the nature of interfacial defects (Nielsen *et al.*, 1989; Baker *et al.*, 1989). Tabuki *et al.* (1992) studied interfacial phenomena in the W-Si system. They found that positrons diffused beyond the positive charge depletion layer towards the Si bulk region and the phase charge during silicidation. Weng *et al.* (1995) investigated the solid state reaction of Co/Ti/Si and Co/Si. They found that the S parameters were sensitive to thin film reaction and crystalline characteristics. Xiong *et al.* (1995) studied defects in a layered Cu/Co/Cu structure and found significant concentration of open-volume defects in all three layers of Cu/Co/Cu sputtered on a Si substrate. The trapping phenomena are often analyzed in terms of the diffusion properties of positrons in the bulk lattice. Applying the positron diffusion model to low-energy positron beam results for a sample of Si with an overlayer of $SiO_2$, Smith *et al.* (1992) determined the interface depth, the width of the depletion region and the charge at the interface.

Jiang *et al.* (1995) reported a theoretical study of the diffusion of positrons to the surface for a semi-infinite medium and a film with a thickness in semiconductors by using a $\delta$-function method. The results can be used to the analysis of experimental data for the overlayer/substrate system that contains one film or several films and a semi-infinite medium. In experiments on semiconductors, there may be an internal electric field near the surface which can influence the positron diffusion. The one-dimensional diffusion equation is given by

$$\frac{\partial n(t)}{\partial t} = D_+ \frac{\partial^2 n(x,t)}{\partial^2 x} - v_d \frac{\partial n(x,t)}{\partial x} - \lambda_{\text{eff}} n(x,t) \,. \tag{9.4.24}$$

This equation differs from Eq. (9.4.16) by the second term for a constant electric field and was the same as that given by Schultz and Lynn (1988). In Eq. (9.4.24), $n(x,t)$ is the positron density as a function of both time and position, $D_+$ is the positron diffusing coefficient, $\lambda_{\text{eff}}^{-1} = \tau_{\text{eff}}$, is the effective lifetime of the positron in a freely diffusing state and $v_d$ is the field-dependent drift velocity. By applying a radiative boundary condition

$$n(0,t) = \beta \frac{\partial n(x,t)}{\partial x} |_{x=0} \tag{9.4.25}$$

a general solution $n(x,t)$ can be obtained. For a film of thickness of $d$, and boundary conditions, $n(0,t) = 0$, $n(d,t) = 0$, and initial conditions at $x = a$, $n(x,0) = \delta(x-a)$, the fractions $f_0$ and $f_d$ of the positrons diffusing to the surfaces at $x = 0$ and $x = d$ respectively can be determined. The results obtained are the same as those of Beling *et al.* (1987) and Novikov *et al.* (1991), but they used different methods and did not give the result for such general cases.

### 9.4.4. *Voids and Cracks*

As the vacancy cluster size becomes larger than 10 Å, the positron lifetime approaches an asymptotic value of (450–500) ps which seems saturated and independent of the size of the cluster being further increased. This is a large void seen by positrons. This has been explained in terms of a positron surface state. Efficient positronium (ps) formation at surfaces was a key factor in facilitating the spectroscopic studies of this basic quantum-electrodynamic system (see Nieminen, 1983). For thermalized positrons, the analysis of trapping phenomena in voids can be based on the solution of diffusion-annihilation equations. The lifetime spectrum is a superposition of annihilations in the bulk, defects, the surface state and ps annihilations. Computer programs for the void and the semi-infinite geometry are available (Nieminen, 1983).

Further applications of the positron-annihilation techniques to the study of crack problems in materials appear to have considerable potential interest. As the cracked metal is loading, a plastic zone always appears at the crack tip. The size and defect structure of the plastic zone are closely related to the resistance of crack propagation, which is referred to as the fracture toughness of materials. The positron-annihilation technique (PAT) can be used to investigate this problem because of its sensitivity to defects in the plastic zone. PAT can determine the plastic zone size, as in the work of Jiang *et al.* (1982) for $\alpha - T_i$ specimens (Fig. 9.8).

The progress of elastic-plastic fracture of materials needs deeper understanding of the defect structure in the plastic zone at the crack tip. Kobayashi

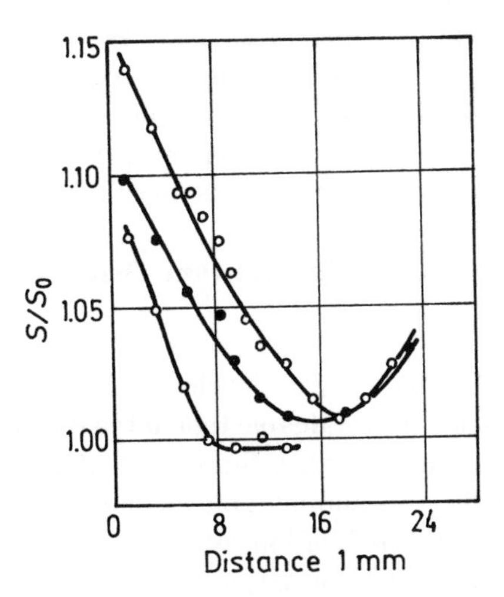


Fig. 9.8. Relative change of S as a function of distance. *l*-Distance from crack tip along direction of crack.

and Ohr (1980) investigated *situ* the distribution in the plastic zone at a crack tip. They found that there is a dislocation-free zone at the crack tip, which is contrary to the original Bilby-Cottrell-Swinden (BCS, 1963) crack dislocation model for the crack. Lung and Xiong (1983) performed calculations on the dislocation-distribution function in the plastic zone. They found that there was a negative-sign dislocation zone near the crack tip (see Sec. 3.6.6 in Chap. 3). Such behaviour was observed near a crack tip in NaCl (Narita and Takamura, 1985), stainless steel (Ohr, 1987) and Al single crystal (Xu *et al.*, 1991). The distribution function for dislocations in the plastic zone is directly related to the calculated results for crack-opening displacement, which is an important parameter for the fracture toughness of materials. Few studies of this problem have been performed to date. A high-intensity positron source focused to produce a very narrow beam may possibly be used for the study of this problem, although this may lead to some technical difficulties. Advances in techniques of positron microbeams seem possible to allow a probe with a diameter of 0.1 $\mu$m and more than $10^6$ positrons per second (see Seijbel *et al.*, 1995). For super high strength steels, the plastic zone size is grossly of the order of a micrometer. Using PAT for studies of many materials is practical. Is there really a dislocation-free zone just ahead of the crack tip? Some workers have suspected that it is an artefact related to the use of a thin specimen in Ohr's experiments. If a positron microbeam is used, then thicker specimens can be studied. Is there also a negative-sign dislocation zone? What is the dislocation shielding effect on the stress field near the crack tip? Is there any antishielding effect on the stress field due to the presence of the negative-sign dislocation? How does the work done in forming this dislocation configuration relate to the fracture toughness of the materials? All these questions should lead to deeper understanding of the elastic-plastic fracture of materials.

## 9.5. *Electron and Positron Momentum Distributions in Solids*

### 9.5.1. *Introduction*

Among the methods for the study of the electronic structure of solids, positron annihilation has proven to be a valuable tool in the investigation of the electron momentum distribution in materials. Since the first observation of the angular correlation between the two annihilation quanta by Beringer and Montgonmery (1942), the positron annihilation technique has developed into a method capable of yielding relevant information on the electronic structure such as Fermi surface dimensions and relevant wave functions (see Mijnarends, 1979). In nondilute disordered alloys, excessive electron scattering due to the short mean free path of the electrons precludes the use of other methods (say, the de Haas-van Alphen effect, the magnetoresistance effect, the $rf$ size effect, etc.) at solute concentrations $\geq 1\%$. Kohn anomalies in X-ray diffuse scattering (Moss *et al.*, 1974) or inelastic neutron scattering (Powell *et al.*, 1968), and Faraday effect (McAlister *et al.*, 1965) can still be used for investigations of the Fermi surface geometry. The Compton effect (see Williams, 1977) is able to provide direct information about the entire electron momentum distribution, but cannot compete in resolution with positron annihilation. For the study of nondilute alloys, positron annihilation is one of the very few useful techniques. It also has some problems. The disturbance that the positron produces on the electron distribution is a point of weakness. The joint electron-positron momentum density depends not only on the electronic structure of the sample, but also on the positron wavefunction, and it is affected by the many-body electron-positron correlation. Fortunately, it has been shown that positron-electron correlations do not shift the discontinuities of the momentum density from the true Fermi surface. The accuracy that can be obtained in the determination of the parameter of the Fermi surface in a sample metal is of the order of $10^{-3}$ a.u. which is quite comparable to de Haas-van Alphen data though more delicate analysis involving band calculation with consideration on many-body perturbations for the electron wavefunctions is required. The second problem is its high affinity to low-density lattice defects, but this problem can be overcome in many cases by careful preparation of the specimens. The vastly increased resolution in momentum space resulting from the measurement of two transverse momentum components (2D ACAR),* designed by Berko and co-workers (1977), instead of one (1D long slit), opened up a new plethora of problems which could be addressed by the positron technique (see Mijnarends, 1979; 1983; Mijnarends and Bansil, 1995). Early spectra of 1D or 2D ACAR permitted a straightforward interpretation mostly in terms of the geometry of Fermi surface. Since the progress of this technique, the complexity of the systems investigated has grown from pure metals, semiconductors and alloys to highly complex systems such as half-metallic ferromagnets, heavy-fermion compounds and high-$T_c$ superconductors.

### 9.5.2. *Positron Implantation and Thermalization*

The positron most frequently is taken from a $\beta^+$ source; in a few laboratories with access to accelerator facilities, positrons from high-energy photon

*1D and 2D ACAR are one-dimensional and two-dimensional angular correlation of annihilation radiation experiments for positron momentum distribution measurements respectively.

materialization are also used. A high-energy positron, emitted by a radioactive source, enters a solid with an initial energy in a range characteristic of the beta spectrum of the radioisotope. The implant energy of the positron is an important variable as it controls the distribution of penetration depths. It is rapidly slowed down in a time of the order of picoseconds by atomic ionization, excitation, positron-electron collisions and positron-phonon interactions, until it reduces (near) thermal equilibrium with the surrounding crystal. The stopping profile of energetic positron from a radioactive source is exponential

$$P(x) = \alpha \exp[-\alpha x]\,,$$

$$\alpha \approx 16 \frac{\rho[g/\mathrm{cm}^3]}{E_{\max}^{1.4}[\mathrm{MeV}]}\,\mathrm{cm}^{-1} \tag{9.5.1}$$

where $\rho$ is the density of the material and $E_{\max}$ the maximum energy of the emitted positron. The most common isotope for positron lifetime and Doppler-broadening exeperiment is $^{22}$Na with $E_{\max} = 0.54$ MeV. For this isotope the characteristic penetration depth $1/\alpha$ is 110 $\mu$m in Si, 50 $\mu$m in Ge, and 14 $\mu$m in $W$. For monoenergetic positrons in the range 0–30 keV, the stopping profile can be described by a derivative of a Gaussian function

$$P(x) = -\frac{d}{dx} \exp[-(x/x_0)^2]\,. \tag{9.5.2}$$

The mean stopping depth is

$$\bar{x} = 0.886 x_0 = A E^n [\mathrm{keV}] \tag{9.5.3}$$

where

$$A \approx \frac{4}{\rho} \mu g/\mathrm{cm}^2\,, \quad n = 1.6\,. \tag{9.5.4}$$

The mean stopping depth varies with energy from 1 nm up to a few $\mu$m. In a solid, the positron energy loss rate in the range 1 MeV $> E_+ >$ 100 keV is about 1 MeV/ps, and from 100 keV to 100 eV it is 100 keV/ps. Below 100 eV the effects are gradually switched on. In metals, calculations indicate that positron momentum distribution rapidly relaxes to the Maxwell-Boltzmann distribution. The thermalization time at 300K is (1–3) ps, i.e. much less than a typical positron lifetime of (100–200 ps). Even at 10K, calculated thermalization times are less than the positron lifetime (Nieminen and Oliva, 1980; Jenson and Walker, 1990). Usually, it is assumed that positrons are thermalized at $t \approx 0$, and that the thermalization time in metals and semiconductors is negligible compared to the positron lifetime. In an insulator with a wide band gap of several eV, phonons may not be effective enough to thermalize positrons during their lifetime; the nonthermal positrons can transverse long distances to reach the surface (Gullikson and Mill Jr, 1986).

At any one time, there is on the average only one positron in the sample due to the low $e^+$ density. The positron is in its ground state at the bottom of the positron conduction band. Its motion obeys the Boltzmann distribution

$$f_+(E_+, T) = (\pi m^* k_B T)^{-3/2} \exp[-E_+/k_B T] \tag{9.5.5}$$

where $E_+$ denotes the positron kinetic energy, $m^*$ its effective mass, and $k_B$ is the Boltzmann constant. With $E_+ = p^2/m^*$, it is clear in Eq. (9.5.5) that depending on temperature and effective mass, the positron displays a small thermal motion which will broaden the momentum resolution function in an angular-correlation experiment. There are also contributions due to various many-body effects and to positron-phonon interaction.

### 9.5.3. *Positron Diffusion*

The thermalized positron, scattered by phonons, diffuses until it annihilates with an electron. During its lifetime, it diffuses over a volume of about $L_+^3$, where $L_+$ is the average diffusion length before annihilation. The motion of the thermalized positron in solids is limited by positron-electron interaction, described by electron-hole pair generation, by positron-lattice interaction and by scattering off impurities. Owing to experimental difficulties, there exist few direct measurements of the positron mobility in solids and the diffusion constant, defined via the Einstein relation

$$D_+ = \mu_+ \frac{kT}{e} \tag{9.5.6}$$

where $\mu_+$ is the mobility, and $e$ the positron charge. In metals, the measurement of positron mobility is obviously extremely difficult, and one has to rely on theoretical estimates. Bergerson *et al.* (1974) have calculated the various contributions in a number of simple metals. Soininen *et al.* (1990) measured the positron mobility $\mu_+$ and positron diffusion constant in high-purity defect-free Mo, Al, Cu, and Ag single-crystal samples in the temperature range 20–1400K with a slow-positron-beam technique. The values of $D_0$ ($D_+ = D_0$ $(\mathrm{T}/300\mathrm{K})^{-\alpha}$) is about 1.0–2.0($\mathrm{cm}^2$/s) where $\alpha$ is the power and is about 0.5. Brusa *et al.* (1995) reported measurements of positron mobility in polyethylene, their measurements were carried out by improving the acquisition and analysis of Doppler-broadening annihilation spectra. The results measured (Soininen *et al.*, 1990) are in agreement with theory where thermal positron motion is limited by scattering from acoustic phonons. The diffusion length $L_+$ is defined by $L_+ = \sqrt{D_+\tau_b}$; $L_+$ is about 1100–1800 Å at 300K. This is the case in high-purity defect-free single-crystals.

In industrial materials, there are many defects and impurity atoms. The diffusion length $L_+$ of the positron will be shorter. Defects and impurity atoms may form fractal structures in the material. The diffusion length of thermalized positrons can be treated as a random walk on fractals (Lung, 1995). The mean square displacement does not obey Fick's law $R^2 \sim t$, but scales with time as $R^2(t) \sim t^{2/d_w}$, where $d_w > 2$ is the fractal dimension of the random walk. $R$ and $t$ are dimensionless quantities. In a simple discrete random walk the walker advances one step in unit time. Now, we have

$$R_f(t) \sim t^{1/d_w} \tag{9.5.7}$$

$$R(t) \sim t^{1/2} \tag{9.5.8}$$

and

$$R_f(t)/R(t) \sim t^{\beta}\,, \quad \beta = \frac{2 - d_w}{2d_w}\,. \tag{9.5.9}$$

In general, $d_w > d_f$, the fractal dimension of the objects, $R_f(t) < R(t)$, due to $d_w > 2$. It has been shown that $d_w = 3$ in a medium that is filled with static trapping sites at a finite concentration both for traps distributed in a d-dimensional Euclidean space or in a fractal space (Havlin and Ben-Avraham, 1987). Results of calculated values of diffusion lengths of thermalized positron in media with traps showed that $R_f(t)$ can be shorter than half of $R(t)$. This is a quite an approximate model calculation. The key parameter is $d_w$. If this parameter can be obtained by measurements or precise calculation, the positron diffusion length can be estimated accurately.

### 9.5.4. *Positron Distribution in Solids*

The spatial distribution of the thermalized positron is not uniform. The density distribution is relatively uniform apart from the ion cores due to the strong repulsion. Using pseudopotential theory, Stott and Kubica (1975) calculated accurate positron wave functions and energies with modest numerical labor. The construction of a single-particle potential for a positron in a metal is simpler than for electron. There is no exchange repulsion. The positron wave

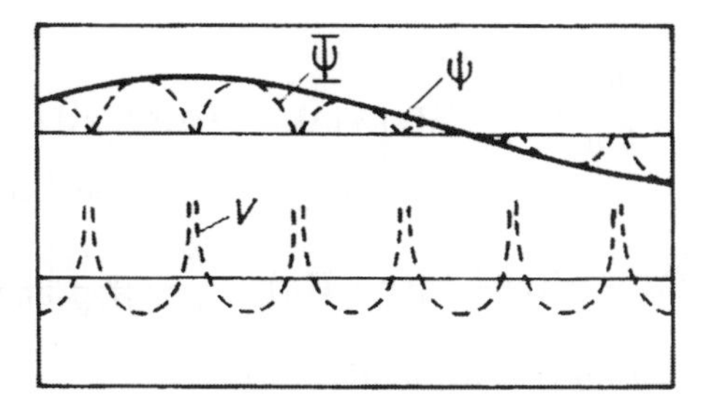


Fig. 9.9. Schematic positron wave function and potential in a perfect metal. $V$ is the full Hartree potential and $\Psi$ the corresponding full wave function. $\psi$ is the pseudo wave function corresponding to the nearly constant pseudopotential.

function for states near the bottom of the lowest energy band is separated into two factors. One reflects the strong repulsion of the positron from the ion core, and the other is a smooth envelope, which is energy dependent, sensitive to the environment and reflects the positron distribution in the interstitial regions and/or between atomic cells in the crystal. Figure 9.9 schematizes this division for a nonzero $\mathbf{k}$ corresponding to a temperature of about 1000K. The envelope satisfies a Schrödinger-like equation with a relatively weak potential term which, e.g., in perfect metals can be attacked with low-order perturbation theory.

The positron wave function for a wave vector $\mathbf{k} = 0$ is factorized as

$$\Psi_{\mathbf{k}}(\mathbf{r}) = U(\mathbf{r} - \mathbf{R})\psi_{\mathbf{k}}(\mathbf{r}) \tag{9.5.10}$$

where $\mathbf{r}$ lies in an atomic cell centered on $\mathbf{R}$, and $U(\mathbf{r})$ is chosen to satisfy the Schrödinger equation with the spherically symmetric single-ion potential $V_a(r)$ with Wigner-Seitz boundary conditions imposed at a conveniently chosen muffin-tin radius. In the cell,

$$\left[-\frac{\hbar^2}{2m}\frac{\partial^2}{\partial r^2} + V_a(r)\right] U(r) = E_{WS} U(r) \tag{9.5.11}$$

$$\frac{\partial U}{\partial r}|_{r=R_0} = 0 \tag{9.5.12}$$

where $R_0$ is the muffin-tin radius. Between the muffin-tin spheres, $U(r)$ can be chosen to be a constant, $U(r) = U(R_0)$. Substituting (9.5.9) into the full Schrödinger equation, one can obtain an equation for the pseudo-wave function

$\psi_{\mathbf{k}}(r)$

$$\left[-\frac{\hbar^2}{2m}\nabla^2 + W(\mathbf{r})\right]\psi_{\mathbf{k}}(\mathbf{r}) = E_0(\mathbf{k})\psi_{\mathbf{k}}(\mathbf{r}) \tag{9.5.13}$$

where

$$W(\mathbf{r}) = \begin{cases} E_{ws} + V(\mathbf{r}) - V_a(\mathbf{r} - \mathbf{R}_n) - \dfrac{\hbar^2}{2m}\dfrac{\nabla U(\mathbf{r})}{U(\mathbf{r})} \cdot \nabla \\ \text{(within the } n\text{th muffin-tin)} \\ V(\mathbf{r}) \\ \text{(outside a muffin-tin)} \end{cases} \tag{9.5.14}$$

and $V(\mathbf{r})$ is the full positron potential. From Eq. (9.5.13), we may see that the pseudopotential $W(\mathbf{r})$ is much weaker than $V(\mathbf{r})$ in the core regions. This method has been used to calculate positron ground-state wave functions, energies and band masses in simple metals and ionic crystals (see Nieminen, 1979; 1983; 1995).

### 9.5.5. *Electron-Positron Correlation in Pure Metals and the Core Effects*

The positron attracts a cloud of electron which screens its charge (see Mijnarends and Bansil, 1995). An extra term $V_{ep}$ describing $e^- - e^+$ correlation to the Coulomb-Hartree potential $V_e$ due to the ions and the electron should be added to describe the mean-field potential felt by the positron

$$V^+(\mathbf{r}) = -V_c(\mathbf{r}) + V_{ep}(\mathbf{r})\,. \tag{9.5.15}$$

This leads to a redistribution of the $e^+$ wave function. Moreover, the pile-up of the electronic wave functions at the position of the positron increases the annihilation probability to a value much larger than that obtained via the overlap of the independent-particle wave functions. This leads to the enhancement factor in Eq. (9.4.1). However, the effects of enhancement on the momentum density are much smaller, but not negligible.

Kahana (1960, 1963) and Carbotte (1966, 1967) presented a theory of correlation, who used a Green function formalism to treat the $e^- - e^+$ and $e^- - e^-$ correlations in a free-electron gas. Boronski and Nieminen (1986) calculated the enhancement factor* as

$$\gamma(r_s) = 1 + 1.23r_s + 0.8295r_s^{3/2} - 1.26r_s^2 + 0.3286r_s^{5/2} + \frac{r_s^3}{6}\,. \tag{9.5.16}$$

*See also the early work of N. H. March and A. M. Murray, Phys. Rev. *126*, 1480 (1962).

Wang *et al.* (1995) used the self-consistent charge discrete variational $X_\alpha$ cluster method (SCC-DVM) to calculate the conduction electron density of metals. Comparing with experimental data on positron annihilation rate, the relation between electron density and positron annihilation rate is obtained:

$$\lambda = 22.57 n_{\mathrm{con}}^{0.72} \tag{9.5.17}$$

and an expression for the enhancement factor is given by

$$\gamma(r_s) = 1.34 r_s^{0.84} + 0.166 r_s^3 \,. \tag{9.5.18}$$

The enhancement factor is also a function of the momentum $p$. The enhancement due to $e^- - e^+$ correlations increases as one approaches the Fermi surface, while, on the other hand, the high momentum tail, present at $|\mathbf{k}| > k_F$ as a result of $e^- - e^-$ correlations, is strongly attenuated. For $k < k_F$, $\gamma(p, r_s)$ is given by

$$\gamma(p, r_s) = a(r_s) + b(r_s)(p/p_F)^2 + c(r_s)(p/p_F)^4 \,. \tag{9.5.19}$$

The interaction of the positron with the correlated electron gas tends to sharpen the discontinuity at the Fermi momentum. A discussion of $e^- - e^+$ and $e^- - e^-$ correlations which takes into account the interaction of the particles with the lattice is extremely complicated. Here, the problem is discussed in terms of a local-density approximation (LDA) which is introduced in Sec. 9.4 (Eqs. 9.4.5 and 9.4.6).

The more tightly bound "core" electrons on real metals lead to a faster annihilation rate than what would be predicted from the conduction electron gas alone. In transition and noble metals, the core contribution is large due to the extrusive $d$ shell.

West (1973) used the electron gas theories by renormalizing the valence electron density $n$ according to the description

$$n_{\mathrm{eff}} = n\left(1 + \frac{\Gamma_c}{\Gamma_v}\right) \tag{9.5.20}$$

to calculate the annihilation with the core electrons. In Eq. (9.5.20), the $\Gamma_c$ and $\Gamma_v$ are the partial annihilation rates with the "core" and "valence" electron respectively. The ratio $\Gamma_c/\Gamma_v$ can be estimated from the angular correlation curves, which consist of a separable Gaussian core electron part and a free electron-like parabola. (Readers may refer also to Stern (1991), Daniuk *et al.* (1991), Berko (1983) and Barbiellini *et al.* (1997).*

*See also J. A. Alonso, J. Jiang, C. W. Lung, J. Y. Wong and N. H. March, An. Fis. (Spain) 93, 136 (1997).

### 9.5.6. *Momentum Density in Crystalline Solids*

The numerous aspects of the technique of band-structure calculation are covered by a number of review articles and books (see Callaway, 1964, Mijnarends, 1979). The band-structure methods most commonly employed in momentum-density calculation are the augmented-plane-wave (APW), the Korringa-Kohn-Rostoker (KKR), and the orthogonalized-plane-wave (OPW) methods. The most transparent one is the OPW method; the KKR method has been extensively used for momentum density computations in metals and compounds as complex as the high-$T_c$ superconductors.

In a periodic crystal of volume $V = N\Omega$, where $N$ denotes the number of lattice sites and $\Omega$ the volume of the unit cell, the positron and electron wave functions are solutions of the Schrödinger equation

$$[-\nabla^2 + V(\mathbf{r})]\psi(\mathbf{r}) = E\psi(\mathbf{r})\,. \tag{9.5.21}$$

Here $V(\mathbf{r})$ is the potential 'felt' by the positron or electron when it moves through the crystal, $V(\mathbf{r})$ is a periodic function of $\mathbf{r}$, and $\psi(\mathbf{r})$ has the Bloch form $\psi(\mathbf{r}) = V^{-1/2}u_{k,j}(\mathbf{r})\exp[i\mathbf{k}\cdot\mathbf{r}]$, where $u_{k,j}(\mathbf{r})$ has the periodicity of the lattice, $\mathbf{k}$ denotes the wave vector of the positron or electron, and $j$ labels the energy bands.

In the KKR formalism the solution of Eq. (9.5.21) is written in the form of an integral equation

$$\psi_k(\mathbf{r}) = \int G(\mathbf{r},\mathbf{r}')V(\mathbf{r}')\psi_k(\mathbf{r}')d\mathbf{r}'\,. \tag{9.5.22}$$

The integration extends over the unit cell of volume $\Omega$, and the Green function $G(\mathbf{r},\mathbf{r}')$ is given by

$$G(\mathbf{r},\mathbf{r}') = \Omega^{-1}\sum_n \frac{\exp[i\mathbf{k}_n\cdot(\mathbf{r}-\mathbf{r}')]}{E-k_n^2} \tag{9.5.23}$$

where $\mathbf{k}_n = \mathbf{k}+\mathbf{K}_n$ and $\mathbf{k}_n = |k_n|$, while the summation is over the reciprocal-lattice vectors. The crystal potential $V(\mathbf{r})$ is assumed to have the muffin-tin form: spherically symmetric inside the (nonoverlapping) muffin-tin spheres with radii $r_i$ and a constant value $V_0$ between the spheres. When the zero of energy is chosen such that $V_0 = 0$, the only nonzero contributions to the integral (9.5.22) come from the interior of the muffin-tin spheres. A trial wave function in the muffin-tin spheres has the form

$$\psi_k(\mathbf{r}) = \sum_L i^l C_L R_l(r,E)Y_L(\mathbf{r}) \tag{9.5.24}$$

where $Y_L$ are the real-spherical harmonics. Substituting $\psi_k(r)$ into integral (9.4.2), one finds $P_0(p)$. Readers should refer to articles and reviews for details (see, for example, Mijnarends and Bansil, 1995).

The KKR method of computing the momentum density presented here can be made into a fast and highly efficient scheme by the extensive use of interpolation in precomputed tables for many of the relevent quantities such as the structure functions and by extensive vectorization to which the method lends itself extremely well.

### 9.5.7. *1D and 2D ACAR Measurements*

A measurement of the deviation from collinearity between two quanta emitted in the annihilation of a positron could provide information about the momentum distribution of electrons. Apparatus for the measurement of the angular correlation of annihilation in one or two dimensions (1D or 2D ACAR) has been developed (see Berko, 1983; Mijnarends, 1979; 1983; Mijnarends and Bansil, 1995).

If both the positron and the electron are at rest, the two photons emitted after annihilation of the pair are at an angle of 180°, with total energy $E_T = 2mc^2 - E_b$, $E_b$ being the binding energy of the electron and the

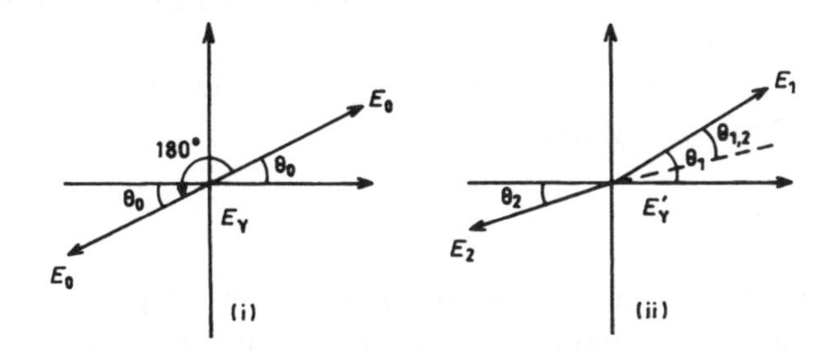


Fig. 9.10. (a) (i) Center-of-mass frame; (ii) laboratory frame.

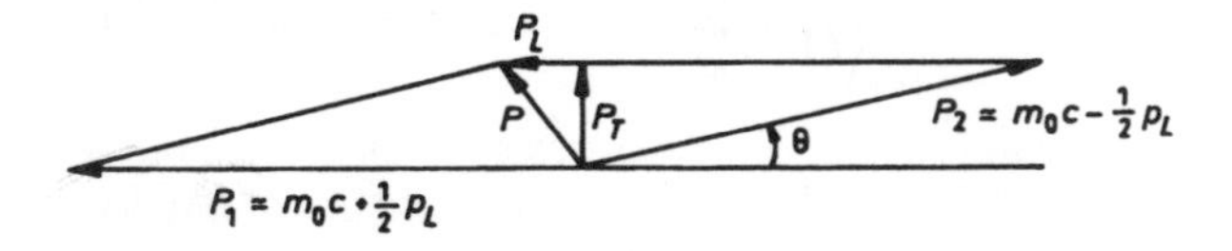


Fig. 9.10. (b) The vector diagram of the momentum conservation in the 2$\gamma$-annihilation process. The momentum of the annihilating pair is denoted by $p$, subscripts $L$ and $T$ refer to longitudinal and transverse components, respectively.

positron in the system. Thermalized positrons have kinetic energy of only about 0.025 eV at room temperature and usually are considered to be at rest. In their center-of-mass frame, the positron annihilates with the moving electron and the two photons emitted are moving strictly into opposite directions. However, in the laboratory frames, the two photons are not anticollinear because of the finite momentum $\mathbf{p}$ of the electron to be conserved, and their energy is Doppler shifted. As illustrated in Fig. 9.10(a), in the laboratory frame, their center-of-mass moves with velocity $\mathbf{v}$ relative to the laboratory frame. Asuming $v_1$, $v_2$, $\theta_1$ and $\theta_2$ to be the velocities and directions of two photons relative to the laboratory frame respectively, then,

$$\tan\theta_1 = \frac{\sqrt{1-\frac{v^2}{c^2}}\sin\theta_0}{\cos\theta_0 - \frac{v}{c}} \approx \frac{\sin\theta_0}{\cos\theta_0 - \frac{v}{c}}$$

$$\tan\theta_2 = \frac{\sqrt{1-\frac{v^2}{c^2}}\sin\theta_0}{\cos\theta_0 + \frac{v}{c}} \approx \frac{\sin\theta_0}{\cos\theta_0 + \frac{v}{c}}$$

$$\theta_{12} \sim \tan(\theta_1 - \theta_2) \approx \frac{\frac{2v}{c}\sin\theta_0}{1 - v^2/c^2} \approx \frac{2v}{c}\sin\theta_0 = \frac{P_T}{m_0 c} \tag{9.5.25}$$

The momentum conservation yields the result (9.5.25), due to $v \ll c$. It is illustrated in Fig. 9.10(b).

The Doppler frequency shift in the energy of the annihilation photons measured in the laboratory system is given by

$$\frac{\Delta\nu}{\nu} = \frac{v_L}{c} \tag{9.5.26}$$

where the longitudinal center-of-mass velocity $v_L$ of the pair equals $P_L/2m_0$. Since the energy of a photon is proportional to its frequency, the Doppler shift at the energy $m_0c^2$ is given by

$$E_{1,2} = \frac{\frac{1}{2}E_T(1 \pm \frac{v}{c}\cos\theta_0)}{\sqrt{1-\frac{v^2}{c^2}}} \sim \frac{1}{2}E'_T\left(1 \pm \frac{v}{c}\cos\theta_0\right)$$

and

$$E'_T \sim E_T = 2E_0 = 2m_0c^2 - \text{binding energy} \sim 2m_0c^2$$

$$E_{1,2} \approx m_0c^2 \pm \frac{cP_L}{2} = E_0 \pm \Delta E\,. \tag{9.5.27}$$

This shows that the line shape of the annihilation radiation reflects the momentum distribution of electrons in metals.

In Fig. 9.10(b), given a typical atomic momentum $\mathbf{p}(|\mathbf{p}| \sim 10^{-2}mc)$, the angle between the two photons deviates from 180° only by a few milliradians. A momentum $\mathbf{p}$ corresponding to one atomic (momentum) unit (i.e. $|\mathbf{p}| = 1$ a.u., $m = 1$, $c = 137$) produces an angular deviation of 7.297 mrad., when $\mathbf{p}$ is perpendicular to the $2\gamma$ axis, and an energy shift of $\pm 1.86$ keV (from 511 keV) when $\mathbf{p}$ is parallel to the $2\gamma$ axis.

The angular correlation between the photons, measured in 1D and 2D ACAR experiments are essentially the one and two dimensional projection of the momentum density distribution $\Gamma(\mathbf{p})$ of the photon pairs.

$$N(p_y, p_z) = \text{const.} \int_{-\infty}^{+\infty} \Gamma(\mathbf{p}) \mathrm{dp_x} \tag{9.5.28}$$

$$N(p_z) = \text{const.} \int_{-\infty}^{+\infty} \int_{-\infty}^{+\infty} \Gamma(\mathbf{p}) \mathrm{dp_x} \mathrm{dp_y} \,. \tag{9.5.29}$$

### 9.5.8. *Examples of Momentum Density Calculations and Experiments*

#### i. Fermi-surface Measurements

Using the high-resolution long-slit apparatus, Stewart's group reported Fermi diameter in the [100], [110] and [111] directions of Li at temperatures just above

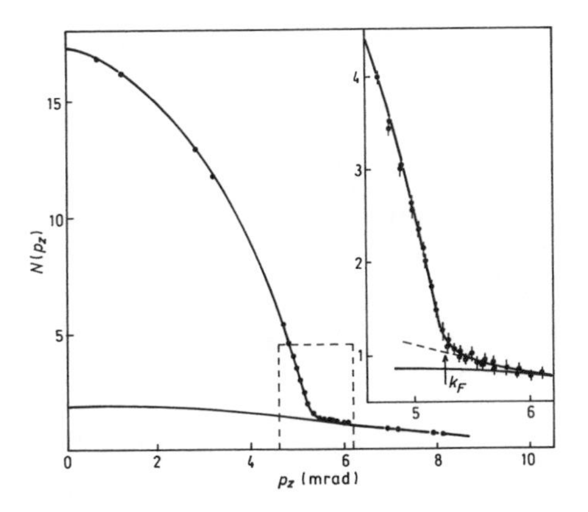


Fig. 9.11. High-resolution 1D ACAR curve in Mg at 4.2K, $p_z$ along the $c$-axis, by Kubica and Stewart (1975); $\Delta p \sim 0.16$ mrad. Inset shows $N(p_z)$ in the region of the Fermi momentum $p_z = k_F$.

the martensitic transition (78K) (Berko, 1983). Their data show about a 2% departure from the free-electron sphere, with small bulges in the [110] direction. Figure 9.11 shows 1D ACAR from a Mg crystal at 4.2K to exhibit the high precision required around the Fermi momentum (Kubica and Stewart, 1975)

ii. 2D ACAR Al results

Figure 9.12 shows $N(p_y, p_z)$ for Al with crystal orientation indicated in the figure (Berko *et al.*, 1977). They obtained good fits with OPW and APW computations.

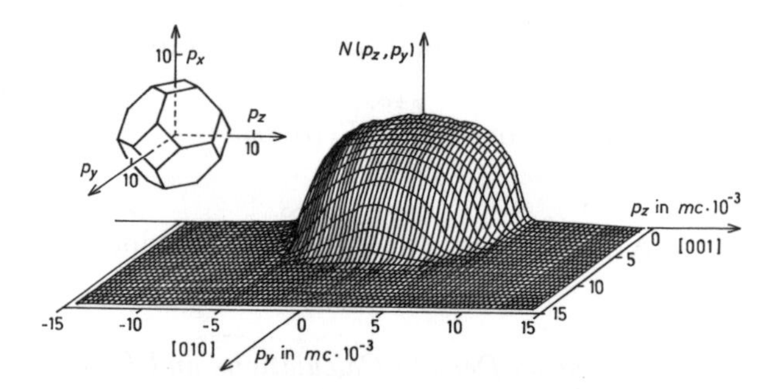


Fig. 9.12. 2D ACAR surface of Al by Berko *et al.* (1977). The orientation of the crystal is illustrated in terms of the Brillouin zone in the inset. Each crossing of lines is an independent measurement. Sample at 100K.

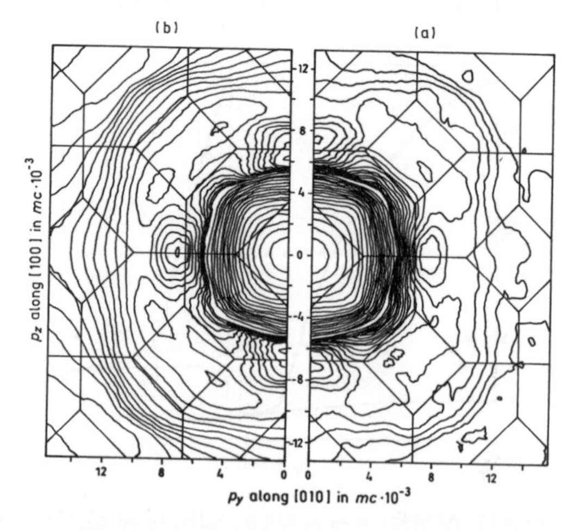


Fig. 9.13. Contour map (Berko, 1983) of the experimental (a) vs. theoretical (APW) (b) $N(p_y, p_z) - R(p_y, p_z)$ for Cu, with $p_x$ along the [100] direction.

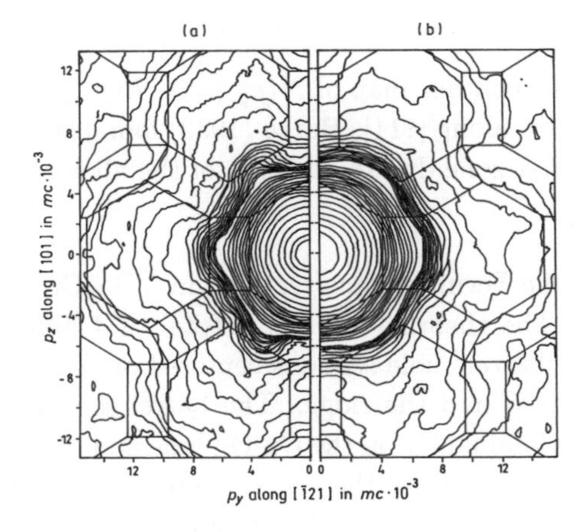


Fig. 9.14. The contour map of $N(p_y, p_z) - R(p_y, p_z)$ for (a) Cu and (b) Cu-30at.% Zn, with $p_x$ along [111]. $R(p_y, p_z)$ is a smooth, rotationally symmetric surface. (Berko, 1983)

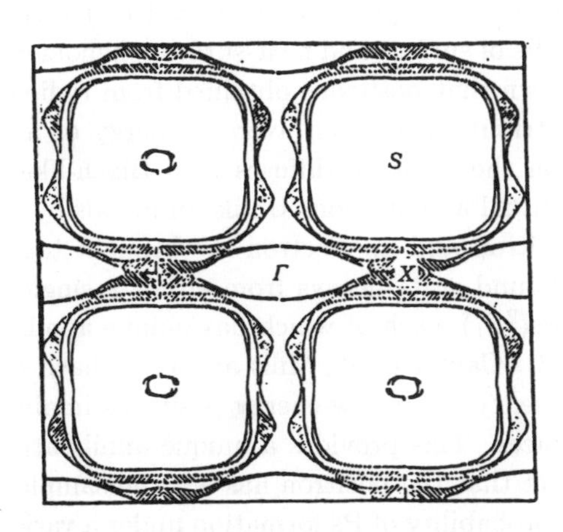


Fig. 9.15. The Fermi-surface sections in $YBa_2Cu_3O_7$, projected onto the $k_z = 0$ plane. The hatched regions indicate all the points in the projected Brillouin zone where there is a Fermi surface at some value of $k_z$ [110]. (after West, 1995)

### iii. Copper

Figure 9.13 shows the contour map of the expermental (a) vs. theoretical (APW) (b), $N(p_y, p_z) - R(p_y, p_z)$ for Cu, with $p_x$ along the [100] direction. (Berko, 1983).

iv. Disorder alloys

Figure 9.14 shows the contour map of $N(p_y, p_z) - R(p_y, p_z)$ for (a) Cu and (b) Cu-30at.% Zn, with $p_x$ along [111]. Berko (1983) found the clear growth of the FS features in the first zone due to addition of the alloy element Zn.

v. Compounds

Figure 9.15 shows the Fermi-surface sections in $YBa_2Cu_3O_7$, projected onto the $k_z = 0$ plane. The theorists predict a total of 4 different Fermi-surface sections: two large "barrel" hole sections centered on $S$, a further and smaller closed-hole pocket at $S$ and the ridge. The barrel sections arise from bands whose states are predominantly on the Cu-O plane and the remaining two from states on the Cu-O chains. Both the general depression around $S$ and the lack of any obvious discontinuities noted by Berko *et al.* (1991) and West (1995) are the effects of the positron preference for the Cu-O chain.

## 9.6. *Experimentation with Low-Energy Positron Beams*

A high flux of monoenergetic positrons ($\triangle E < 1$ eV) with variable energy is most useful as a probe of surface and near-surface phenomena ($< 10^4$ Å). From beta-decay it is known that positrons obtained from radioactive sources have a broad energy distribution with an average energy of a few hundred keV. After thermalization the positron diffuses in a Bloch-like state in a defect-free metal until eventual annihilation. Unlike in metals by direct annihilation, in vacuum, if a positron and an electron are brought together, positronium (Ps) forms. This bound state decays from either a singlet state, p-Ps($^1S_0$), or triplet state, o-Ps($^2S_1$), each of which has unique annihilation characteristics (see Lynn, 1983). Canter *et al.*, Mills and Lynn have shown that Ps does form with high efficiency, when low-energy positrons impinge on metallic and semiconductor surfaces. This provides a unique annihilation signature which enables one to detect that the positron has left the sample. One can use this result to study the probability of Ps formation under a variety of experimental conditions with respect to sample temperature, incident-positron energy and angle (relative to the surface normal), surface conditions and detect concentration. The dominant processes in the interaction of thermalized positrons at a metallic surface have been shown to be (i) localization of the positron in a surface state (annihilation or Ps state). (ii) direct re-emission of the positron from the metal into the vacuum (iii) re-emission into the vacuum as Ps (iv) reflection of the positron wave from the surface potential back into the metal. Other processes are less probable.

### 9.6.1. *Positron Diffusing Back to Metal Surface*

The probability of the positron diffusing back to the surface is dependent on the implantation depth beneath the surface as well as the positron diffusion constant ($D_f$) and the effective annihilation rate in the lattice. The diffusion behaviour of a thermalized positron is determined mainly by scattering from acoustic phonons.

Positron and Ps interactions with a surface affords a new quantitative method to examine surfaces. The positron's sensitivity to lattice defects can provide unique information on defects residing near a surface or at interfaces. Central to these measurements is an accurate determination of the fraction (F) of those positrons which form Ps while leaving the surface of the sample. The reader is referred to the survey by Lynn (1983) for details related to the determination of the Ps fraction.

### 9.6.2. *Positron Trapping at Monovacancies Near a Surface*

Lynn *et al.* (see Lynn, 1983) express the diffusion length $L_+$ in terms of an energy, $E_0$ of the incident-positron; then $L_+ = \sqrt{D\tau_{\text{eff}}} = AE_0^n$. After some calculation, the Ps fraction $F$ is given by

$$F = \frac{f_0}{1 + (E/E_0)^n} \tag{9.6.1}$$

where $f_0$ is the branching ratio which includes the term found from the radiative boundary condition. In this derivation no effect from defect trapping or localization in the surface state has been included (see Lynn, 1983).

In the work of Kreuzer *et al.* (1980) a series of rate equations including positron trapping and detrapping at thermally generated vacancies in the bulk, trapping in the surface state, direct Ps formation and annihilation from these states has been given and solved in terms of the experimentally varied parameters, namely $F$ vs. incident-positron energy and sample temperature.

Figure 9.16 represents the Ps fraction for Al(110) as a function of incident-positron energy at various sample temperatures. The effect of positron trapping at thermally generated monovacancies can be observed at implant energies $>$ keV and at higher sample temperatures by the large decrease in the fraction of the positrons which diffuse back to the surface and form Ps, or by the change in the curvature demonstrated by solid lines in Fig. 9.16.

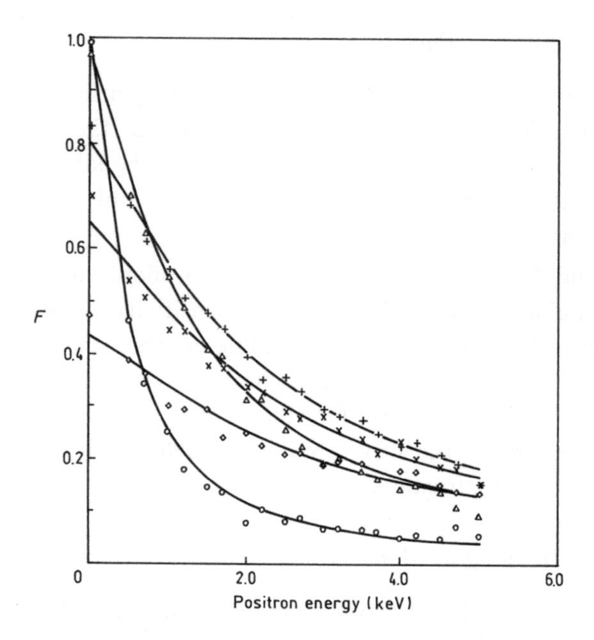


Fig. 9.16. The Ps fraction for Al(100) is shown as a function of incident-positron energy at various sample temperatures (K): ◇ 323, × 448, + 498, △ 648, ○ 873. The solid lines are the best fit of Eq. 9.6.1 with $n = 1.35$ through the data. No impurities were observed with AES* measurements and a sharp LEED pattern was observed. The statistical-error bars are approximately the size of the data points. *AES is Auger electron spectroscopy. (After Lynn, 1983)

### 9.6.3. *Defect Profiling with Positron-Beams*

The investigation of defects close to the surface by monoenergetic positron beams has been applied to a wide variety of problems, such as radiation damage, thin films and epitaxial layers. Within a decade, there have been many studies of Doppler-broadened annihilation radiation (DBAR). Various computer codes have been developed to evaluate the defect profile from the observed DBAR data (see Schultz *et al.*, 1990).

At München and Tsukube, pulsed positron beams have been developed, which enable also positron lifetime (PLT) studies of defect profiles close to the surface. Pulsed positron beams provide much more detailed results than with continuous beams. A brief review by Kögel (1995) was presented on the 10th International Conference on Positron Annihilation, (Proc. ICPA-10). Studies on natural oxide layers on V, Nb, Ta; defects close to the surface in hydrogen-implanted graphite; in nitrogen-implanted and fatigued titanium and in laser-shocked nickel have been performed. Puska *et al.* (1997) studied this problem two years later.

## 9.7. *Summary*

In investigating defects, positron annihilation is an important tool to study vacancies, dislocations, grain boundaries and surfaces in metals. Compared with other atomic-scale probes, the positron technique has a firm theoretical basis though its results are not always easy to interpret.

Positron-defect interactions are well established in metals. The trapping rate is related to the defect concentration, whereas the annihilation characteristics of trapped positrons give information on the nature and geometry of the defects. The positron lifetime is a powerful parameter; the trapped-positron lifetime reflects the three-dimensional open volume of a defect.

The quantitative interpretation of positron annihilation experiments is coupled to progress in describing the electronic structure. Because of its indirect nature, the positron probe requires a solid theoretical framework and good computational method, both of which are now available. The technique is best used in conjunction with other condensed-matter probes and powerful computational methods.

The electronic structure of defects is the deeper layer intrinsic quality of the mechanical property of materials. With the knowledge of positron-defect interaction, the relationship of the phenomenon (mechanical property) with the essence (electronic structure) will not doubt be understood more deeply in the future.

# Chapter 10

# Stretched Chemical Bonds, Electron Correlation and Extended Defect Propagation

In the earlier Chaps. (2–9), defects and mechanical properties of metallic materials including fracture have been discussed together with the elastic strain field of dislocations (Chap. 3), phonon processes (Chap. 5), electronic structure (Chap. 6), interatomic forces (Chaps. 7 and 8) and positron annihilation studies on defects (Chap. 9). For the metallic materials considered in this book, the plastic deformation processes always occured in degradation and failure of materials. These processes are now studied in relation to other physical processes (electric,* thermal, and etc.). Studies on electronic structure and interatomic forces are important for deeper understanding of mechanical properties of metallic materials (see Chap. 2). Even in the quantitative analysis of a fracture surface with the concept of fractals (Chap. 4), interatomic forces and electronic structure are also necessary intrinsic ingredients (see Sec. 4.17.3). Below, particular emphasis will be placed on chemical bonding in relation to electronic structure and mechanical properties (see also Sec. 8.6).

## 10.1. *Roughness and Toughness of Metals and Metallic Alloys*

As Mandelbrot and coworkers (1984) have pointed out, fracture surfaces are self-affine and exhibit scaling properties on two or three decades of length scales (see, for example, Daguier *et al.*, 1996). The roughness index found in early studies lay around 0.8 and it was conjectured that this could be a 'universal value' i.e. material-independent and also not sensitive to the fracture mode

*For example, D. Schaible *et al.* (Phil. Mas. Lett. *78*, 121, 1998) have discussed the relation between residual resistance and plasticity of high purity NiAl single crystal.

(Bouchaud *et al.*, 1990). Later work has shown that fracture toughness, while not correlated with the above roughness index, is related to relevant length scales measured on the fracture surfaces (E. Bouchaud and J. R. Bouchaud, 1994).

For the metallic materials considered in this book, significantly smaller exponents are determined through scanning tunneling microscopy (see Appendix A2.5), i.e. for length scales of the order of nanometers. Thus, the roughness index for fractured tungsten (regular stepped region) is reported by Milman *et al.* (1993) to be $\sim 0.6$, while that for the semimetal graphite is near to 0.5 (Milman *et al.*, 1994). Also, low cycle fatigue experiments on steel yield a roughness index $\sim 0.6$ (McAnulty *et al.*, 1992). Subsequently, it was demonstrated that a different roughness index appropriate to small length scale could be observed with scanning electron microscopy (SEM) (Bouchaud and Navéos, 1995) which can associated with 'quasistatic' fracture regime. While this smaller roughness index was found from SEM to be $\sim$ 0.4-0.5, subsequent experiments using atomic force microscopy have been reported on a $Ti_3Al$-based alloy, for which the fracture surface has been studied by both the atomic force microscope and the scanning electron microscopy (Appendix 2.5) and SEM (Dagiuer *et al.*, 1996). Results from the two techniques agree quantitatively. Two fracture regimes were observed by Daguier *et al.* and it was demonstrated that the roughness index characterizing the regime of small length scales was 0.5. These same workers note that the length scales fractal domain extends over five decades of length scales.

Models have emerged which distinguish two regimes, such as mentioned above. There is a crossover, at some particular length scale, from 'quasistatic' to dynamic behavior. Daguier *et al.* (1996) note that the existence of these two regimes, with a crossover length decreasing with increasing crack velocity, has also been observed in the molecular dynamics simulations carried out by Nakano *et al.* (1995). For a lattice dynamical model of crack propagation, see Sec. 10.5.4.

Of particular relevance in the present context is that, as Daguier *et al.* (1996) note, the 'quasistatic' regime has only been observed in metallic materials (see Plouraboué *et al.*, 1996). Hence Daguier *et al.* (1996) infer that plasticity might be important for the formation of this regime, either due to different fracture mechanisms which are indeed involved within the plastic zone, or because plastic dissipation may reduce crack propagation velocities in the region of small length scales.

What Daguier *et al.* (1996) have convincingly demonstrated is first that the results from AFM are quantitatively in agreement with standard SEM. By simultaneous use of the two techniques, they have been able to observe the 'universal' fracture regime, with a roughness index of 0.8, over about five decades of length scales in their $Ti_3Al$-based metallic alloy. Importantly, at sufficiently small length scales, a regime has been found, extending over some three decades, and this has permitted a roughness index of 0.5 to be reliably extracted.

Two fracture regimes observed by Daguier *et al.* (1996) strongly support the concept of multirange fractals introduced in Chap. 4 since the geometrically scaling found much earlier (see Sec. 4.14). Their results also indicate the importance of plasticity in the fracture of metallic materials.

One may well enquire why many sorts of materials have been shown to exhibit scaling properties over only two or three decades of length scale? The reason may be that materials usually possess complicated microstructures, some of which form fractal structures which may influence the roughness index of the main fractal structures considerably. According to the concept of multirange fractals introduced in Chap. 4, different fractal structures in various ranges of length scales may partially overlap one another. The overlapping (or crossover) regions cause the data to deviate from a linear plot (see Fig. 4.15). From this point of view, the effect of microstructure in the description of fractured surfaces by fractals cannot be ignored.

The debate on the universality and specificity of roughness index is still going on (see Milman *et al.*, 1993; Hansen *et al.*, 1993) at the time of writing. However, we believe there is not by any means a real conflict between current viewpoints. The generality, even in the same universal class (say dynamic instability, complex mode crack propagation, intergranular cracking etc. i.e. less dependent of material type) may lie low in the specificities of materials. As we have explained in Sec. 4.17.1, the measured fractal dimension is the synthesis of various elementary fractal structures including material dependent and less dependent ("universal") fractals. In Sec. 4.14, and with particular reference to Fig. 4.16, if two fractals overlap each other in the entire range of length scale, the synthesized effective fractal dimension would lie between these two. The concrete value depends on the fraction of population of the two mechanisms in the material. This effect may reduce the specificity of materials. Let us take an example (see Lung, 1998). If we have two samples of materials A and B with specific fractal dimensions 1.50 and 1.26 respectively, the difference between them is 0.24. This lies outside the experimental error. Now, we have some universal mechanisms to create fractal structures which are less material-dependent [say dynamic instability (see Sec. 4.18.5), oscillatory propagation of a slant crack (see Sec. 4.18.7) and evolution induced catastrophic model (see Sec. 4.18.6)]. Assuming their common fractal dimension value to be 1.1 (see Bai *et al.*, 1994, and Sec. 4.18.7), the synthesized effective fractal dimensions would be $\sim 1.3$ for sample A and 1.18 for sample B. The difference between them is then reduced to 0.12. We recall that range of values of D is between 1-2, which is twice that of the roughness index, 0.5-1. Then, the difference between A and B is reduced further to $\sim 0.06$. This appears to lie within the experimental error. It seems that the fractal dimension of fractured surfaces will not correspond to a universal value.

Moreover, from Eq. (4.17.5)

$$D = 1 - \frac{\ln[G_{Ic}(D, \varepsilon_n)/G_{Ic}(1,1)]}{\ln(\eta/L_0)} \tag{10.1.1}$$

Even with a large change in the sorts of materials considered, the relative change of logarithmic values of the ratio $G_{Ic}(D, \varepsilon_n)/G_{Ic}(1,1)$ might be small (in brittle fracture case, $G_{Ic}(1,1) = 2\gamma_s$, the specific surface energy). If we assume the denominator does not change very much, the observed values of D being in a narrow range can be understood in a semiquantitative way.

In short, the less material dependent mechanism may lie low in the phenomena of various specificities. At large length scales ($r > 0.1 - 1.0$ $\mu$m, see Daguier *et al.*, 1996), the fractal dimension has weaker dependence on the material; and at smaller length scales ($r < 0.1$ $\mu$m), the fractal dimension is more strongly dependent on the material.

The results observed by Daguier *et al.* (1996) strongly support the importance of plasticity in the fracture of metallic materials. Milman *et al.* (1994) showed the material-structural-dependent character of the roughness index (or local fractal dimension). Both of these groups arrived at the same conclusion that for metallic materials considered in this book, dislocation processes are important.

Since the pioneering studies of Vitek and coworkers (see Vitek, 1994, 1995), there have been many investigations by computer simulation of dislocation core structure by means of N-body potentials. It seems true to say that interatomic forces are a decisive factor in dislocation core structures. It also seems that much progress in studies of mechanical properties with electronic structural theory is to be expected though some difficulties for a realistic representation of an extended defect such as a dislocation are still awaiting a decisive solution at the time of writing.

## 10.2. *Perfect Crystal Properties: Elastic Constants and Melting Points*

### 10.2.1. *Bonding Energies of Metals*

Parallel to many numerical computer *ab initio* calculations, Rose and Shore (1995; 1991) studied the qualitative picture of the trends in the cohesive energies of elemental metals and elastic constants of the transition metals from a uniform electron gas. Their model includes the additional electron-ion interaction (i.e., that part not accounted for by jellium) by adding an electron potential that is constant inside pseudojellium and zero outside. Their pseudojellium model requires an adjustable parameter (fixed by the chemical potential) and gives reasonably good agreement for the surface properties and cohesive energies of the simple metals. The only input parameter is the average electron density. They identify the crucial variables that determine the bonding energies of the metals as the size of the atom and the electron density at the cell boundary. Paxton (1995) and collaborators (1991) used the local density approximation to density functional theory to reexamine the 'theoretical strength' of metals. They calculated ideal twin stress in five b.c.c. transition metals, and in Ir, Cu and Al.

### 10.2.2. *Melting Energies*

Based on analyses of data of melting points, of 14 kinds of metals, Lung and Wang proposed a qualitative electronic structural criterion for the strength of atomic bonding (Lung and Wang, 1961). This involved a function $f(k)$ which is, in essence, a product of a group velocity, $dE/dk$, which depends on direction, and density of states. $f(k)$, to be physical is to be evaluated at the Fermi momentum. Relevant data from Lung and Wong (1961) is collected in Tables 10.1 to 10.3. Note that group velocity is directionally dependent (see below). Tables 10.2 and 10.3 show that $f(k)$ reflects the strength of atomic bonding.

This present criterion for the strength of atomic bonding, synthesizes two factors, group velocity and $n_{\text{eff}}$ instead of previous criteria which consider one

Table 10.1. The electronic structure and atomic bonding (Lung and Wang, 1961).

| elements | melting points (°C) | $m^*/m$ (related to density of states) | $n_{eff}$ (related to group velocity) | $N_a$ ($\times 10^{-24}$) | $E_F$ (eV) | $\sigma/M\Theta^2$* ($\times 10^2$) |
|---|---|---|---|---|---|---|
| Cu | 1084° | 1.012, 1.5 | 0.37 | 0.0842 | 7.1, 7.0 | 9.1 |
| Ag | 960° | 0.992 | 0.89 | 0.0586 | 5.51 | 12.4 |
| Au | 1063° | 0.994 | 0.73 | 0.0589 | 5.51 | 8.1 |
| Zn | 419.5° | 0.8–0.9 | – | 0.0662 | 11 | 6.1 |
| Cd | 320.9° | 0.75 | 2.4 | 0.0463 | | 4.5 |
| Hg | −38.9° | 1.8–2.2 | 2.1 | 0.0407 | | 3.4 |
| Be | 1300° | 0.46 | – | 0.1229 | 14.8, 13.5, 13.8 | 2.0 |
| Mg | 650° | 1.33 | – | 0.0431 | 7.3, 9.0, 6.2 | 8.1 |
| Fe | 1535° | 12 | – | 0.0847 | 4.4 | 1.14 |
| Co | 1480° | 14 | – | 0.0890 | 5.8 | 1.7 |
| Ni | 1455° | 18 | – | 0.0913 | 4.7 | 1.9 |
| Li | 180° | 1.4 | 0.55 | 0.0463 | 4.2, 4.1, 3.7 | 12.9 |
| Na | 97.8° | 0.98 | 0.87, 1.1 | 0.0254 | 3.5, 3.0, 2.5 | 24 |
| K | 63.6° | 0.93 | 0.97, 0.75 | 0.0134 | 1.9 | 15.9 |

* $\sigma$ is the specific electrical conductivity, $m$ is ionic mass and $\Theta$ is Debye temperature.

Table 10.2

| $f(k)_{k=k_{max}}$ | melting point |
|---|---|
| Ag > Au > Cu | Cu > Au > Ag |
| Hg > Cd > Zn | Zn > Cd > Hg |
| Mg > Be | Be > Mg |
| Ni > Co > Fe | Fe > Co > Ni |

Table 10.3

| $f(k)_{k=k_{max}}$ | melting point |
|---|---|
| Zn > Cu | Cu > Zn |
| Cd > Ag | Ag > Cd |
| Hg > Au | Au > Hg |

factor only, e.g. density of energy states, Fermi energy of conduction electrons. It is worthwhile to pointing out that progress in positron annihilation, photon emission and Compton scattering techniques may provide information on momentum distribution of electrons, and effective number of free electrons. It is to be hoped that this qualitative criterion can be checked with new experimental results. Using this qualitative criterion, one can explain the reason why crystals usually glide along the highest density direction in the highest density crystal plane. Let us see what will happen on the Brillouin zone when a shear strain occurs. The lattice vector after a shear strain $\gamma$ is given by

$$\mathbf{a}_1 = a\begin{pmatrix}1\\0\\0\end{pmatrix}, \quad \mathbf{a}_2 = a\begin{pmatrix}0\\1\\0\end{pmatrix}, \quad \mathbf{a}_3 = d\begin{pmatrix}\gamma\\0\\1\end{pmatrix} \tag{10.2.6}$$

where $d \neq a$, and the reciprocal lattice vectors are given by

$$\mathbf{b}_1 = \frac{1}{a}\begin{pmatrix}1\\0\\-\gamma\end{pmatrix}, \quad \mathbf{b}_2 = \frac{1}{a}\begin{pmatrix}0\\1\\0\end{pmatrix}, \quad \mathbf{b}_3 = \frac{1}{d}\begin{pmatrix}0\\0\\1\end{pmatrix} \tag{10.2.7}$$

where $\gamma = \frac{x}{d}$, and the first Brillouin zone is composed by the following eight planes according to the conditions $\mathbf{k} \cdot \mathbf{h}_b = \pi h_b^2$,

$$k_x - \gamma k_z = \pm\frac{\pi}{a}(1+\gamma^2)$$

$$\frac{1}{a}k_x + \frac{1}{d}k_z - \frac{\gamma}{a}k_z = \pm\left[\frac{\pi}{a^2} + \frac{\pi}{d^2}(1 - \frac{\gamma d}{a})^2\right]$$

$$k_y = \pm\frac{\pi}{a}$$

$$k_z = \pm\frac{\pi}{d}\,.$$

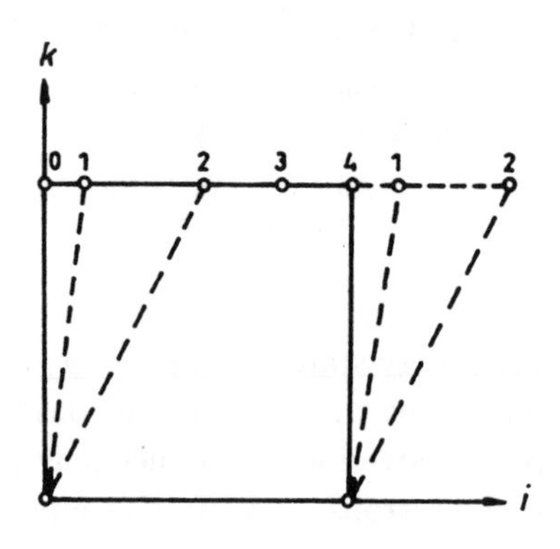


Fig. 10.1(a)

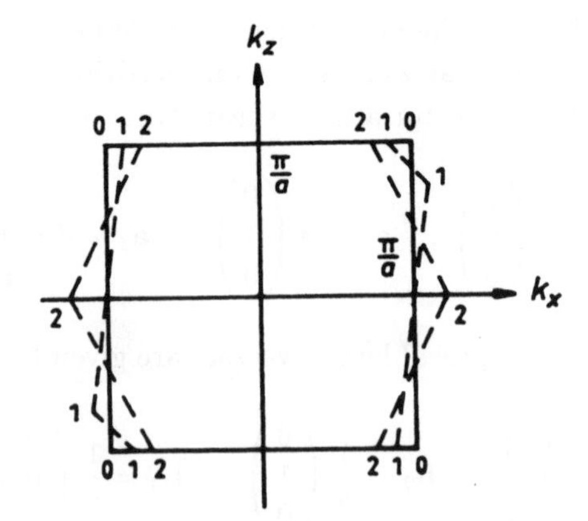


Fig. 10.1(b)

Assuming that $a = d$, Fig. 10.1(a) shows the primitive cell of a simple cubic lattice in two dimensions before and after deformation; 10.1(b) shows the first Brillouin zone of the simple cubic lattice in two dimensions before and after deformation. The shape of the first Brillouin zone has been changed by shear deformation. The distribution of state of electrons in $k$-space might be changed under transitions in $k$-space to a new equilibrium state due to interactions with the Brillouin zone boundary.

Now, we discuss the case of $a \neq d$. Assuming that atoms will glide when they displace to a critical value, $\beta a$ (i.e. $\gamma = \frac{\beta a}{d}$). We analyse the degree of the distortion of the Brillouin zone shape. We calculate the interaction between $x - Z$ plane and planes showed by the first and second equations of the first Brillouin zone above.

$$\begin{aligned} k'_x &= \pm\frac{\pi}{a}\left[1 + \beta(1-\beta)\frac{a^2}{d^2}\right] \\ k'_Z &= \pm\frac{\pi}{d}(1-2\beta)\,. \end{aligned} \tag{10.2.1}$$

The ratios of $k'_x/k_x$ and $k'_Z/k_Z$ are given by

$$\begin{aligned} \frac{k'_x}{k_x} &= 1 + \beta(1-\beta)\frac{a^2}{d^2} \\ \frac{k'_Z}{k_Z} &= 1 - 2\beta\,. \end{aligned} \tag{10.2.2}$$

Equation (10.2.2) reflects the distortion of Brillouin zone. The value of $\frac{k'_Z}{k_Z}$ is independent on $a/d$, and however, $\frac{k'_x}{k_x}$ is dependent on $a/d$. We may see:

(i) If $a$ is smaller, $\frac{k'_Z}{k_Z}$ is nearer to unity. That means smaller distortion of the Brillouin zone. The atom is easy to glide along the direction in the line of closest packing ($a$ is smallest).

(ii) If $d$ is larger, $\frac{k'_Z}{k_Z}$ is nearer to unity. That also means smaller distortion of the Brillouin zone. The slip takes place along crystallographic planes of the closest packing ($d$ is the largest).

Li and Lin (1982) calculated the second and third order elastic constants of seven simple metals (Li, Na, K, Cs, Al and Pd) with the one-parameter Heine-Abarenkov model potential with the Hubbard-Sham dielectric function. The two adjustable parameters are determined from the experimental data of the elastic constant $C_{44}$ and the lattice constant of the crystal at 0K and zero applied pressure. The calculated results are in good agreement with experiments. Again, Li (1986, 1987) calculated the third order elastic constants and the pressure derivatives of the second order elastic constants of noble metals (Cu, Ag, Au) at 0K with the model of pseudopotential and the overlap energy of noble metals. This shows that the results for gold are closer to the experimental data than those of other authors.

### 10.2.3. *Interatomic Forces in Solid Solution of Transition Metal Alloys*

In transition metals with partially filled d-band the attractive part of the total energy originates principally from the formation of this band (Friedel, 1969; Pettifor, 1983). Within the tight binding framework the total energy of the system can be divided into a repulsive pair-potential and an attractive bond energy part (Finnis *et al.*, 1988; Sutton *et al.*, 1988). For a chosen form of the hopping integral the bond energy of the system can be found. This method has to be carried out in the **k**-space representation and thus the system must be three-dimensionally periodic. Several approximate schemes have been developed in which the total energy is evaluated in the real space (Carlsson, 1990).

Pettifor *et al.* (Pettifor, 1981; Aoki and Pettifor, 1994) advanced the bond-order potentials, which are one such scheme calculated in coordinate space. The concept of bond order is equal to the difference between the numbers of electrons in the bonding and antibonding states corresponding to the respective orbital. In the framework of bond order potentials, for a given configuration of atoms, it can be done numerically using the recursion scheme. Aoki and Pettifor (1994) used the bond order potentials to study d-bonded transition metals.

Another different approach by Cyrot-Lackmann (1967), and Ducastelle and Cyrot-Lackmann (1970), uses the scalar moments referred to its centre of gravity. A real space description of the energy of transition metals that includes the dependence on angles between pairs of bonds, has been proposed by Carlsson (1990). If only a finite number of moments are taken into consideration and a certain approximation to the shape of the local density of states (LDOS), the bond energy can be evaluated.

For solid solutions of transition metal alloys, Dehjar *et al.* (1957) and Gou *et al.* (1962) made some qualitative analyses. They discussed the strength of interatomic forces in transition metal alloys. They analysed experimental data on sublimation energy, melting point, elastic modulus, coefficient of thermal expansion, bulk modulus, Debye temperature and activation energy of diffusion of Fe-C, Fe-Mo, Fe-W, Fe-V, Fe-Mn, Fe-Ni, Fe-Cr, Ni-Mo, Ni-Mn, Ni-Fe, Ni-Zr, Ni-Nb, Ni-Cu, Co-Cr, and Co-Ni alloys. Gou *et al.* (1962) found that Cr, Mo, W, Mn, and Nb increase the interatomic forces in Fe-alloys, but Ni decreases it. The role of V is not clear. They also found that Cr, Mo, W, Mn, Nb, Fe, and Zr increase the interatomic forces in Ni-alloys, but Cu decreases it. They pointed out that Cr may strengthen atomic forces in Co-alloys but Ni decreases it. Gou *et al.* (1962) summarized that for Fe based alloys, the strength of interatomic forces would increase if elements on the left of iron in the Periodic Table are added to form solid solutions in general, and would decrease if elements on the right is added. Furthermore, Gou *et al.* (1962) drew the conclusion that elements having more unpaired electrons than the solvent metal will increase the interatomic force and that elements having less unpaired electrons than the solvent metals will decrease the interatomic force. Moreover, Gou *et al.* (1962) pointed out that there will be a strengthened or weakened atomic cluster formed when a second element atom is added in the base metal. If the second element increases the interatomic force, the strengthened atomic cluster will become more and more as the concentration of the second element increases. Thus, the macroscopic properties will increase. The concept of clusters they proposed in 1960's may be one of the earliest attempts to study clusters after Taylor *et al.* (1933) on the growth of a crystal by adding of successive atoms to a bcc lattice for up to eight sodium atoms and up to five copper atoms (de Heer *et al.* 1987). Another interesting phenomena is the non-monotonic change of properties in extremely dilute solid solution (0.001%–0.1%at.): Borovski (see Guo *et al.*, 1962) found that it is a general phenomenon in transition metal alloys, say Fe-W, Fe-Cr, Fe-Mo, Fe-Al alloys and etc. Based on some experimental data on Al-Ag, Al-Cu, Al-Zn, Al-Mg; Cu-Zn, Cu-Ga, Cu-Ge, and Cu-Cs alloys, Gou *et al.* pointed out that it is not only a general phenomenon in transition metal alloys but also in non-transition metal alloys. Then, it is universal. They explained this phenomenon as follows: As a second element (or impurity atom) is added into the base metal, the interatomic force is strengthened by two mechanisms: One is the unpaired electrons and the other is the polarization of the second element. The latter effect increases the atomic force till the polarization spheres of the impurity atoms overlap each other and then compensate a part of this effect. The maximum effects appear when the polarization spheres are just in contact with each other.

### 10.2.4. *Bonding Characteristic of $Ni_3Al$*

$Ni_3Al$ and its alloys have aroused interest in potential high temperature structural applications, but their brittleness is the most formidable obstacle to fabrication and use. Using discrete variational $X\alpha$ (DFT-$X\alpha$) method, Meng *et al.* (1993, 1994) studied a cluster of monocrystalline $Ni_3Al$. Their calculations show that Ni-Ni bond is very similar to Ni-Al bond. The ductility of monocrystal is due to the similarity of Ni-Ni bond to Ni-Al bond. The brittleness of polycrystal is attributed to the weakening of directional covalent bonding at grain boundaries. The addition of ternary elements (Pd, Ag, Cu, and Co) substituting for the Ni sites leads to the increase of delocalized bonding electrons, the decrease of the covalent bond directionality and then the increase of the ductility.

Aoki and Izumi (1977) have shown that an increase in ductility of $Ni_3Al$ could be obtained by B doping. Liu *et al.* (1989, 1990a, 1990b, 1992) and Wan *et al.* (1992a) have demonstrated that the elongation and fracture of $LI_2$-type ordered intermetallic compounds are commonly very susceptible to the test atmosphere. Losses of tensile ductility have been observed for a number of polycrystals with $LI_2$-type structure which were injected by cathodic hydrogen charging or exposed to hydrogen gas. Wan *et al.* (1992b) have reported that the ductility of $Ni_3Al$ with low boron content (120 wppm) was sensitive to environmental embrittlement in the presence of water vapour. Wang *et al.* (1992) calculated the localized electronic structure of boron-impurity-vacancy (B-V-B) complex in Ni for four models with different configurations by use of the multiple-scattering $X\alpha$ method. Results are presented for the total energy, density of states and local charge transfer. Comparisons are made among the four models. The calculations indicate a strong binding between the boron atom and the vacancy. Boron-containing alloys with a B-V-B complex show a strong potential for charge redistribution. Deng *et al.* (1995) using positron annihilation technique studied the behaviour of hydrogen in $Ni_3Al$ alloys as well as the interaction of hydrogen and boron with defects in $Ni_3Al$ alloys. Their results demonstrate that the more the hydrogen atoms segregate to defects, the shorter the mean lifetime. The magnitude of the decrease of $\tau$ is about 10 ps in B-free $Ni_3Al$ and only 4–5 ps in B-doped $Ni_3Al$. This indicates that the filling effect of hydrogen into defects in $Ni_3Al$ alloys is related to B content. B atoms which sink into defects such as vacancies, dislocations and grain boundaries effectively impede the entrance of protons and/or hydrogen atoms into the defects. More valence electrons are presented in the region of defects. This gives some increase of positron mean lifetime. The increase in ductility of $Ni_3Al$ obtained by B doping involves the contributions of both effects: B strengthening the metallic bonding cohesion in grain boundaries and B suppressing hydrogen embrittlement.

### 10.2.5. *The Empirical Electron Theory of Solids (EET) and The Brittleness of σ-Phase in Fe-Cr Alloys*

Based on Pauling's electron theory of metals, Yü (1978, 1981) proposed an empirical electron theory of solids and molecules. The empirical electron theory separates the charge distribution in crystals into three kinds of electrons, the dumb pair electrons,* the covalent bond valence electrons and the lattice valence electrons, and modifies the relation between the number of equivalent bonds and the experimental covalent bond lengths. He pointed out that the dumb pair electrons as well as the covalent bond electrons are made of *d* electrons, whereas the lattice electrons are made of *s* and *p* electrons. He reported that more than a thousand crystalline and molecular structures together with some of their related experimental data and experimentally verified theoretical information, were analyzed for 78 elements of the first six periods in the Periodic table. To the first order of approximation, the author reported that the results appear satisfactory.

Chen *et al.* (1980) calculated the number of dumb electrons, covalent bond valence electrons and lattice valence electrons in $Ni_3Al$ with the empirical electron theory and compared with that calculated from Cooper's (1963) structure diagram of the equidensity curves on surface (110) of outer electron distribution in $Ni_3Al$ (both experimental and theoretical). The calculated results are listed in Table 10.4. From Table 10.4, one may see that the result calculated by the EET agrees well with the result from the experimental data in the first order of approximation whereas that from the theoretical curve by Cooper (1963) is quite different due to its basis of free electron theory for transition metal alloys. On the other hand, Wang *et al.* (1981) applied the EET to the interpretation and analyses of Cu and Ag charge density distribution diagrams obtained by Fong *et al.* (1975) from the bond structures with the pseudopotential method. It was shown that the results by EET and energy band theory

*Y. Guo, R. Yü, R. Zhang, X. Zhang and K. Tao (J. Phys. Chem. B. *102*, 9, 1998) offer the explanation that 'the so-called dumb pair electrons represent either a bonding and an antibonding electron, whose resultant bonding power is mutually cancelled by each after and whose spins are opposite ... It can also represent a pair of nonvalence electrons of opposite spins, which sinks deeply down to the atomic orbit.

Table 10.4. Comparison of distribution in $Ni_3Al$ of the outer electrons by the EET with that from Cooper's data.

| Number of electrons | Dumb electrons pairs | Covalent electrons | Lattice electrons | Outer total electrons |
|---|---|---|---|---|
| E.E.T | 2.860 | 9.710 | 0.430 | 13.000 |
| exp.(Cooper) | 3* | 9.229 | 0.401 | 12.630 |
| Th.(Cooper) | 7.932 | no | no | 13.066** |

* It is an estimated value.

** a number of valence electrons, 5.134 has been added.

agree with each other. Yuan (1985) applied the EET to analyse the brittleness of $\sigma$-phase in Fe-Cr alloys. It was found that there are only four weak bonds, $n_p$ and $n_T$ ($n_p = 0.1584$; $n_T = 0.0946$) connecting atoms between two parallel (001) planes. He reported that it is due to the inhomogeneous distribution of the covalent electrons, and that this is the physical origin of the brittleness of Fe-Cr $\sigma$-phase crystals.

## 10.3. *Morse Potentials and crss of Iron Single Crystals*

The transition-state theory has been applied widely and has had considerable success in treating thermally activated process, especially in describing the movement of point defects. This theory has been also applied widely to the thermally activated movement of dislocations. For a review of these topics see Dorn (1968). Then, the problem of dislocation motion is simplified to calculate the energy of dislocation core structure. Schöck (1980) discussed the limitations of transition-state theory and dynamical theory. He pointed out that the essential assumption of the transition-state theory were that at the saddle point the modes of the activated complex should need two counter contrary requirements. The first one is that the system interacts strongly enough with the temperature bath and stays long enough to acquire thermal equilibrium; and the second one is that the system interacts weakly enough with the temperature bath that a positive velocity is not reversed. The problem, we think, is whether the deformation process is quasi-stationary or not. Conrad and Sprecher (1989) discussed the forces opposing dislocation motion and pointed out that the motion of dislocations in a metal falls into two general categories: (i) At high stresses or high velocities, where the waiting time $t_w$ at a structural obstacle is either zero or considerably less than the running time $t_r$ between the obstacles, the dislocation velocity in this regime is given by a drag mechanism. The drag coefficient $B$ is given by

$$B = \tau^* b/v \tag{10.3.1}$$

where $\tau^* = \tau - \tau_i$; $\tau$ is the applied stress and $\tau_i$ is the long-range internal stress resulting from all crystal defects. (ii) At low stresses or low-velocities where $t_w > t_r$, the thermal activation plays a role in overcoming the obstacle. The average dislocation velocity $v$ is given by

$$v = v_0 \exp\{-[\triangle G(\tau^*)/kT]\} \tag{10.3.2}$$

where $\triangle G(\tau^*)$ is the Gibbs free energy. Three cases of dislocations under high frequency vibration exist: (1) ultrasonic wave propagation (2) high temperature creep and (3) the intermediate case of normal plastic deformation which is more complicated than cases (1) and (2).

Table 10.5 shows the relative values of $t_w(\sim \gamma^{-1})$ and $t_r(\sim v^{-1})$ of dislocation lines for normal tension test and creep in comparison with atomic diffusion. $P_T$(tension) $\ll P_c$(creep) $\ll P_d$(diffusion). The running time $t_r$ is too short for the tension test case. It seems not sufficient to assume that the dislocation line stays long enough to acquire thermal equilibrium in tension test case.

For avoiding the assumption of thermal equilibrium, the thermal activation was described by so-called "dynamical" theory of thermal activated ion developed by some authors. A rather simplified dynamical treatment of unpinning of dislocations can be made when the action of thermal forces was considered. This was first analysed by Leibfried (1957). He considered a dislocation

Table 10.5

| | velocity $\bar{v}$ ($b$/sec) | vibration frequency $\gamma$ (1/sec) | $p = \gamma/\bar{v}(b^{-1})$ |
|---|---|---|---|
| Tension test (disl.) | $\sim 10^4$ | $\sim 10^{10}$ | $10^5$ |
| Creep (disl.) | $\sim 1$ | $\sim 10^{10}$ | $10^{10}$ |
| Diffusion (atom)* | $\sim 1$ | $\sim 10^{12}$ | $10^{12}$ |

* C in $\alpha$-Fe at 40°C.

segment of length $L$ pinned at the ends in the string approximation. Its shape can be represented by a set of linear oscillators. At the centre, the dislocation segment is pinned by a force per unit length $k = -\tau_0 y$ acting over a segment of length $l \ll L$. It has been shown that the maximum displacement is essentially the amplitude of the lowest mode, whereas the higher ones make very little contribution. He obtained

$$\langle y_0^2 \rangle_{av} \approx 2kT/(\pi^2 E/L + 2\tau_0 l) . \tag{10.3.3}$$

The corresponding frequency of the lowest mode is approximately

$$\nu_0 = \nu_D [b^2/L^2 + 4\tau_0 l/\pi^2 \mu L]^{1/2} . \tag{10.3.4}$$

The basic assumption is that a jump takes place when an atom is displaced a critical distance from its equilibrium position.

Similar to this approach, Lung *et al.* (1964, 1965) considered mainly the stretched chemical exchanging bonds with the moving dislocation centre dangling atoms as the pinning force. The formation and disruption of the bonds resist the motion of dislocations. The stress needed to overcome this resistance is the frictional stress. One bond between the dislocation centre atom and the impure atom has been considered as the representative bond to simplify the treatment. This seems to be reasonable. Wang *et al.* (1993) calculated the electronic structure of edge dislocation in iron and showed that the charge redistribution of the dislocation core caused by the local inhomogeneous deformation forms a stronger interaction between the interstitial impurity atom (carbon, nitrogen, etc.) and the edge dislocation than the elastic interaction. Moreover, from Eqs. (1.1.1) and (1.1.2), we may have seen that the interaction energy of a pair of univalent ions is similar to that of having summed the repulsive and attractive interactions with nearest neighbours. Their differences are only between the coefficients. This is quite straight forward for ionic bonding, the effects of more distant ions are usually taken into account by a geometrical correction factor called Madelung constant, which depends on the details of the crystal structure. This concept has been used by Wang *et al.* (1989) to calculate the interatomic potential in transition metal Ni. Let us borrow this concept for a simplified approximate assumption. For covalent bonds, one has no trouble such as with inequivalent ions. They chose a Morse potential function as the interaction between the paired atoms.

$$u_c = D\{e^{-2\alpha(r-r_0)} - 2e^{-\alpha(r-r_0)}\} . \tag{10.3.5}$$

Here $r_0$ is the distance between the dislocation centre atom and the impurity atom under equilibrium condition, while under the action of applied stress, it will be changed to $r$. $D$ is $u_c$ under equilibrium condition, and $\alpha$ is the parameter of the Morse function. These parameters can be calculated from the quantum theory of molecules.

Figure 10.2 shows the schematic figure of the configuration of the edge dislocation core in a simple cubic lattice with an impurity atom just below the dislocation core atom. Like the Peierls model, they chose the edge dislocation model for simplicity. This analysis will not lose the generality of screw and other complicated dislocation cores.

$$\bar{r} - r_0 = \{\overline{[r_0^2 + (x+u)^2]}\}^{1/2} - r_0 \sim \frac{x^2 + \overline{u^2}}{2r_0}\,.$$

Then Eq. (10.3.5) may be written in the following simplified form

$$u_c = D\{e^{-2Ax^2 - 2BT} - 2e^{-Ax^2 - BT}\} \tag{10.3.6}$$

and $A = \frac{\alpha}{2r_0}$, $B = (\frac{\alpha}{2r_0})(\frac{\overline{u^2}}{T})$. The critical resolved shear stress may be written as

$$\sigma_c = S^{-1}\left(\frac{\partial u_c}{\partial x}\right)_{\max} \tag{10.3.7}$$

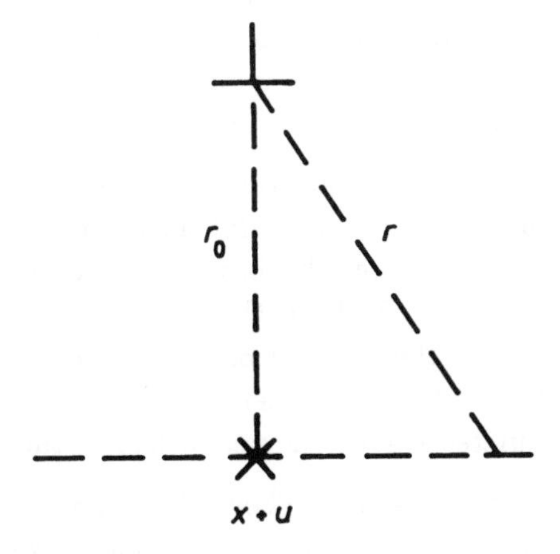


Fig. 10.2

and $S = [\frac{(c_i n)^{\frac{1}{3}}}{b}]^{-1}$ is a constant with the dimension of area, $c_i$ is the atomic concentration of interstitial, $n$ is the number of atoms per unit volume (containing all solute and solvent atoms), and $b$ is the Burgers vector, $T$ is the absolute temperature. The value $x$ of $(\frac{\partial u_c}{\partial x})_{\max}$ is determined by $\frac{\partial^2 u_c}{\partial x^2} = 0$ at any given temperature, i.e.

$$BT = -\ln \frac{2Ax_c^2 - 1}{4Ax_c^2 - 1} - Ax_c^2 \,. \tag{10.3.8}$$

From Eqs. (10.3.6) and (10.3.7), one obtains

$$\sigma_c = 2\left(\frac{x_c}{r_0}\right)\left\{e^{-2Ax_c^2 - 2BT} - e^{-Ax_c^2 - BT}\right\}\alpha D S^{-1} \,. \tag{10.3.9}$$

In Chap. 2, we have explained that the approximation of Eq. (10.3.7) in appropriate range of temperature may be expressed as $\sigma(T) = \sigma^*(0)\exp(-BT)$. This expression has been compared with Lung *et al.*'s experiments on Mo-single crystals and exhibited good agreement. This expression is also similar to later experiments on the temperature dependence of the plastic flow-stress of covalent crystals (Suzuki *et al.*, 1995).

According to quantum theory of molecules and thermal vibration theory of dislocations by Leibfried (1957), the parameters, $A$, $B$, $\alpha$, $D$ can be calculated in simple cases. Thus $\sigma_c$ at any temperature of this model may be calculated from Eqs. (10.3.8) and (10.3.9). However, in general, for common metals, especially for transition metals, it is much more complicated. It is not practical to calculate the absolute values of critical stress for comparison with experimental results. To overcome this difficulty, Lung *et al.* (1966) determined some of the parameters empirically and using this method they have compared the curve of their model with experiments in a semi-quantitative sense.

(1) Within a certain temperature range, by means of the asymptotic relationship $\ln \sigma_c(T) = \ln \sigma_c(0) - BT$ in Chap. 2 and the least square method, the slopes $B$ on the $\ln \sigma - T$ plot have been determined. They are $1.17 \times 10^{-2}(\mathrm{K}^{-1})$ in the case of Edmondson, ($\alpha$-iron) and $8.7 \times 10^{-3}(\mathrm{K}^{-1})$ in the case of Cox *et al.* ($\alpha$-iron).
(2) Assuming that a certain experimental value of crss is fitted with the theoretically calculated result, one may calculate the curve of this model.

Figure 10.3 shows the results. Then, the curves can be calculated by Eqs. (10.3.8) and (10.3.9) and are in good agreement with the experimental results of iron single curve in the whole range of temperatures.

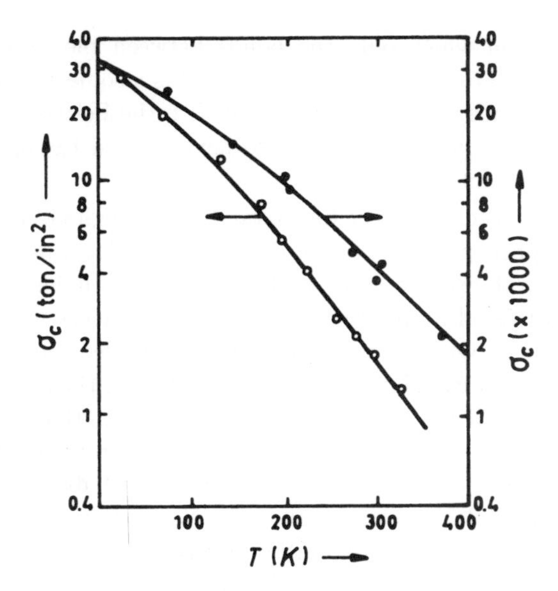


Fig. 10.3. $\sigma_c - T$ curve of iron single crystals (black lines are theoretical curves). (From Lung *et al.* (1996) ∘ – Edmondson; • – Cox *et al.*

This agreement shows the behaviours of Morse potential in the crss of iron single crystal especially in the extreme low temperature range though it is not rigorous to know whether the plastic mechanism is changed or not. Anyway, the agreement shows another possibility — that Morse potential is one of the factors deciding the crss beginning to deviate from the usual temperature dependence (exponential relationship) at the temperature ($\sim$ 200K), higher than that of twinning formation (140K) (Louat and Hutchison, 1962). The authors emphasize that this is not in contradiction with the available computer simulations on dislocation core structures introduced in Sec. 10.1. (i) The computer simulation emphasizes the atomistic core structure and energy of dislocations of different kinds of metals even at high temperature with molecular dynamics method, not the temperature dependence of crss under the *same glide mechanism*. (ii) Morse potential model is based on the experiments on Mo-single crystals which are selected to have the orientation in such a way that (110) [111] remained to be the only slip system throughout the whole testing temperature range. Thus, this model excludes the complicated cases of dislocation core configuration transformations from the glissile into sessile form with increasing temperature, such as in the $LI_2$ structure crystals.

The treatment in which only the bonds between the dislocation centre atoms and impurity atoms have been considered is a somewhat oversimplified model. However, if one compares this model with the Peierls model, one may see that it is as rough as Peierls model to have treated the upper and lower parts of crystals as continuous media.

The treatment needs a deeper understanding through electronic structure calculations. In principle, this can be done with the local density functional theory. However, at present, it is used routinely in studies of the stability of crystal structures. Its application to extended defects is still limited at the time of writing. Qualitative analysis based on electronic theory is still worthwhile. If we can calculate the local $N(E)$ at the Fermi surface and the $N_{\text{eff}}$, we may characterize the local atomic bonding strength at defect core. We know that $N_{\text{eff}}$ has directionality. $N_{\text{eff}}$ would be anisotropic at the dislocation core and thus it describes the different behaviours of dislocation glide and climb. Positron annihilation can measure the $N_{\text{eff}}$ at the dislocation (Deng *et al.*, 1995) and the momentum distribution of the electrons (see Chap. 9).

Computer simulation with empirical atomic potentials still affords a realistic approach to this problem. However, at the time of writing, calculations on the structure and energy of the dislocation core at different stages of dislocation movement in various ranges of finite temperatures are lacking.

The Morse potential model may predict the anomalous yield stress in $LI_2$ alloys in the following way (Fig. 10.4).

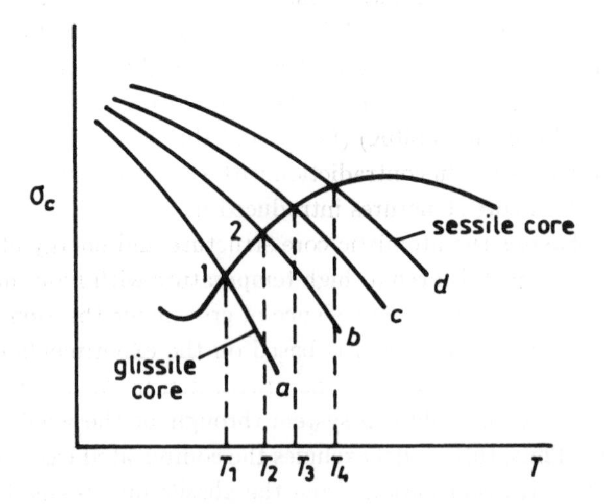


Fig. 10.4. Anomalous yield stress versus temperature.

If the dislocation core structure were not change at $T_1$ or so, the $\sigma_c$ vs $T$ relationship would be along curve $a$ and then, at $T_1$ the $\sigma_c$ would be at point 1. As the temperature increases to $T_2$, the dislocation core structure changes partly to be sessile. If the dislocation core structure does not change at $T_2$ or so, the $\sigma_c$ vs $T$ relationship would be along curve $b$ and then, at $T_2$ the $\sigma_c$ would be at point 2. Repeating this argument again, we obtain curve 1 2 3 4 as the anomalous yield stress in $LI_2$ alloys. This explanation is not in contradiction with electron microscope observations by Hirsch (1992), but is another kind of "dynamical" model to explain the temperature dependence of yield stress in bcc metal alloys rather than thermal activation mechanism.

The interaction between phonons and a dislocation has been calculated by many workers (Maradudin, 1958; Tewary, 1969; Ninomiya, 1970; Boyer and Hardy, 1971; Tewary and Bullough, 1971; (see also Bullough and Tewary, 1979)). Following the above studies, Ohashi *et al.* (1978) used the double-time Green function technique to work out thermal vibrations of a dislocation. An expression for the mean-square displacement of the dislocation is obtained. The results were used to calculate the root-mean-square displacement of dislocations in fcc metals. In their calculations, they did not consider the dissociation of the perfect dislocation into Shockley partials. However, the authors argued that the effect of the dissociation of the perfect dislocation does not affect the results obtained for low temperature because at low temperatures the phonons of long wavelength are dominant. At high temperatures the phonons of short wavelengths become important and the atomic configurations around the dislocations must be considered. The results have not been used to bcc metals. It might be the same way to obtain the root-mean-square displacement of dislocations provided the dissociation into partials is not being considered. Their interesting finding is that at low temperatures ($T > 0.1 \times \theta_D$, the Debye temperature), mean-square displacements of both edge and screw vary linearly with temperature. Then the B value in Eq. (10.3.6) would be constant ($\frac{\langle u^2 \rangle}{T} \sim$ const.). This conclusion seems the same as the result from experiments on Mo single crystals by one of the present authors and his co-workers (see Chap. 2). In those experiments, they selected Mo single crystals according to their orientations being such that all their slip systems are $\{110\}$ $[111]$. The reason why the behaviour of Mo single crystals is like that of fcc metals is not clear. It seems that the temperature dependence of bcc metals is like that of fcc metals if samples are selected such that their slip systems are the same in the entire range of temperatures.

Based on Lung *et al.*'s analysis (1997), in Sec. 2.1.2, we pointed out that the approximate exponential relationship between crss and temperature is common even for bcc simple crystals provided that the slip plane and direction are kept the same. Using double-time Green function, Wang *et al.* (1998) calculated the thermal vibration of a dislocation line. With Morse potential model, Wang *et al.* (1998) calculated values of crss of Al, Cu, Fe, and Mo single crystals. The calculated results showed that the temperature dependence of crss of many metals obeys the approximate exponential low.

Using the Morse potential model, one can also explain the influence of the concentration of solute atoms in metals on the temperature dependence of yield stress of bcc metals. Jiang *et al.* (1984, 1989) calculated the $B$ values in Eq. (2.2.1) of various concentrations of carbon atoms on the temperature dependence of yield stress of $\alpha$-Fe. They extended the method by Leibfried (1954) in which the interaction of impure atom with the dislocation line was simplified to be frictional stress parameter rather than a Morse potential. It may be a reasonable approximation when the temperature is low, and $\overline{u^2}$ is small enough. Table 10.6 shows the data calculated in comparison with experimental results measured by He *et al.* (1966) with internal friction method for determination of the concentration of carbon solute atoms in $\alpha$-Fe solid solution.

Based on the elasticity theory and considering the lattice modification for the dislocation core, Wang *et al.* (1993) studied the relaxed atomic structure of the dislocation core with a self-consistent calculation by using the molecular dynamics. From the modified atomic coordinates and by use of the recursion method, they calculated the electronic structure of the dislocation core. They found that the edge dislocation results in the splitting of degenerate states, and the movement of Fermi level. The local density of states in the dislocation core diffuses markedly in energy space, which reflects the perturbation of the dislocation. The neighboring environment around the atoms in the core affects the local density of states sensitively.

Table 10.6

| Exp. | Cal | $C(\times 10^6)$ |
|---|---|---|
| 11.00 | 10.49 | 66 |
| 7.58 | 8.29 | 300 |
| 6.54 | 6.40 | 630 |
| 6.16 | 6.03 | 720 |

It seems that by combining this theoretical calculation on the *local* density of states in the dislocation and the measurement of *local* $N_{\text{eff}}$ at the defect by positron annihilation (Deng *et al.*, 1995), the mechanical behaviour of the dislocation would be understood more deeply in terms of electronic properties.

## 10.4. *Plastic Deformation of L1₂ Ordered Alloys*

In the context of the relation between mechanical properties and atomistic (and finally electronic!) structure, we summarize below the study by Tichy *et al.* (1986) on the plastic deformation of $L1_2$ ordered alloys.

This subject has been studied extensively, both experimentally and theoretically (see reviews of Stoloff and Pope, 1966; Liang and Pope, 1977; Pope and Ezz, 1984; Pope and Vitek, 1984). Often, these investigations had as one focal point a very prominent feature of the plastic properties of these alloys, namely the peak in the dependence of the yield stress, observed in a number of different $L1_2$ alloys at high temperatures. The widely accepted mechanism for this anomalous increase of the flow stress involves cross-slip of screw dislocations from (111) planes, where they are mobile, to (010) planes, where they are assumed immobile (Kear and Wilsdorf, 1962; Thornton *et al.*, 1970) and to act as obstacles to moving dislocations. The driving force for the cross-slip is provided by the anisotropy of the antiphase boundary (APB) energy, which is lowest for (010) planes (Flinn, 1960).

It is not our purpose here to go into calculational details. But we summarize the main conclusions of the study of Tichy *et al.* (1986).

An important deduction from these atomistic studies of screw dislocations in $L1_2$ alloys in which no stable APBs exist on $\{111\}$ planes is that these dislocations are always sessile. Following Tichy *et al.* (1986) we note that the cores of the $\frac{1}{2}[\bar{1}01]$ superpartials bounding an APB on the (010) plane are spread entirely outside this plane into two $\{111\}$ planes intersecting along the $[\bar{1}01]$ direction. The core of the $\frac{1}{3}[\bar{1}\bar{1}2]$ superpartial bounding a superlattice intrinsic stacking fault (SISF) on the (111) plane is split, into an edge part $\frac{1}{6}[1\bar{2}1]$ and a screw part $\frac{1}{2}[\bar{1}01]$. The core of the latter is again spread into the two intersecting $\{111\}$ planes. Hence it appears from the work of Tichy *et al.* (1986) that, in the alloys they consider, the core of the $\frac{1}{2}[\bar{1}01]$ screw dislocation always has a strong tendency to spread equally into the two most densely packed planes, (111) and $(1\bar{1}1)$, intersecting along the $[\bar{1}01]$ direction. The reason proposed by Tichy *et al.* (1986) is that the complex stacking fault (CSF) is also unstable, and thus the usual planar splitting of $\frac{1}{2}[\bar{1}01]$ dislocations into Shockley partials is not possible. Tichy *et al.* (1986) in this context contrast the situation with that in pure fcc metals and fcc — based ordered alloys. In these cases, $\frac{1}{6}\langle 112\rangle$ type stacking faults on $\{111\}$ planes are stable and therefore a glissile form of $\frac{1}{2}\langle 101\rangle$ screw dislocations always exists.

Tichy *et al.* (1986) emphasize that very different plastic behaviour is to be anticipated for $L1_2$ alloys with stable and unstable APBs and CSFs on $\{111\}$ planes, respectively. This is due to the markedly different character of screw dislocations in these two alloy classes. In the former, a glissile configuration of screw dislocations always exists, even if it is only metastable and energetically less favorable than the sessile configuration formed by the transformation of the core to produce a ribbon of APB on the (010) plane. Tichy *et al.* (1986) note that such a glissile configuration results when superdislocations split into $\frac{1}{2}\langle 101\rangle$ superpartials on $\{111\}$ planes, and these further dissociate into the Shockley partials. The low-temperature behaviour can then be anticipated to be like that of pure fcc metals.

If a more stable sessile configuration exists, however, the glissile dislocations may transform into sessile ones at high temperatures which will result in anomalous yield behaviour (Tichy *et al.*, 1986; see also Paidar *et al.*, 1984). On the other hand, as Tichy *et al.* point out, when the APB is not stable on $\{111\}$ planes, the low-temperature plastic behaviour of these alloys will be like that of pure bcc metals in which no glissile forms of screw dislocations can occur. Screw dislocations can move only via thermal activation; Tichy *et al.* presume via the formation of double kinks. The flow stress will then rapidy increase with decreasing temperature. Moreover, due to the anisotropy of the distribution of core displacements a complex slip geometry and deviations from Schmid's law can be anticipated, as in bcc metals. A detailed theory of the thermally activated motion of sessile screw dislocations has been given by Tichy *et al.* (1986), but we must refer the interested reader to their paper for details.

## 10.5. *Breaking-Bond Models of Propagation of Extended Defects*

The Celli-Flytzanis (CF) (1970) lattice dynamical model of the propagation of a screw dislocation in a solid has been referred to already in Sec. 2.2.1. It was stressed there that in contrast to thermally activated motion referred to above, the CF model is appropriate to high velocity propagation. The model is crucially about bond-breaking, and therefore we first summarize briefly the electron theory of the stretching of the simplest chemical bond; namely that in $H_2$ molecule in free space.

### 10.5.1. *Stretched Chemical Bond in $H_2$ and Electron Correlation*

We present below an argument concerning the crucial role of electron-electron repulsion in the ground-state of the $H_2$ molecule in free space as we stretch the chemical bond. The theory presented below, though not the criterion for bond-breaking proposed, goes back to Coulson and Fischer (1949).

These workers considered two H nuclei, $a$ and $b$, at internuclear separation $R$. They then formed asymmetric orbitals: $\phi_a + \lambda\phi_b$ centred dominantly on $a$: $0 < \lambda \leq 1$, and $\phi_b + \lambda\phi_a$ centred likewise on $b$. Here $\phi_a$ and $\phi_b$ are atomic orbitals (to be definite, say hydrogenic atom 1s wave functions centred on nuclei $a$ and $b$ respectively). Coulson and Fischer (1949) then formed the spatial ground-state variational wave function $\Psi(1,2)$ (to be multiplied by the usual anti-symmetric spin wave function):

$$\Psi(1,2) = N[\phi_a(1) + \lambda(R)\phi_b(1)][\phi_b(2) + \lambda(R)\phi_a(2)] \tag{10.5.1}$$

with $N$ the normalization factor; 'parameter' $\lambda$ to be determined by minimizing the expectation value of the Hamiltonian of the $H_2$ molecule with respect to $\lambda(R)$ at each value of $R$.

The finding of Coulson and Fischer is then simply stated. It $R_{\text{equil}}$ denotes the bond length of the free space equilibrium molecule, such a minimization procedure led to the exact value unity for $\lambda$ in the range $R < 1.6R_{\text{equil}}$. But putting $\lambda = 1$ in Eq. (10.5.1) regains the usual 'molecular orbital' wave function for $H_2$, in the 'linear combination of atomic orbitals (LCAO)'. This is the regime in which the electrons embrace both nuclei a and b equally.

However, for $R > 1.6R_{\text{equil}}$, Coulson and Fischer demonstrated with the variational wave function (10.5.1) that $\lambda(R)$ decreased very rapidly from unity, with a discontinuity in its slope at $R = 1.6R_{\text{equil}}$, and as $\lambda \to 0$ in Eq. (10.5.1) one regains $\phi_a(1)\phi_b(2)$ which is one half of the symmetrized Heitler-London (valence bond) wave function. The interpretation of this is that, when the chemical bond in $H_2$ is stretched to 1.6 times its equilibrium length, electron-electron repulsion 'drives the electrons back on to their own atoms'. The Coulson-Fischer type of approach is the forerunner of what quantum chemists refer to as generalized valence bond theory.

We turn immediately to the relevance of this simple example of the essential role electron correlation eventually plays when a chemical bond is stretched to the cleavage force needed to separate two pieces of crystal, discussed at same length in Chap. 5 and Appendix 5.1.

### 10.5.2. *Cleavage Force in Directionally Bonded Solids*

The first example, probably the best available at the time of writing, concerns the covalently bonded solid, silicon, studied by Matthai and March (1997): see also Chap. 5. These workers used the coordination-dependent potential of Tersoff (1989), which was fitted to available density functional calculations using the local density approximation for the exchange-correlation potential $V_{xc}(\mathbf{r})$ (see Secs. 6.9 and 8.6).

The main point we wish to make here is that Matthai and March (1997), in the example of the free space $H_2$ molecule discussed immediately above, identified the separation $R = 1.6R_{\text{equil}}$ with the 'breaking of the chemical bond' in $H_2$, brought about by electron correlation.

In their Si study, they found that the maximum of the cleavage $F(z)$ (see Sec. 5.2) occurred at an additional interplanar spacing of $\sim 0.7a$, with $a$ the equilibrium interplanar separation between the two parallel crystal planes being pulled apart. They interpreted the maximum of $F(z)$, at position $z_{\text{max}} \sim 0.7a$ as the position to be identified with 'bond-breaking'. Of course, this is a 'criterion' again, as for $H_2$ in free space above, but now for bonds in crystalline Si. These bonds in Si seem comparable then in their 'elasticity' with the $H_2$ molecule in free space.

Subsequent work by Osetsky *et al.* (1998) on the cleavage of Fe, using an empirical N-body potential has given a less 'elastic' directional bond in this material but it remains to be seen how sensitive this result is to the details of the interatomic force field employed.

To conclude this brief discussion of cleavage force $F(z)$, it was stressed in Chap. 5 that for $z \gg z_{\text{max}}$, $F(z) = c/z^3$ and that in a metal $c$ can be related to the electronic plasma frequency $\omega_p$. Of course, this quantity $\omega_p$ is entirely due to long-range Coulombic correlations between electrons, adding credence to the view that the cleavage force as one goes to even larger separations, is very dependent on electron correlation. This discussion of stretched chemical bonds, and bond-breaking, leads us naturally into a discussion of bond-breaking models for the description of the propagation of extended defects in solids.

### 10.5.3. *Steady-State Propagation of Screw Dislocation in a Bond-Breaking Model*

The Celli-Flytzanis model is a model based on bond-breaking: it was discussed qualitatively already in Chap 5, Sec. 5.2.4. We set out below the equation of motion of this model and then consider explicitly the 'almost continuum' limit as in the subsequent work of March *et al.* (1998).

Celli and Flytzanis (1970) treat the steady-state propagation of a straight screw dislocation in uniform motion in a cubic lattice in which nearest-neighbour interactions are assumed. The only component of the displacement, denoted below by $s_{m,n}$, is parallel to the dislocation axis, $m$ and $n$ being half-integers lablling, atomic positions in a lattice. Their 'bond-breaking' model can then be stated as follows. The force between near neighbours is assumed to be a linear function of the relative displacement $D$, that is

$$F(D) = -A(D - \nu b)\,, \tag{10.5.2}$$

$b$ denoting the lattice spacing in the direction parallel to the dislocation (the magnitude of Burgers vector) and $\nu$ being an integer such $|D - \nu b| < b/2$. Celli and Flytzanis note that $D$ exceeds $b/2$ only for atom pairs across the slip plane, before the dislocation passes two given atoms and the connecting bond breaks.

One may then write the equation of motion as

$$\begin{aligned} -M\left(\frac{d^2 s_{m,n}}{dt^2}\right) &+ A(s_{m+1,n} + s_{m-1,n} + s_{m,n+1} + s_{m,n-1} - 4s_{m,n} \\ &= Ab\Theta[m - (vt/a)][\delta_{n,\frac{1}{2}} - \delta_{n,-\frac{1}{2}}] \end{aligned} \tag{10.5.3}$$

In Eq. (10.5.3), $M$ denotes the atomic mass, $v$ is the constant velocity of the dislocation, while $a$ is the lattice constant in the plane perpendicular to the dislocation. $\Theta(z)$ in Eq. (10.5.3) is the step function

$$\Theta(z) = \begin{cases} 1 & \text{for } z > 0 \\ 0 & \text{for } z < 0\,, \end{cases} \tag{10.5.4}$$

and finally the transverse sound velocity $c$ employed below is given by

$$c^2 = \left(\frac{Aa^2}{M}\right) \tag{10.5.5}$$

(1) Replacement of equation of motion (10.5.3) by differential equation

Celli and Flytzanis (1970), using Fourier analysis, provide a complete solution of Eq. (10.5.3). Here, following March *et al.* (1998), attention below will be focussed entirely on replacing Eq. (10.5.3) by a differential equation (see

Eq. (10.5.13)). Let us proceed by first quoting the result already given by Celli and Flytzanis for the continuum limit:

$$s_{m,n}(t) = \frac{b}{2} + \frac{b}{2\pi}\tan^{-1}\left\{\frac{\left[na\left(1-\frac{v^2}{c^2}\right)^{\frac{1}{2}}\right]}{ma - vt}\right\} \tag{10.5.6}$$

with $c^2$ as in Eq. (10.5.5).

To motivate physically the construction of a differential equation obeyed by Eq. (10.5.6), let us first replace the discrete quantities $na$ and $ma$ by continuous variables $x$ and $y$ respectively (compare Eq. (2.2.1), with obvious minor differences). Since the displacement $s(x, y, t)$ is then a function of $y - vt$, a 'wave' equation for propagation with speed $v$ is a natural starting point, for $s_0(x, y - vt)$. However,to see how $s_0$ must be modified to obtain $s$, March *et al.* (1998) invoke energy balance considerations as follows.

Celli and Flytzanis (1970) impose on their discrete lattice dynamical model the condition that, on average, there be a balance between the work done per unit-time by the applied stress and the power lost through lattice phonons emitted by the moving dislocation. Then the net energy flow beyond the $n$th atom row is found to be

$$P(n,t) = \frac{Mc^2}{a^2}\sum_m (s_{m,n+1} - s_{m,n})\frac{ds_{m,n}}{dt} \tag{10.5.7}$$

Equation (10.5.7) can be replaced in the 'almost continuum' limit (March *et al.* (1998) invoke the Euler-Maclaurin summation formula) by

$$P(x,t) = \frac{Mc^2}{a^2}\int_{-\frac{a}{2}}^{t+\frac{a}{2}} dy \frac{\partial s}{\partial x}\frac{\partial s}{\partial t} \tag{10.5.8}$$

This result (10.5.8) lies at the heart of the treatment of March *et al.* (1998). Briefly, it motivates the substitution of the point nonlinearity of the interatomic force on the right-hand side of Eq. (10.5.3) by the radiation force due to the emission of waves. The physical assumption which underlies the development below is that the radiation force has the same damping effect as the emission of waves. But now the loss of discreteness means that the equilibrium positions must be put in as asymptotic boundary conditions, consistent with the Celli-Flytzanis limit of Eq. (10.5.6). The above considerations, plus the form of the integrand in Eq. (10.5.8) for the power $p$ motivate the assumption of March *et al.* (1998) that the appropriate (solitary wave) equation satisfied by the limit (10.5.6), namely $s(x, y, t)$ in the notation introduced above, has the form

$$\frac{\partial^2 s}{\partial x^2} + \frac{\partial^2 s}{\partial y^2} - \frac{1}{v^2}\frac{\partial^2 s}{\partial t^2} + C\frac{\partial s}{\partial t}\frac{\partial s}{\partial x} = 0 \tag{10.5.9}$$

Substituting Eq. (10.5.6), with continuum replacements $na \to x$, $ma \to y$, into Eq. (10.5.9), one readily obtains the constant $C$ as

$$C = \frac{4\pi}{bv}\left[1 - \frac{v^2}{c^2}\right]^{\frac{1}{2}} \tag{10.5.10}$$

For reasons that will emerge clearly below, it will be valuable in gaining further insight into the nature of the solitary wave Eq. (10.5.9) to define

$$B = \left[1 - \frac{v^2}{c^2}\right]^{\frac{1}{2}} \tag{10.5.11}$$

and hence $1/v^2$ entering the 'acceleration term, $(\partial^2 s/\partial t^2)$ in Eq. (10.5.9) can be replaced by

$$\frac{1}{v^2} = \frac{1}{c^2} + \frac{B^2}{v^2} \tag{10.5.12}$$

Thus one is led to the main result of March *et al.* (1998), namely to rewrite Eq. (10.5.9) as

$$\left[\frac{\partial^2 s}{\partial x^2} + \frac{\partial^2 s}{\partial y^2} - \frac{1}{c^2}\frac{\partial^2 s}{\partial t^2}\right] - \left[\left(\frac{B}{v}\right)^2\frac{\partial^2 s}{\partial t^2} - \left(\frac{4\pi B}{bv}\right)\frac{\partial s}{\partial t}\frac{\partial s}{\partial x}\right] = 0 \tag{10.5.13}$$

March *et al.* (1998) stress that, in this solitary wave Eq. (10.5.13), the Celli-Flytzanis continum limit (10.5.6) makes both the square brackets exhibited in Eq. (10.5.13) identically zero, as is readily verified by direct insertion. The solitary wave solutions for fixed $x$ are found to be indeed very localized in $y$.

### 10.5.4. *Unified Treatment of Lattice Dynamical Models of a Crack and of a Screw Dislocation*

Following their establishment of the solitary wave Eq. (10.5.13) describing the displacement $s(x, y, t)$ of a propagating screw dislocation in the 'almost continuum' limit of the Celli-Flytzanis lattice dynamical model, March *et al.* (1998)

have proposed a unified treatment of the above problem and crack propagation* as described in the (now two-dimensional) model of Slepyan (1981).

The argument of March *et al.* (1998) proceeds from the displacement $s(x, y, t)$ given in Eq. (10.5.6) for the screw dislocation propagation. Noting that this displacement is a function of the single variable, $\xi$ say, defined by

$$\xi = \frac{Bx}{y - vt} \tag{10.5.14}$$

these workers differentiate $s(\xi)$ to find

$$\frac{\partial s(\xi)}{\partial s} = \frac{\text{constant}}{(1 + \xi^2)} \tag{10.5.15}$$

Multiplying both sides of Eq. (10.5.4) by $(1 + \xi^2)$ and differentiating once more yields

$$(1 + \xi^2)\frac{\partial^2 s}{\partial \xi^2} + 2\xi\frac{\partial s}{\partial \xi} = 0 \tag{10.5.16}$$

It is next noted by March *et al.* (1998) that this Eq. (10.5.6) is a special case of Legendre's equation

$$(1 - x^2)\frac{\partial^2 Q_\nu(x)}{\partial x^2} + 2x\frac{\partial Q_\nu(x)}{\partial x} + \nu(\nu + 1)Q_\nu = 0 \tag{10.5.17}$$

This linear Eq. (10.5.17) has a further linearly independent solution $P_v(x)$ which will be utilized below.

However, returning to the motion of the screw dislocation, it turns out that $s(\xi)$ in Eq. (10.5.6) is related to $Q_0(x)$ in Eq. (10.5.17) by

$$s(\xi) = \text{constant } Q_0(i\xi) \tag{10.5.18}$$

where $i = \sqrt{-1}$.

(1) Relation to crack propagation in model of snapping brittle-elastic bonds.

March *et al.* (1998) have further stressed that the crack propagation treatment of Slepyan (1981), in which he analyzed, again by a 'breaking-bonds' model (see especially Fig. 1 of Marder and Liu, 1993) cracks in simple two-dimensional lattices, has an asymptotic solution which is also contained within Legendre's Eq. (10.5.17). The physical solution this time is in fact $P_{\frac{1}{2}}(i\xi)$. Using an

*See also F. S. C. Ching, J. S. Langer and H. Nakanishi (Phys. Rev. Lett. *76*, 1087 (1996)).

integral representation of $P_{\frac{1}{2}}(i\xi)$, March *et al.* demonstrate that, for large $\xi$, the Slepyan solution for the crack displacement, proportional to $\xi^{-1/2}$, is regained from the Legendre function $P_{\frac{1}{2}}$

To conclude this section, we first emphasize, as noted by March *et al.* (1998), that while it is the solitary wave Eq. (10.5.13) which gives insight into the screw dislocation propagation, it is the linear Legendre Eq. (10.5.17) which unifies the 'bond-breaking' models of extended defect propagation: both screw dislocation and crack. In particular, the displacement for crack propagation can be constructed explicitly from the screw dislocation displacements. This is satisfying in that previous workers have noted that a crack can be represented as a pile-up of screw dislocations. Further points to be stressed here are (i) while the phonons emitted by a moving stress dislocation are embodied in Eq. (10.5.13), a moving crack emits dislocations and thus the latter point needs a more complete study; (ii) the Rayleigh wave velocity should play a role in crack propagation descriptions and this requires further investigation, and (iii) in a metal, propagation of a crack should take account of the fact that volume plasmons, with electronic plasma frequency $\omega_p$, are converted to surface plasmons as the propagation proceeds. Such aspects have been considered already in connection with surface energies of metals in the early work of Schmit and Lucas (1972); see also Lang, (1983) but they need treating in the study of crack dynamics. Again, as with bond stretching and breaking, electronic correlation is involved, but now crucially the long-range Coulombic electronic correlations.

## 10.6. *Grain Boundaries (GB),* Plastic Behaviour and Fracture*

Zhou *et al.* (1990) studied the microprocess of deformation and fracture for pure and bismuth-segregated tilt bicrystals of copper. Using the MD method with the atoms interacting via the empirical N-body potential proposed by Finnis and Sinclair (1984), they found that for pure $\sum 33$ bicrystal, the deformation is mainly due to the glide of partial dislocations generated from the GB structural units where the GB dislocation exists. The ductile fracture is attributed to the dislocation emission, which leads to vacancy generation and void coalescence. The bismuth segregation weakens the atomic bonds between copper atoms in the vicinity of GB. Under the action of external load, the weakened bonds break and lead to formation of microcracks. Finally, the brittle fracture

*Z. F. Zhang and Z. G. Wang (Phil. Mag. Lett. *78*, 105, 1998) have made experimental observations which show the effect of GBs on cyclic deformation behaviour.

takes place along the binding weakening region. Peng *et al.* (1992) and Zhou *et al.* (1992), again studied the selective bismuth segregation and the micro-process of fracture for the three [101] tilt copper bicrystals $\sum 9(2\bar{1}2)$ 38.94°, $\sum 11(\bar{3}\bar{2}3)$ 50.48° and $\sum 33(\bar{5}\bar{4}5)$ 58.99° by the MD method. The results show that the Bi segregation and the fracture behaviour of the Cu-Bi bicrystal are strongly dependent on the grain boundary structure. The severe intergranular brittle fracture that happens in the $\sum 9$ bicrystal is mainly caused by the breaking of weakened Cu-Cu bonds, which is related to the highly concentrated Bi segregation at the GB region. In the case of the $\sum 11$ bicrystal, the segregation of Bi atoms at the GB shows an inhomogeneous distribution characteristic, as though the fracture is intergranular but with a large amount of shear deformation. The transgranular fraction that appears in the $\sum 33$ bicrystal is related to the low concentration of the Bi atoms along the GB and in the grains. Chen *et al.* (1992) calculated and showed that the structure of the grain boundary component of nano-$\alpha$-Fe appears to have an atomic distribution of short-range order.

The plastic behaviour of solids is of a high degree of "complexity". Plastic deformation is the growth of dissipative structures and is a nonlinear, far-from-equilibrium situation. The natural approach to the problem of the emergence of new patterns of dislocations is in terms of the bifurcation theory to chaos. In the past decade, it has become more evident that plastic processes are well organized* spatial, temporal, or spatio-temporal structures arising out of chaotic states. It would be impossible to introduce these new developments in the framework of this book.

Several texts (Nicolois and Prigogine, 1977; Haken, Synergetics, 1983a, 1983b and etc.) provide an introduction to this subject. In addition, the proceedings of Workshops on Non-Linear Phenomena in Materials Science I(1988)–II(1992) (Eds. G. Martin and L. P. Kubin) and III(1995) (Eds. G. Ananthakrishna, L. P. Kubin and G. Martin) provide a valuable collection of research and review papers describing the studies in this field. Readers are recommended to consult these sources for full details of this important area.

---

*Self-organization and annealed disorder in fracture is treated, for example, by G. Caldarelli *et al.* (Phys. Rev. Lett. 77, 2503, 1996).

# Appendix

## A2.1. *High-Resolution Electron Microscopy (HREM): Use to Study Grain Boundaries*

The experimental image from HREM provides an accurate measure of the grain-boundary rigid-body displacements. It also gives information on the internal structure of the grain boundary within the resolution and projection limits of HREM.

*Example of $\sum = 5$, (310) [001] Grain Boundary in NiAl*

The experimental HREM image, reproduced from Fonda *et al.* (1995), (their Fig. 2), of the $\sum = 5$, (310) [001] grain boundary in NiAl, is shown in Fig. A2.1. The two grains were imaged by Fonda *et al.* along their common [001] directions which, for the B2-ordered NiAl, is along columns of pure Ni and Al. This image displays a repeat distance of $10d_{1\bar{3}0}$ (9.1 Å) along the grain boundary. Analysis of this image shows (Fonda *et al.*, 1995) that there is an asymmetry produced by a $\frac{1}{2}d_{1\bar{3}0}$ (0.46 Å) rigid-body translation of the top grain towards the right, with a small grain-boundary expansion within the experimental resolution (about 0.2 Å). The way this experimental HREM data can be fruitfully combined with atomistic structure calculations using N-body empirical potentials developed for the NiAl phase is discussed in the main text.

## A2.2. *Scanning Acoustic Microscope*

Wuri *et al.* (1995) have constructed a scanning acoustic microscope (SAM) which is suitable for transmission measurements in addition to the usual reflection mode. They were concerned, in the example they utilize, with a cubic single crystal of GaAs, with two polished (100) surfaces, which they examined

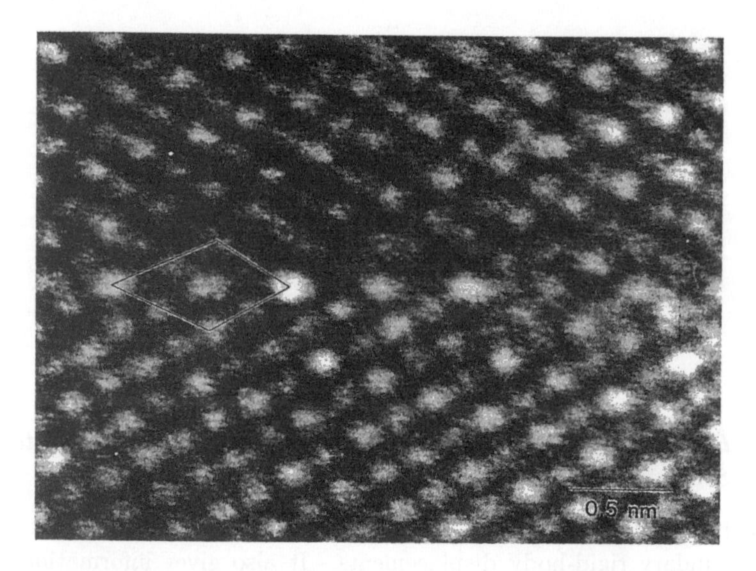


Fig. A2.1. HREM image of the $\sum = 5$, (310) [001] grain boundary in NiAl. The thickness is about 58 Å and the defocus about −650 Å.

with their SAM in transmission mode. They emphasize that the images of the transmitted ultrasound amplitude measured at frequencies of 350–400 MHz contain sufficient information about the acoustic anisotropy of the crystal to determine the elastic constants $C_{11}$, $C_{22}$ and $C_{44}$ numerically. They achieved this by an iterative fitting procedure which correlated simulated images with the one measured by the SAM.

They stress that their technique appears to be universally applicable to virtually all kinds of single crystals. The practical point they emphasize is that, in comparison with plane wave ultrasound (Truell *et al.*, 1969), it avoids the laborious and sometimes damaging work of cutting and polishing crystals in several different orientations, since their technique allows use of one single crystal orientation for a determination of all elastic constants.

## A2.3. *Quartz-Crystal Microbalance: Application to Study of Atomic-Scale Friction*

Krim *et al.* (1991) have used a quartz-crystal microbalance (QCM) to study atomic-scale friction.

Their QCM considered of a single crystal of quartz which oscillates in transverse shear motion with a quality factor $Q$ near to $10^5$. This was mounted horizontally within a vacuum chamber which was then plunged into a liquid-nitrogen bath.

They utilized both Au and Ag surfaces, and observed sliding-friction effects in Kr monolayers undergoing solidification on these noble metal surfaces.

## *Results*

Their experiments (Krim *et al.*, 1991) were carried out on 'smooth' and 'rough' Au and Ag surfaces, whose surface areas were found by nitrogen-adsorption

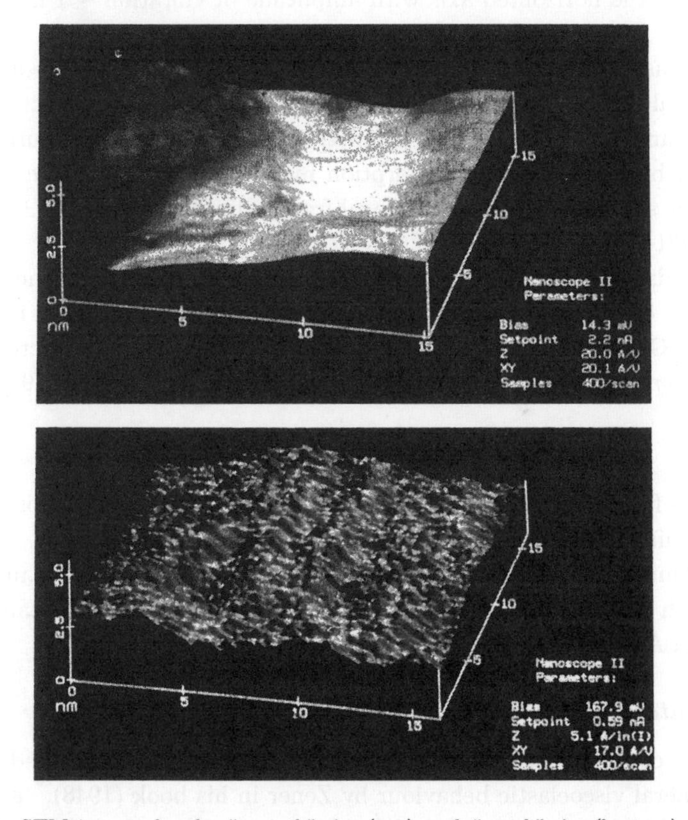


Fig. A2.2. STM images for the "smooth" Au (top) and "rough" Ag (bottom) substrates utilized for these studies. Each is the surface of a film which has been evaporated onto the surface of a quartz oscillator which translates back and forth along the horizontal axis with amplitude of vibration $\sim 1$ nm. (From Krim *et al.* 1991).

measurements. Their finding was that the systems Kr | Au and Kr | Ag showed practically no difference in their sliding behaviour, provided the interface morphologies were directly comparable. They argue that this is consistent with the fact that the van der Waals force which binds a Kr atom to a Au surface differs only by some 10% of that of a Ag surface ($\sim 2 \times 10^{-5}$ dynes; see also Rauber *et al.*, 1982).

Figure A2.2 (Fig. 1 from Krim *et al.*, 1991) shows scanning tunnelling microscope (STM) images for the 'smooth' Au (upper part of figure) and the 'rough' Ag (lower part) substrates. Each shows the surface of a film which was evaporated on to the surface of a quartz oscillator, which translates back and forth along the horizonted axis with amplitude of vibration $\sim 1$ nm (Krim *et al.*, 1991).

The rough Ag surface studied by Krim *et al.* (1991) were 8- and 5-mhz commercial quartz crystals. These can be characterized by respective surface areas 1.5 and 3.5 times that of a geometrically flat plane. These surfaces were indicated by liquid nitrogen adsorption measurements and analysis of STM data to be self-affine fractals with local fractal dimension $D = 2.3 \pm 0.1$ (Krim *et al.*, 1991).

The adsorbed film of Kr produces shifts in both the frequency ($f$) and amplitude of vibration, which are simultaneously recorded as a function of pressure. Gas-phase corrections were performed and then a characteristic film slip time ($\tau$) was obtained from the relation (Krim and Widom, 1988)

$$\delta(Q^{-1}) = 4\pi\tau\delta f$$

where the film-substrate frictional force is assumed to be directly proportional to the sliding speed (Stokës law: see Krim *et al.*, 1991).

To summarize the results of Krim *et al.* (1991), solid films slide more easily on smooth Au than do liquid films, while the reverse situation obtains for the 1.5 rough Ag substrate.

## A2.4. *Internal Friction*

Since the concept of anelasticity behaviour of solids was separated from the more general viscoelastic behaviour by Zener in his book (1948),* much literature on this subject has accumulated and showed its successful potential for

*Zener C. *Elasticity and Anelasticity of Metals* (Univ. of Chicago Press, Chicago, Illinois, 1948).

application to the understanding of atomic structure and dynamical behaviour of materials. For example, the Snoek peak has been used to determine the concentration of carbon atoms in the solid solution of $\alpha$-iron. Kê's pendulum (1947) is still the simplest technique in principle to study the low frequency internal friction in metals, just as Kê has used it to study the grain boundary more than 40 years ago. The technique has been developed more recently, such as the automatic inverted torsion pendulum. electromagnetically excited forced vibration pendulum. During the eighties, Kê *et al.* (1989) discovered a new internal friction peak which is different from the old one and which is associated with the 'bamboo' boundaries. Nowick (1967) has studied internal friction originating both in dislocations and in point defects.

Ultrasonic measurements provide information about the symmetry of defects, can often be used at low defect concentrations where defect interactions can be neglected, and permit the determination of properties of individual defect species even when several different types of defects are present simultaneously in the sample. These techniques are also well known for their usefulness in the study of dynamic dislocation effects. Measurement of the temperature dependence of the yield stress at constant strain rate have often been found by theoretical curves based on different assumed force-distance, but the measurements are not sufficiently precise to distinguish between the theories or to determine the temperature law. Granato (1990) gave a review on defect tunnelling in ICIFUAS-9. Internal friction has been applied to many topics in relation to mechanical properties of metals, such as dislocation kinks (Bordoni, 1949; Seeger, 1956), grain boundaries (Kê, 1990b), interaction of dislocations with point defects (see Kê, 1990a), martensite transformation (Wang *et al.*, 1990), grain growth (Lung, 1960), creep (Kong, 1990) and other mechanical properties (see Kê, 1990a).

Takeuchi (1995) asserts that one of the controversial issues over a long period of time in crystal plasticity is to magnitude of the Peierls stress in fcc metals. Bordoni (1949) discovered a low temperature internal friction peak of relaxation type in defined fcc metals (now known as the Bordoni peak). The mechanism giving rise to the Bordoni peak was proposed by Seeger (1956) in terms of the thermally-activated kink-pair formation on dislocations lying in the Peierls potential (see also the review by Fantozzi *et al.*, 1982). Using this interpretation one can estimate the Peierls stress for the motion of dislocations; Takeuchi (1995) estimates this stress, from the Bordoni peak, to be $\sim 10^{-3}$ in units of the shear modulus G.

But Takeuchi (1995) has noted that, in an apparently more straightforward way, one can make an alternative estimate of the Peierls stress by extrapolating the critical resolved yield stress to absolute zero of temperature. Mechanical tests at He temperatures for high-pure fcc metals show that the extrapolated critical resolved shear stresses are of the order of $10^{-5}$ G. Thus, as Takeuchi (1995) emphasizes, these two estimates of the Peierls stress in fcc metals differ by 2 orders of magnitude.

Takeuchi (1995) presents arguments that the Bordoni peak observed in annealed and deformated fcc metals is to be interpreted as due to the slide of undissociated dislocations, which is controlled by thermally activated kink-pair formation on the undissociated dislocations.

The Peierls stress for undissociated dislocations can be anticipated to be much larger than for the dissociated dislocation for the same total Burgers vector.

Thus, the conclusion of Takeuchi (1995) is that the Bordoni peak in fcc metals is due to the motion of undissociated dislocations. A more recently discovered lower temperature peak in Al (Kosugi and Kino, 1989; Kosugi and Kino, 1993) is attributed by Takeuchi (1995) to the motion of dissociated dislocations. Takeuchi also points out that the macroscopic yielding of crystals at helium temperatures is governed by the glide of dissociated dislocations and hence the Peierls stress estimated from the mechanical test for fcc metals is inconsistent with that estimated from the Bordoni peak. He proposes an experiment, namely to observe by transmission electron microscopy whether the network dislocations in annealed high-purity Cu are dissociated or not, but, to the knowledge of the present authors, this has not yet been carried out at the time of writing.

### A2.5. *Atomic Force Microscope*

The atomic-force microscope (AFM) provides a method for measuring ultra-small forces between a probe tip and an electrically conducting or insulating surface. The AFM is a combination of the principles of the scanning tunneling microscopy (STM) and the stylus profilometer (SP). It incorporates a probe of 30 Å and a vertical resolution less than 1 Å. It has been used for topographical measurements of surfaces on the nanoscale. With the STM, the atomic surface structure of conductor is well resolved. With the SP 3D images of surfaces with a lateral resolution of 1000 Å and a vertical resolution of 10 Å are recorded (Binnig *et al.*, 1986).

AFM images are obtained by measurement of the force on a sharp tip created by the proximity to the surface of the sample. This force is kept small and at a constant level with a feedback mechanism. When the tip is moved sideways it will follow the surface contours to be a trace.

Subsequent modifications of the AFM led to the development of the friction-force microscope (FFM) designed for atomic-scale and microscale studies of friction. This instrument measures frictional forces with an optical beam deflection AFM (Meyer, 1990).

## A3.1. *Disclinations*

Disclinations (rotation dislocations) are the topogical concepts which help in a description of broken symmetries of directional media. Disclinations are described in references (Friedel, 1964; Kléman, 1980).

### A3.1.1. *The Volterra Process in a Continuous Medium*

Consider a closed line $l$ and cut the crystal along a surface $\sum$ bounded by $l$; displace the two lips $\sum_1$ and $\sum_2$ (Fig. A3.1) of the cut surface one relative to the other by an amount

$$\mathbf{d}(\mathbf{r}) = \mathbf{b} + 2\sin\frac{1}{2}\Omega\mathbf{v} \times \mathbf{r}\,. \tag{A3.1.1}$$

Here, $\mathbf{b}$ is a translation that preserves the symmetry of the medium; $\Omega$ is a rotation that likewise preserves the symmetry. The displacement may be any $\mathbf{b}$, or $\Omega$, or a combination of the two as Eq. (A3.1.1).

In the most anisotropic elastic medium, some rotation dislocations are forbidden and some allowed rotation dislocations are quantized, the possible rotations are all the rotations of multiples of $\pi$ around any axis. Less anisotropic elastic media can have axes of rotation of $\frac{1}{2}\pi$, $\frac{2\pi}{3}$, ... or axes of revolution.

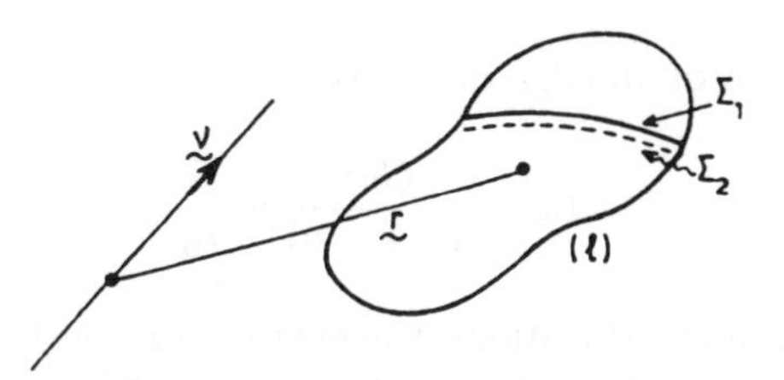


Fig. A3.1. The Volterra process.

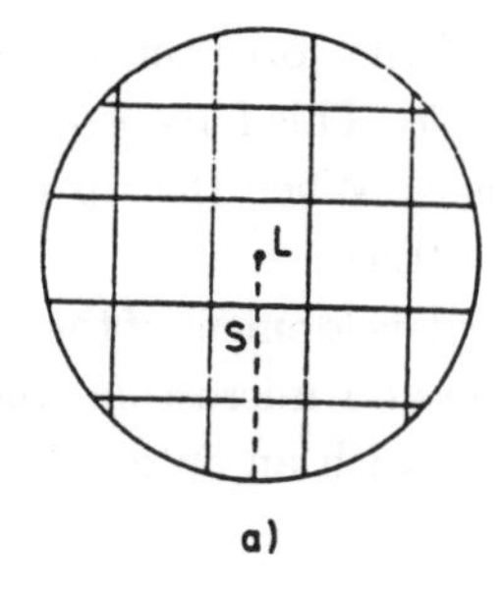


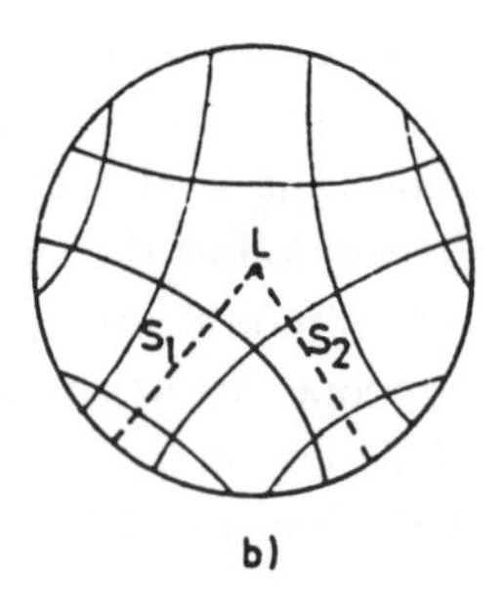


Fig. A3.2. Wedge disclination in a solid lattice; (a) perfect lattice and (b) dislocated lattice.

### A3.1.2. *Disclinations in Crystals*

Strong distortions should exist on the line, except when the line is along the axis of rotation. Figure A3.2 shows the wedge disclination in a cubic crystal ($\Omega = \frac{1}{4}\pi$). It is defined by the condition that the rotation vector $(\Omega, \nu)$ is along the line.

The notion of Burgers vector is particularly well adapted for the translation dislocations. An extension has been adapted to rotation dislocations. In many cases, the use of this extended Burgers circuit does not add any new information, because rotation dislocations are equivalent to a distribution of translation ones. With some exceptions, the use of normal Burgers circuits is sufficient.

### A3.1.3. *Energy of a Disclination*

The approximate value of the energy of wedge disclinations was calculated by Huang and Mura (1970) (also see Lihaqiaov and Hayinov, 1989).

$$E_\Omega \simeq \frac{G\Omega^2 R^2}{16\pi(1-\nu)} \quad (r_0 \ll R)\,. \tag{A3.1.2}$$

Comparing with that of an edge dislocation

$$E_b \simeq \frac{Gb^2}{4\pi(1-\nu)} \ln \frac{R}{r_0} \tag{A3.1.3}$$

we may find that unlike dislocations, the energy per unit length of disclinations depends on $R$ strongly. Even for small crystals ($R \sim 10^{-3}$ cm), the energy per unit length of disclinations is much larger than that of dislocations even if

$\Omega \sim 1^\circ$ and $R \sim 10^{-3}$ cm. This is the reason why disclinations rarely exist in common crystals.

## A3.2. *Dislocation Interactions*

### A3.2.1. *Interaction between Two Dislocation Loops*

If dislocation loop 1 is created while loop 2 is present, the stresses originating from loop 2 do work $-W_{12}$, when $W_{12}$ is the interaction energy between the two loops. When the loops are composed of straight line segments, the total interaction splits into interactions between the individual segments. Although an isolated dislocation segment has no physical meaning, formal expressions for segment-segment interaction enable us to determine by summation the total interaction with any piecewise straight dislocation array.

It has been shown that the interaction energy between two dislocation loops is

$$\begin{aligned} W_{12} = &-\frac{\mu}{2\pi}\oint_{c_1}\oint_{c_2}\frac{(\mathbf{b_1}\times\mathbf{b_2})\cdot(d\mathbf{l_1}\times d\mathbf{l_2})}{R} \\ &+\frac{\mu}{4\pi}\oint_{c_1}\oint_{c_2}\frac{(\mathbf{b}\cdot d\mathbf{l_1})(\mathbf{b_2}\cdot d\mathbf{l_2})}{R} \\ &+\frac{\mu}{4\pi(1-\gamma)}\oint_{c_1}\oint_{c_2}(\mathbf{b_1}\times d\mathbf{l_1})\cdot\mathbf{T}\cdot(\mathbf{b_2}\times d\mathbf{l_2}) \end{aligned} \tag{A3.2.1}$$

where $\mathbf{T}$ is a tensor with components

$$T_{ij} = \frac{\partial^2 R}{\partial x_i \partial x_j} \tag{A3.2.2}$$

$\mathbf{b_1}$ and $\mathbf{b_2}$ are Burgers vectors of loop $1(c_1)$ and $2(c_2)$ respectively. $\mathbf{R} = \mathbf{r} - \mathbf{r}'$ and $d\mathbf{l}$ and $d\mathbf{l}'$ are along dislocation lines $\boldsymbol{\xi}_1$ and $\boldsymbol{\xi}_2$ (Fig. A3.3).

### A3.2.2. *Force Produced by an External Stress Acting on a Dislocation Loop*

The force produced by an external stress acting on a dislocation loop can be calculated in a simple way. Let $\underline{\sigma}$ denote the stress tensor in the medium, excluding the self-stress of the dislocation loop under consideration. The stress does work for creating the loop

$$W = \int_A -\mathbf{b}\cdot(\underline{\sigma}\cdot d\underline{\mathbf{A}})\,. \tag{A3.2.3}$$

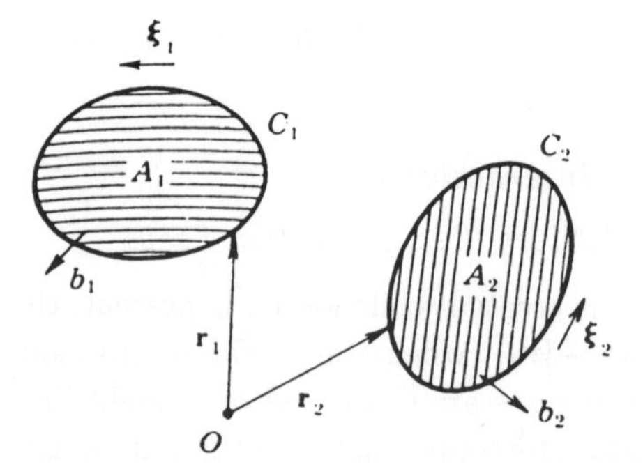


Fig. A3.3. Two dislocation loops within the same elastic continuum.

The line element $d\mathbf{l}$ of the loop is displaced by some distance, and the area $A$ changes by increments $\delta\mathbf{r} \times d\mathbf{l}$ and the stress $\underline{\sigma}$ does additional work

$$\begin{aligned}\delta W &= \oint_c d\mathbf{F} \cdot \delta\mathbf{r} \\ &= -\oint_c \mathbf{b} \cdot [\underline{\sigma} \cdot (\delta\mathbf{r} \times d\mathbf{l})] \\ &= -\oint_c (\mathbf{b} \cdot \underline{\sigma}) \cdot (\delta\underline{\mathbf{r}} \times d\mathbf{l}) \\ &= \oint_c [(\mathbf{b} \cdot \underline{\sigma}) \times d\mathbf{l}] \cdot \delta\mathbf{r}\,. \end{aligned} \tag{A3.2.4}$$

Then

$$d\mathbf{F} = (\mathbf{b} \cdot \underline{\sigma}) \times d\mathbf{l} \tag{A3.2.5}$$

Equation (A3.2.5) can be used to determine the interaction force between dislocation segments.

### A3.2.3. *Interaction between Two Parallel Straight Dislocations*

Consider two screw dislocations parallel with the $z$-axis, and with Burgers vectors $\mathbf{b_1}$ and $\mathbf{b_2}$. These two segments can be considered to be segments of two closed loops (Fig. A3.4). Only these two segments are close enough to contribute appreciably to Eq. (A3.2.1). The end effects, the other interaction terms have a negligible effect on the interaction energy as a function of $R$, the distance between two parallel dislocation lines, in the limit of $L \gg R$. For the

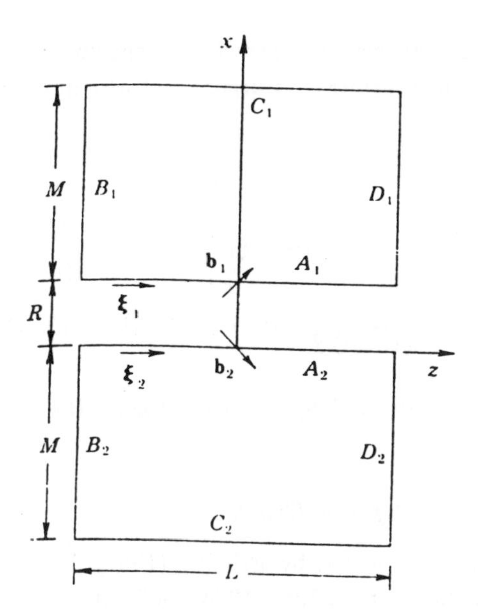


Fig. A3.4. Two coplanar dislocation loops.

special case, $\mathbf{b_1} = \mathbf{b_2} = (0, 0, b_z)$ the only term that varies with $R$ is

$$\frac{W(A_1 - A_2)}{L} = \frac{\mu b_z^2}{2\pi} \ln \frac{2L}{R} . \tag{A3.2.6}$$

The interaction energy can be expressed as the increase in the energy of the system when the dislocations are brought together from some value $R_a$ to a separation $R$:

$$\frac{W_{12}}{L} = -\frac{\mu b_z^2}{2\pi} \ln \frac{R}{R_a} . \tag{A3.2.7}$$

The interaction force per unit length is

$$\frac{F}{L} = -\frac{\partial(W_{12}/L)}{\partial R} = \frac{\mu b_z^2}{2\pi R} . \tag{A3.2.8}$$

In the case $\mathbf{b_1} // \mathbf{b_2}$, $|\mathbf{b_1}| \neq |\mathbf{b_2}|$,

$$\frac{F}{L} = -\frac{\partial(W_{12}/L)}{\partial R} = \frac{\mu b_1 b_2}{2\pi R} . \tag{A3.2.9}$$

Screw dislocations of like sign ($b_1 b_2 > 0$) repel, while those of unlike sign ($b_1 b_2 < 0$) attract. It is a radial force.

Dropping end-effect terms, an expression for the interaction energy between parallel dislocations with arbitrary Burgers vectors can be derived,

$$\begin{aligned}\frac{W_{12}}{L} &= -\frac{\mu(\mathbf{b_1}\cdot\xi)(\mathbf{b_2}\cdot\xi)}{2\pi}\ln\frac{R}{R_a} \\ &\quad -\frac{\mu}{2\pi(1-\nu)}[(\mathbf{b_1}\times\xi)\cdot(\mathbf{b_2}\times\xi)]\ln\frac{R}{r_a} \\ &\quad -\frac{\mu}{2\pi(1-\nu)R^2}[(\mathbf{b_1}\times\xi)\cdot\mathbf{R}][(\mathbf{b_2}\times\xi)\cdot\mathbf{R}]\,. \end{aligned} \tag{A3.2.10}$$

This equation was first derived by Nabarro (1952). Here, we cite from Hirth and Lothe (1982).

## A3.3. *Inclusion Model for a Crack*

Inclusion theory, first suggested by Eshelby (1957), has been applied in the studies of fracture (Zhang *et al.*, 1981; 1983; 1985; 1989), mechanical behaviour of composite materials (Wu and Chou, 1982; Zhang and Chou, 1985).

The inclusion model considers the crack as an elliptic inclusion in 2-dimensional system with the elastic constant being zero (Fig. A3.5). Assuming the applied stress is along the $y$-axis, ($\sigma_{22}^A = p$), the strain induced is given by

$$\begin{aligned}\varepsilon_{22}^A &= \frac{p}{2G(1+\nu)}\,, \quad \varepsilon_{11}^A = \varepsilon_{33}^A = -\frac{\nu p}{2G(1+\nu)} \\ \varepsilon^A &= \varepsilon_{11}^A + \varepsilon_{22}^A + \varepsilon_{33}^A = \frac{(1-2\nu)p}{2G(1+\nu)}\end{aligned} \tag{A3.3.1}$$

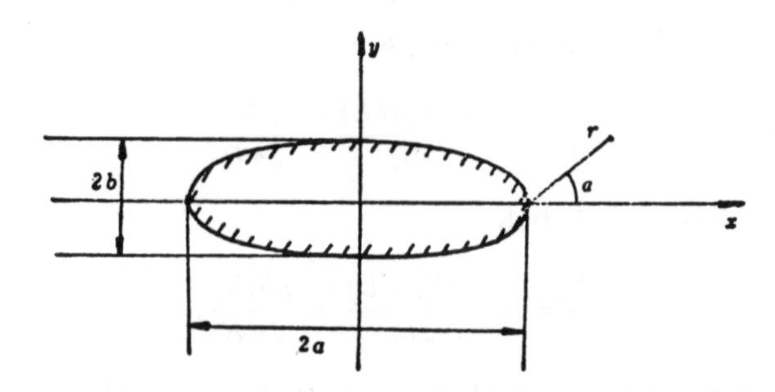


Fig. A3.5. The elliptic hole.

where $G$ is the shear modulus and $\nu$ is the Poisson's ratio. Solving a series of equations, the eigenstrains of the elliptic hole under the unidirectional extension are given by

$$\begin{aligned}
\varepsilon_{11}^{A} &= -\frac{p}{2G}(1-\nu)\left(1-\frac{\nu^2}{1-\nu^2}\right) \\
\varepsilon_{22}^{A} &= \frac{p}{2G}(1-\nu)\left(\frac{2}{\beta}+\frac{1}{1-\nu^2}\right) \\
\varepsilon_{33}^{A} &= -\frac{p}{2G}(1-\nu)\cdot\left(\frac{\nu}{1-\nu^2}\right)
\end{aligned} \tag{A3.3.2}$$

where $\beta = b/a$, the ratio of the two axis of the elliptic hole. In the limiting case of $\beta \ll 1$, Zhang and Zhe (1981) calculated the approximate values of the stress field near the crack tip,

$$\sigma_{xx}^{c} = \frac{G}{1-\nu}\left(\frac{\pi a}{2\pi r}\right)^{1/2}\cdot\beta(\varepsilon_{22}^{*}-\varepsilon_{11}^{*})\cos\frac{\alpha}{2}\left[1-\sin\frac{\alpha}{2}\sin\frac{3\alpha}{2}-\frac{2(\varepsilon_{11}^{*}+\nu\varepsilon_{33}^{*})}{(\varepsilon_{22}^{*}-\varepsilon_{11}^{*})}\right]$$

$$\sigma_{yy}^{c} = \frac{G}{1-\nu}\left(\frac{\pi a}{2\pi r}\right)^{1/2}\beta(\varepsilon_{22}^{*}-\varepsilon_{11}^{*})\cos\frac{\alpha}{2}\left[1+\sin\frac{\alpha}{2}\sin\frac{3\alpha}{2}+\frac{2(\varepsilon_{11}^{*}+\nu\varepsilon_{33}^{*})}{(\varepsilon_{22}^{*}-\varepsilon_{11}^{*})}\right]$$

$$\tau_{xy}^{c} = \frac{G}{1-\nu}\left(\frac{\pi a}{2\pi r}\right)^{1/2}\beta(\varepsilon_{22}^{*}-\varepsilon_{11}^{*})\sin\frac{\alpha}{2}\left[\cos\frac{\alpha}{2}\cos\frac{3\alpha}{2}-\frac{2(\varepsilon_{11}^{*}+\nu\varepsilon_{33}^{*})}{(\varepsilon_{22}^{*}-\varepsilon_{11}^{*})}\right] \tag{A3.3.3}$$

Equation (A3.3.3) is the same form as stress field near a crack tip when $\varepsilon_{11}^{*} + \nu\varepsilon_{33}^{*} = 0$. The stress intensity factor in this theory may be defined as

$$K_1 = \frac{G}{1-\nu}(\pi a)^{1/2}\cdot\beta(\varepsilon_{22}^{*}-\varepsilon_{11}^{*})\cdot\left[1+\frac{2(\varepsilon_{11}^{*}+\nu\varepsilon_{22}^{*})}{(\varepsilon_{22}^{*}-\varepsilon_{11}^{*})}\right]. \tag{A3.3.4}$$

If we substitute the limiting value of Eq. (A3.3.2) when $\beta \to 0$ to Eq. (A3.3.3), the stress intensity factor can be obtained as

$$K_1 = p(\pi a)^{1/2}. \tag{A3.3.5}$$

This is the well known form in fracture mechanics. The advantage of this model is that only solutions of series of algebraic equations are needed, which is simpler than solving integro-differential equations in the crack dislocation model.

## A3.4. *Coincidence Site Lattice (CSL) and Notion of Displacement Shift Complete (DSC) Dislocations*

Following again Vitek (1994, 1995), to construct a coincidence site lattice (CSL: see Brandon *et al.*, 1964) one can consider two interpenetrating ideal lattices which fill the whole of space but are misoriented with respect to one another. For certain relative orientation, in the cubic case, a three-dimensional lattice of coinciding lattice point is formed. Such a lattice is the CSL and can be characterized by the reciprocal density of the coincidence sites, $\sum$; this quantity being the ratio of the crystal lattice sites to the coincidence sites found in a unit volume (for further information the reader is referred to Warrington and Grimmer, 1974).

Coupled with the CSL is the important notion of dislocation shift complete (DSC) dislocations (Bollman, 1970; see also Vitek, 1994, 1995). The Burgers vector of such dislocations is any vector joining the lattice points of either of the interpenetrating crystals involved in the CSL. Though the origin of the CSL may thereby be shifted, displacement by such a vector of one of the lattices recreates the pattern of the interpenetrating lattices. Though these dislocations are not, in general, lattice dislocations, for a given $\sum$ the corresponding DSC dislocations may be present in grain boundaries (Vitek, 1995) since the structure far from them will not be altered by their presence. However (King and Smith, 1980), DSC dislocations may introduce steps into the boundaries. As Vitek (1995) points out, a boundary with the misorientation slightly deviating away from a coincidence $\sum$ can be regarded as the boundary with the misorientation corresponding to this value of $\sum$ containing a network of relevant DSC dislocations which accommodate the additional misorientation, just as the lattice dislocations do in the case of low angle boundaries. Their content in a boundary satisfies the Frank Eq. (3.7.2), where the angle $\theta$ is now the additional misorientation away from the coincidence $\sum$. For further analysis of geometrical concepts relating particularly to dislocations and to grain boundaries the reader is referred to Sutton (1984; see also Sutton and Balluffi, 1995).

## A4.1. *The Sierpinski Carpet*

The initiator is a square. At the first stage, the square is divided into nine smaller squares. The middle one is deleted. Eight of the remaining smaller squares are divided into smaller squares again. This process is repeated infinitely. Then, (Fig. A4.1)

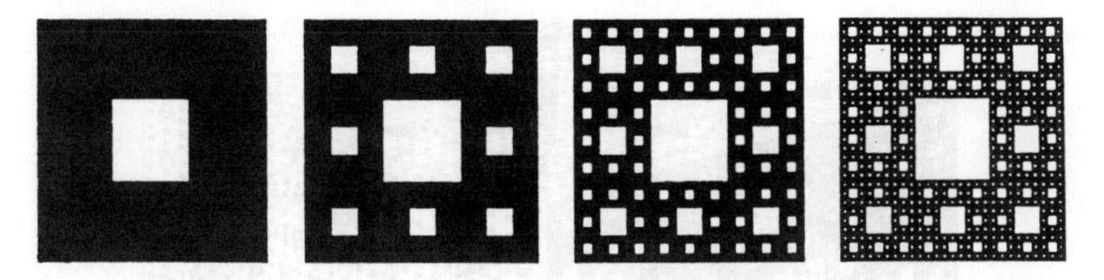

Fig. A4.1. Construction of the Sierpinski carpet. The initiator is a square and the generator (shown on the left-hand side) is made of $N = 8$ squares. They are obtained by contractions of ratio $r = \frac{1}{3}$. The right-hand side of the figure shows the fourth construction state. The similarity dimension is $D = \ln 8/\ln 3 = 1.89\ldots$.

$$N = 8$$
$$r = 1/3$$
$$D = \frac{\ln 8}{\ln 3} \sim 1.89\,.$$

The total area of the fractal is $A_F(\varepsilon_n) = \varepsilon_n^{2-D}$. The total area of the residual is $A_R(\varepsilon_n) = 1 - \varepsilon_n^{2-D}$. It is not a power-law relationship, but the generation distribution (or the area created in one generation) given by

$$A_{R,n+1} - A_{R,n} = \varepsilon^{n(2-D)}(1 - \varepsilon^{(2-D)}) = \text{const. } \varepsilon^{n(2-D)}$$

is a power-law relationship with a fractal dimension which is the same as that of the original fractal.

This model can be used for fragmentation (Turcotte, 1992). Most rock has a natural porosity. This porosity often provides the necessary permeability for fluid flow. A spatial formation of the Sierpinski carpet is the Menger sponge. The Menger sponge can be taken as a simple model for a porous medium. For the Menger sponge, the fractal dimension is (Fig. A4.2)

$$D = \ln 20/\ln 3 = 2.727\,. \tag{4.2.48}$$

The porosity of the nth-order Menger sponge, or the residual set of it, is not a power-law (fractal) relation though the density of the $n$th-order Menger sponge is. Since the ratios $(1/r)$ of above models should be positive integers, the distributions of fragments are discrete. Actual distributions of fragments are statistically continuous.

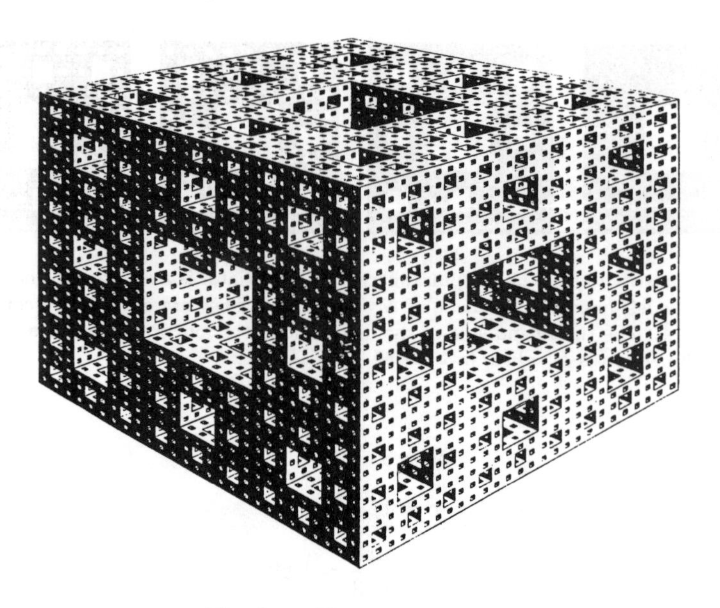

Fig. A4.2. The Menger sponge.

## A5.1. *Phonon Calculation of Cleavage Force $F(z) = Az$ for Small Separation*

Following Kohn and Yaniv, let us calculate the displacements $u_1$ and $u_0$ ($= -u_1$) when forces $F$ (and $-F$) per unit area are acting on the lattice planes $m = 1$ (and $m = 0$). They therefore study the equations of motion of the planes under the influence of a general force $F_m$:

$$M\ddot{u}_m = \sum_{\ell} A(l)u_{m,l} + F_m\,, \tag{A5.1.1}$$

$M$ being the mass of a lattice plane per unit area, given in turn by $\rho a$ where $\rho$ is the density and $a$ denotes the interplanar spacing. Finally the quantities $A(l)$ represent the interplanar force constants.

Writing the harmonic modes of vibrations of the planes in the form of displacements

$$u_m = \exp(\mathrm{ikma} - \omega t)$$

where the square of the angular frequency $\omega(k)$ is

$$\omega^2(k) = \frac{-1}{M}\sum_{l} A(l)\exp(\mathrm{ikla}) \tag{A5.1.2}$$

one now Fourier analyzes $F_m \to F(k)$ and $u_m \to u(k)$. Substitution into Eq. (A5.1.1) then yields

$$u_m = \sum_k \frac{F(k)}{M\omega^2(k)} \exp(\text{ikma})\,. \tag{A5.1.3}$$

For the case where $F_m = F(\delta_{m,1} - \delta_{m,0})$ and imposing periodic boundary conditions after $N$ planes one finds, after Kohn and Yaniv (1979):

$$u_m = \frac{F}{NM} \sum_k \frac{1}{\omega^2(k)} [\exp(ik\{m-l\}a - \exp(\text{ikma}))]\,. \tag{A5.1.4}$$

In particular, one then finds the displacement $u_1$ of the plane $m = 1$ to be given by

$$u_1 = \frac{Fa}{\pi M} \int_{-\pi/a}^{\pi/a} \frac{\sin^2(ka/2)}{\omega^2(k)} dk\,. \tag{A5.1.5}$$

Since $z = (u_1 - u_0) = 2u_1$ we have

$$F = \left(\frac{1}{4}\rho a \frac{1}{\omega_o^2}\right) z \tag{A5.1.6}$$

which is the result of Kohn and Yaniv (1979) when $\omega_o^2$ is found from

$$\frac{1}{\omega_o^2} = \frac{a}{2\pi} \int_{-\pi/a}^{\pi/a} \frac{\sin^2(ka/2)}{\omega^2(k)} dk\,. \tag{A5.1.7}$$

Hence the coefficient $A$ in $F(z) = Az$ for small $z$ is determined by knowledge of the bulk phonon spectrum of the solid. Kohn and Yaniv compare their result with that of Zaremba (1977). For the case when only nearest-neighbour interplanar forces are significant, there is complete accord between the two values of $A$. However, when interactions are included between planes beyond near neighbour, then $A \le A_{\text{rigid}}$. Kohn and Yaniv (1979) find that in typical cases $A$ is some 20 to 40% below $A_{\text{rigid}}$.

## A5.2. *Interstitial Formation Energy in Relation to Vacancy Formation Energy in Alkali Metals*

In Appendix 6.1, we present an electron theory argument, with neglect of ionic relaxation, which, in the end, relates the vacancy formation energy $E_v$ in close-packed metals (e.g. fcc metal) to phonon energies.

We shall see below that a similar formula can be obtained, though from a wholly different starting point, for open bcc metals. Flores and March (1981) proposed a theory showing how the long-range ionic displacements round a vacancy stem from both the elastic long-wavelength limit and from the effect of the topology of the Fermi surface, known as the Kohn anomaly (see Kohn, 1959; Jones and March, 1985). However, when attention is focused on the alkali metals Na and K, with almost-spherical Fermi surfaces, it can be reasonably assumed that the elastic displacements dominate. A model of 'complete relaxation' around a vacancy in then shown by Flores and March (1981) to relate formation energy $E_v$, bulk modulus $B$ and atomic volume $\Omega$ through

$$E_v = \text{constant} \times B\Omega \tag{A5.2.1}$$

which is equivalent to the empirical Mukherjee relation (A6.16) given below.

### A5.2.1. *Application of 'Complete Relaxation' Model to Relate Vacancy and Interstitial Formation Energy in Alkali Metals*

The above model was elaborated by Flores and March to relate vacancy and interstitial formation energies in the nearly free-electron alkali metals Na and K. Their basic argument was that the interstitial with the lowest formation energy would be that in which the relaxed near-neighbour configuration most closely resembles the local relaxation round the vacancy. Then the ratio of interstitial to vacancy formation was calculated. The outcome of this study was to argue that the formation of the interstitial in Na is only some hundreds of an electron volt higher than that for the vacancy.

The major step to be taken, beyond the vacancy study outlined above, is to use the complete-relaxation model of the self-interstitial in a variety of symmetry-allowed configurations in a bcc lattice. Flores and March presented results for 4 different interstitial configurations. The most interesting case, as already anticipated above, arises for a configuration which has close similarity to the relaxed configuration round the vacancy. What can then be calculated is the ratio of interstitial to vacancy formation energies. The result (see Eq. (A5.2.2)) below, can be expressed in terms of a force ratio $w$. For this relaxed interstitial configuration, $w$ is certainly greater than unity. One then must adopt some value for $w$ in order to evaluate the ratio of formation energies. As already indicated, a reasonable choice leaves the interstitial formation energy only some hundredths of an eV higher than that of the vacancy. Of course the force ratio $w$ will eventually need to be calculated from an accurate representation of the force field in such crystals. But the work of Flores and March strongly suggests that the interstitial energy in Na, say, is already determined to better than 0.05 eV. Since it was demonstrated conclusively by Brown *et al.* (1971) that the migration energy of a self-interstitial was exceedingly small in a bcc lattice, for purely geometrical reasons, it is clear that any realistic discussion of diffusion mechanisms in Na and K must pay serious attention to the contribution via interstitial sites (see also March and Pushkarov, 1996).

### A5.2.2. *Relaxation Round Self-Interstitials*

Flores and March considered several configuration for self-interstitials (Fig. 53, Alonso and March, 1989): the octahedral, the tetrahedral and the crowdion configuration, shown in Fig. A5.1(a), (b) and (c) respectively. However, in each of these cases the interstitial formation energy turned out to be well above the corresponding vacancy value. The fourth configuration these workers studied was therefore the relaxed interstitial shown in Fig. A5.1(d). It is assumed that this interstitial is formed by introducing one atom, say A′, and displacing the atom A in such a way that both A and A′ are placed along a line parallel to the cube edge.

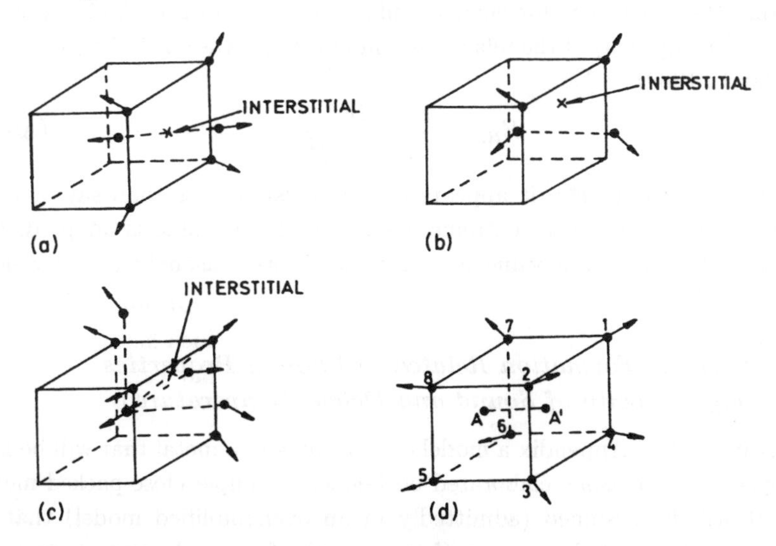


Fig. A5.1. (a) Relaxation around octahedral interstitial in Na. (b) Relaxation around tetrahedral interstitial in Na. (c) Relaxation around crowdion in Na. (d) Relaxation around split interstitial in Na. Redrawn after Flores and March.

The point to be emphasized is that this is now a configuration very similar to that of the vacancy discussed earlier. In fact, following Flores and March, it can be assumed that forces are applied to the atoms labelled 1–8 in Fig. A5.1(d) (Fig. 53(d) of Alonso and March book), in order to increase the crystal volume by one atomic cell. In the vacancy calculation sketched above, one could also regard forces to be applied, in order to decrease the crystal volume by this same value. Once there is room to accommodate an additional atom, this is introduced, while at the same time atom A is moved to its final position. It must be stressed here that the energy spent in this final process of moving atom A and introducing atom A′ is small in Na.

It then turns out that the energy of formation of the self interstitial in the configuration shown in Fig. A5.1(d) is essentially the energy spent in increasing the size of the crystal by one atomic volume. Although, at first sight it might appear that this is equal, in the present model, to the vacancy formation energy, there is an important difference to account for. For a vacancy, the symmetry is such that the forces applied to the eight atoms are directed along the diagonals. However, the symmetry is different for the interstitial, these being indicated in Fig. A5.1(d). It is this changed symmetry which, in the model under discussion, is responsible for a different energy for the interstitial. Denoting the force ratio for energy configuration (d) in Fig. A5.1, by $w$, the formation energy $E_{RI}$ of the relaxed self-interstitial turns out to be (see Alonso and March, 1989)

$$E_{RI} = \frac{3(1+2w^2)E_v}{(1+2w)^2}\,. \tag{A5.2.2}$$

From various arguments, it appears that $w$ must be less than say 4, which indeed seems to be a very extreme case. But if this value is adopted, then $E_{RI} \simeq 1.2E_v$ whereas a value $w = 2$ seems more reasonable, thus yielding $E_{RI} = 1.08E_v$.

## A6.1. *Vacancy Formation Related to Phonon Properties (e.g. Velocity of Sound and Debye Temperature)*

We discuss in this Appendix a model of a vacancy in a metal that will be most appropriate, for reasons elaborated on below, to simple close-packed metals. Here, it will be assumed (admittedly in an oversimplified model) that the vacancy can be treated as a perturbation on the free conduction electron gas.

The simplest approach is to use the linearized Thomas-Fermi (TF) relation between density and potential, discussed in the main text. If $\rho_0$ is the mean conduction electron density before the vacancy is introduced, then the inhomogeneous electron density $\rho(r)$ in the presence of the vacancy is related to the perturbing potential $V(r)$ caused by removal of an ion from the bulk metal by (with $k_F$ the Fermi wave number):

$$\rho(r) - \rho_0 = \frac{q^2}{4\pi} V(r) : \quad q^2 = \frac{4k_F}{\pi a_0}, a_0 = \frac{\hbar^2}{me^2} . \tag{A6.1}$$

In a metal of valence $z$, one can model the vacancy, before screening, as a point charge $-ze$ and screening converts the bare Coulomb potential energy felt by a conduction electron at distance $r$ from the vacancy, namely $ze^2/r$, to the screened Coulomb form

$$V(r) = \frac{ze^2}{r} \exp(-qr) . \tag{A6.2}$$

Let us proceed to estimate the vacancy formation energy $E_v$ by considering the change in the one-electron energy levels caused by the above self-consistent perturbation potential energy $V$. Since one is assuming a free electron gas model, the unperturbed conduction electron wave functions are simply $\vartheta^{-1/2} \exp(i\mathbf{k} \cdot \mathbf{r})$ and the energy change $\Delta\epsilon_{\mathbf{k}}$ of the state of wave vector $\mathbf{k}$ is evidently, from first-order perturbation theory

$$\Delta\epsilon_{\mathbf{k}} = \vartheta^{-1} \int \exp(-i\mathbf{k} \cdot \mathbf{r})\, V(\mathbf{r}) \exp(i\mathbf{k} \cdot \mathbf{r})\, d\mathbf{r} . \tag{A6.3}$$

This is clearly independent of $\mathbf{k}$ and therefore summing over the occupied states, replacing the summation by an integration with the usual factor for the density of states, leads to the total change $\Delta E$ in the one-electron eigenvalue sum as

$$\Delta E = \rho_0 \int V(\mathbf{r})\, d\mathbf{r} . \tag{A6.4}$$

Using either the semiclassical screened Coulomb potential, or the improved wave theory potential of March and Murray (1960) in the above Eq. (A6.4), one finds the same final result

$$\Delta E = \frac{2}{3} z E_F \tag{A6.5}$$

with $z$ the valence and $E_F$ the Fermi energy. We want to stress that the formula (A6.5) is strictly a perturbation theory in the potential $V$, having strength proportional to $z$ (we will display the inadequacy of Eq. (A6.5) as it stands for polyvalent metal vacancies below). But before proceeding to that, we must note an important correction that must be made to Eq. (A6.5) before using it to estimate the vacancy formation energy $E_v$ even for small $z$. This correction is because, when an atom is removed from the bulk metal and placed on the surface, the crystal volume is increased by one atomic cell size in the present model of no relaxation (appropriate only therefore to close-packed metals). This lowers the kinetic energy of the conduction electrons by an amount first calculated by Fumi (1955) to be $(2/5)zE_F$. Subtracting this value from Eq. (A6.5) yields the approximation

$$E_v = (4/15)zE_F \,. \tag{A6.6}$$

Of course, as already stressed, this perturbative argument could, at very best, only be applicable for small valence $z$, i.e. for say Cu, Ag or Au (see Fig. A6.21; March, 1990), when experimental values of $E_v$ are scaled in units of the valence times the free electron Fermi energy and plotted against the valence. The constant 4/15 in Eq. (A6.6) fits reasonably on to this data as the extrapolation back to $z \to 0$ is made. Such a figure makes it abundantly clear that Eq. (A6.6), though valid for sufficiently small $z$, must only be used, and then for semiquantitative purposes, for the monovalent metals.

However, while the above makes it plain that the Fermi energy entering Eq. (A6.6) is an inappropriate unit in terms of which to measure $E_v$ for polyvalent metals, let us next combine this equation with the so-called Bohm-Staver formula for the velocity of sound, $v_s$, in a metal (March, 1966).

*Derivation of Velocity of Sound in Simple Metals*

The above formula for $v_s$ is derived as follows. First, let us write a formula for the plasma frequency of the ions, taken to have mass $M$ and carrying charge $ze$ with $z$ as usual the valence. If one denotes the ionic density by $\rho_i$, then this ionic plasma frequency is given by the usual Langmuir expression

$$\omega_{\text{plasma}}^{\text{ions}} = \left\{ \frac{4\pi\rho_i(ze)^2}{M} \right\}^{1/2} . \tag{A6.7}$$

But next one notes that (i) $\rho_i z$ is the conduction electron density $\rho_0$ and (ii) Eq. (A6.7) represents an 'optical' rather than the desired acoustic mode.

The reason for this becomes clear from the above discussion of the screening of the vacancy, represented by charge $-ze$. Applying the same argument to the screening of ions, around which, of course, in a metal, electrons must pile-up to screen out long-range electric fields which cannot exist in such a conductor, one can write that the bare Coulomb potential of an ion, namely in **k**-space $4\pi ze^2/k^2$, must be screened according to Eq. (A6.2):

$$4\pi ze^2/k^2 \to 4\pi ze^2/(k^2 + q_{TF}^2)\,. \tag{A6.8}$$

Thus, in the long wavelength limit $k \to 0$ appropriate to sound waves

$$ze \to zek^2/q_{TF}^2 \tag{A6.9}$$

and rewriting Eq. (A6.7) by analogy yields

$$\left(\frac{4\pi\rho_0 ze}{m}\right)^{\frac{1}{2}} \to \left(\frac{4\pi\rho_0 ze}{m}\right)^{\frac{1}{2}} \frac{k}{q_{TF}} \tag{A6.10}$$

which evidently converts an 'optical mode' into the desired acoustic branch, to yield a dispersion relation as $k$ tends to zero:

$$\omega = v_s k \tag{A6.11}$$

where the velocity of sound $v_s$ is given by

$$v_s = \left(\frac{zm}{3M}\right)^{\frac{1}{2}} v_F \tag{A6.12}$$

with $v_F$ the Fermi velocity: $mv_F = \hbar k_F : q_{TF}^2 = 4k_F/\pi a_0$, $a_0$ being the Bohr radius $a_0 = \hbar^2/me^2$. Equation (A6.12) is the so-called Bohm-Staver formula.

## A6.2. *Relation of Vacancy Formation Energy to Debye Temperature*

The above formula will now allow the vacancy formation energy $E_v$ to be rewritten in terms of a characteristic 'phonon energy', rather than the Fermi energy $E_F$ which we have seen to be an inappropriate unit for other than small $z$. Squaring Eq. (A6.12) and rearranging yields almost immediately

$$Mv_s^2 = \frac{zm}{3} v_F^2 = \frac{2}{3} zE_F \tag{A6.13}$$

and hence, eliminating $zE_F$ from Eq. (A6.6) for $E_v$ in favour of $Mv_s^2$ leads to the result

$$E_v = \frac{2}{5} Mv_s^2\,. \tag{A6.14}$$

At this point, it is useful to bring this result into contact with an empirical relation proposed by Mukherjee (1965) in which $E_v$ is related rather directly to the Debye temperature $\theta$. Using the elementary model for the phonons in an isotropic solid, as set out for example in the book by Mott and Jones (1936), the Debye temperature $\theta$ is given in terms of $v_s$ by

$$\theta = \frac{v_s}{\Omega^{1/3}} \frac{h}{k_B} \left( \frac{3}{4\pi} \right)^{1/3} \tag{A6.15}$$

with $\Omega$ the atomic volume. Utilizing this equation in Eq. (A6.13) yields (March, 1966)

$$\theta = \frac{h}{k_B} \left\{ \frac{3}{4\pi\Omega} \right\}^{1/3} \left\{ \frac{2E_v}{3\alpha M} \right\}^{1/2} . \tag{A6.16}$$

For the predicted value $\alpha = 4/15$, this formula (A6.16) is not quite quantitatively in agreement with the empirical relation proposed by Mukherjee (1965). But changing $\alpha$ from 4/15 to 1/6 brings Eq. (A6.16) into good agreement with Mukherjee's empirical formula.

The important conclusion of this Appendix is that the vacancy formation energy $E_v$ in close-packed metals is most fundamentally related to a characteristic phonon energy. By eliminating $zE_F$, the obvious weaknesses of using first-order perturbation theory based on a free-electron gas model are avoided and it is then to be emphasized that one is led to a quantitatively useful estimate of the vacancy formation energy by very simple electron plus phonon theory.

## A7.1. *Validity of Electrostatic Model for Interaction Energy between Test Charges in a Fermi Gas*

In the main text, the semiclassical linearized Thomas-Fermi (TF) method was used to derive the interaction energy between test charges 1 and 2, at distance $R$. In the course of that derivation, it was assumed that one could simply calculate the screened potential due to test charge 1 at the site of charge 2 and then simply multiply by that charge.

Below the justification of this procedure is given, following Alfred and March (1957).

We consider below the two-centre problem posed above, but still using

$$\nabla^2 V = q^2 V . \tag{A7.1.1}$$

If the test charges, 1 and 2, create potentials $V_1$ and $V_2$ when present alone in the Fermi gas, then the total potential $V$ of the two-centre problem is, because of the linearity of Eq. (A7.1.1)

$$V = V_1 + V_2 \tag{A7.1.2}$$

i.e. just the superposition potential of the two charges at the specific separation $R$.

The interaction energy between the test charges may now be obtained from (A7.1.2). It will make zero contribution to the kinetic energy difference between the test charges at infinite separation and at distance $R$.

We are now in a position to calculate the changes in both kinetic and potential energies when we bring the charges together. We may write down the explicit contributions as follows (Alfred and March, 1957).

(i) The interaction energy between the charge $ze$ on centre 1 and the perturbing potential, say $V_2$, due to the other.

(ii) The interaction energy between the displaced charge (in linearized TF theory) $(q^2/4\pi e^2)V_1$ round the first charge and the potential $V_2$ due to the second.

(iii) The change in kinetic energy.

For equal test charges $ze$, these terms are evidently given by

(i) $z^2e^2 \exp(-qR)/R$
(ii) $-(q^2/4\pi e^2)\int V_1 V_2 \, d\mathbf{r}$
(iii) $(-q^2/4\pi e^2)^2(E_F/3\rho_o)\,[\int(\{V_1+V_2\}^2 - V_1^2 - V_2^2)\,d\mathbf{r}] = (q^2/4\pi e^2)\int V_1 V_2 \, d\mathbf{r}$

Equation (iii) follows directly from the earlier calculation of the kinetic energy change in Eq. (A7.1.1). Thus, it can be seen that the cancellation between (ii) and (iii) above is complete and one is left with the desired result

$$\Phi(R) = z^2e^2 \exp(-qR)/R \tag{A7.1.3}$$

obtained here from first-principles by calculating the difference between the total energy of the metal when the charges are at infinite separation and when they are brought together to separation $R$. Clearly, this procedure is all to be carried out in the Fermi sea of constant density.

*Separate Changes in Kinetic and Potential Energy*

The kinetic energy in Thomas-Fermi theory, for inhomogeneous density $n(\mathbf{r})$ can be obtained from phase space arguments (see e.g. March, 1992) as

$$T = c_k \int \rho(\mathbf{r})^{5/3}\, d\mathbf{r} \tag{A7.1.4}$$

where $c_k = (3h^2/10m)\,(3/8\pi)^{2/3}$. It is convenient to measure kinetic energy changes relative to the unperturbed Fermi gas state with kinetic energy $T_o$ and density $\rho_o$:

$$T - T_o = c_k \int \left[\rho(\mathbf{r})^{5/3} - \rho_o^{5/3}\right] d\mathbf{r} \tag{A7.1.5}$$

and writing the displaced charge $\rho(\mathbf{r}) - \rho_o = \Delta\rho$ one has for small $\Delta\rho$ from Eq. (A7.1.5) the result

$$T - T_o = E_F \int \Delta\rho\, d\mathbf{r} + \frac{E_F}{3\rho_o} \int (\Delta\rho)^2 + O(\Delta\rho^3) \tag{A7.1.6}$$

with $\rho_o$ and $E_F$ related in the usual free-electron fashion. We now observe that the first term on the right-hand side for $T - T_o$ involves simply the normalization condition for the displaced charge.

Corless and March (1961) have subsequently carried out the calculation for $\Phi(R)$ in the wave theory and have again demonstrated the cancellation of the corresponding wave theory contributions to (ii) and (iii) above. Thus, from the screening of a single test charge discussed in the body of the text, one finds the (now asymptotic) result

$$\Phi(R) \sim A\frac{\cos(2k_F R)}{R^3} : R \text{ large}. \tag{A7.1.7}$$

Modifications of this result occur when the test charges are replaced by ions as in real metals. But there are still oscillations of wavelength $\pi/k_F$, related to the de Broglie wavelength of electrons at the Fermi surface. But $\cos(2k_F R) \to \cos(2k_F R + \propto)$ with a non-zero phase shift $\propto$ and the amplitude $A$ must also be appropriately modified from the result of the linearized theory.

The introduction of a weak pseudopotential into the Corless-March (1961) treatment has been considered by Ziman (1964).

## A7.2. *Derivation of Dielectric Function of a High Density Fermi Gas*

We take as starting point the equation of March and Murray (1960). In atomic units $m = \hbar = 1$ this reads

$$\nabla^2 V = \frac{2k_F{}^2}{\pi^2} \int V(\mathbf{r}') \frac{j_1(2k_F|\mathbf{r}-\mathbf{r}'|)}{|\mathbf{r}-\mathbf{r}'|^2}\, d\mathbf{r}' . \tag{A7.2.1}$$

Introducing the Fourier transform $\tilde{V}(\mathbf{k})$ of the potential, and taking the (test) charge being shielded as having magnitude unity for convenience, we find the bare Coulomb potential has Fourier transform $\tilde{V}(\mathbf{k}) = -4\pi/k^2$ (while the screened Coulomb potential has $\tilde{V}(\mathbf{k}) = -4\pi/(k^2+q^2)$).

In the wave theory corresponding to Eq. (A7.2.1) it will be demonstrated below that $q^2$ in the screened Coulomb potential case will be replaced by a quantity intimately related to the Fourier transform of the wave factor $j_1(2k_F r)/r^2$ appearing on the right-hand side of Eq. (A7.2.1).

Substituting

$$V(\mathbf{k}) = \int V(\mathbf{r}) \exp(i\mathbf{k}\cdot\mathbf{r})\, d\mathbf{r} \tag{A7.2.2}$$

in Eq. (A7.2.2), we find almost immediately

$$\begin{aligned}\int (-k^2)\tilde{V}(\mathbf{k}) \exp(i\mathbf{k}\cdot\mathbf{r}) d\mathbf{k} = {} & \frac{2k_F{}^2}{\pi^2} \int \tilde{V}(\mathbf{k}) \exp(i\mathbf{k}\cdot\mathbf{r}') \\ & \times \frac{j_1(2k_F|\mathbf{r}-\mathbf{r}'|)}{|\mathbf{r}-\mathbf{r}'|^2}\, d\mathbf{r}' d\mathbf{k} \\ & + 4\pi \int \exp(i\mathbf{k}\cdot\mathbf{r})\, d\mathbf{k}\end{aligned} \tag{A7.2.3}$$

where the second term on the right-hand side of this equation accounts for the unit point charge placed at the origin (charge density equal to the Dirac delta function $\delta(r)$ with constant Fourier components therefore). Now the integral over $\mathbf{r}'$ required in the second term can be performed by introducing the difference vector $\mathbf{R} = \mathbf{r}' - \mathbf{r}$ through the function $J(k_F, k)$ defined by

$$J(k_F, k) = 2k_F{}^2\pi^2 \int \exp(i\mathbf{k}\cdot\mathbf{R}) \frac{j_1(2k_F R)}{R^2}\, d\mathbf{R} . \tag{A7.2.4}$$

This definite integral can be evaluated and after some calculation one finds

$$J(k_F, k) = \frac{2k_F}{\pi} + \left(\frac{2k_F{}^2}{\pi k} - \frac{k}{2\pi}\right) \ln\left|\frac{k+2k_F}{k-2k_F}\right| . \tag{A7.2.5}$$

Returning to Eq. (A7.2.3) this reads, by introducing the function $J$:

$$\begin{aligned}\int (-k^2)\tilde{V}(\mathbf{k}) \exp(i\mathbf{k}\cdot\mathbf{r})\, d\mathbf{k} = {} & \int \tilde{V}(\mathbf{k}) \exp(i\mathbf{k}\cdot\mathbf{r}) J(k_F, k)\, d\mathbf{k} \\ & + 4\pi \int \exp(i\mathbf{k}\cdot\mathbf{r})\, d\mathbf{k} .\end{aligned} \tag{A7.2.6}$$

From Eq. (A7.2.6), we have then the desired solution

$$\tilde{V}(k) = \frac{-4\pi}{k^2 + J(k_F, k)}\,. \tag{A7.2.7}$$

As anticipated above, the role of $q^2$ in the screened Coulomb Fourier component is taken over by $J$, with its characteristic singularity at the Fermi sphere diameter $k = 2k_F$. Contact between the two treatments (semiclassical and wave) is readily established by taking the long wavelength limit $k \to 0$ in the equation for $J(k, k_F)$, when one finds

$$J(k_F, k = 0) = \frac{4k_F}{\pi} = q^2\,. \tag{A7.2.8}$$

Using the definition of the dielectric function $\epsilon(k)$ in Eq. (7.1.32) of the main text, the so-called Lindhard dielectric function follows.

It is now known that away from the high density limit $k_F \to \infty$, exchange plus correlation correlations become significant. The reader is referred to the review article by Singwi and Tosi (1981) and the book by March and Tosi (1984) for details.

## A7.3. *Pair Potential for Be Metal from Density Functional Theory*

In the main text, the density functional theory (DFT) of pair potentials was set out and applied to Na metal. At a density appropriate to liquid Na just above its melting point, it proved possible to bring this electron theory pair potential $\phi(r)$ into contact with that obtained by inverting the liquid structure factor $S(k)$ — the so-called diffraction potential.

The purpose of this Appendix is to record related electron theory work on $\phi(r)$ for Be metal. Unfortunately, because of the toxic nature of liquid Be, there seems to be no accurate diffraction data for $S(k)$ at the time of writing.

As for the analogous case of Na, the input data into the calculation on Be of Perrot and March was (a) atomic number 4, (b) mean electron density of conduction electrons and (c) a local density approximation for exchange plus correlation (see Holas and March, 1995, for an exact formula for the exchange-correlation potential in terms of low-order density matrices).

The basic building block is then the total screening charge $Q(R)$ around the divalent ion $Be^{2+}$ embedded in an initially uniform Fermi gas of electrons. For perfect screening, as must obtain in the conductor, $Q(R) \to 2$ or $R \to \infty$ and its detailed form is shown in Fig. 4 of PM(b). Using the DFT for

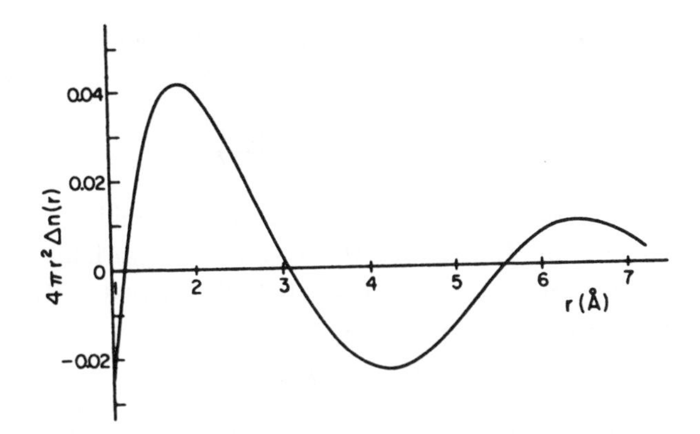


Fig. A7.1. Radial valence density $4\pi r^2 \Delta n(r)$ for Na nucleus embedded in the center of a cavity in originally uniform electron liquid. For graphical accuracy, this is the same for either choice of $V_{xc}(r)$. See also Fig. 7.5 for screening charge around $K^+$ ion.

$\phi(r)$ set out in the body of Chap. 7, PM(b) obtained the result shown in their Fig. 5. It turns out that the first repulsive hump after the initial minimum is predicted to be very large in this case. Some experimental check of this $\phi(r)$ for Be is obviously very desirable: just as the displaced charge $\Delta n(r)$ for $Na^+$ shown in Fig. A7.1 has been checked in PM(a).

## A7.4. *Relation of Vacancy Formation Energy, in Units of Thermal Energy at Melting, to Departures from Joule's Law*

In a hot solid, near its melting temperature $T_m$, the vacancy, formation energy $E_v$ can be obtained by taking out an atom from the bulk solid and placing it on the surface (compare Appendix, for simple metals). This leaves a 'localized hole', where the atom has been plucked out. Now consider that, at melting, the localized hole no longer exists and therefore let us compare the change in internal energy $E$ of the liquid at volume $\vartheta$ and at volume $\vartheta + \Omega$, with $\Omega$ denoting the atomic volume. This assumes no atomic relaxation around the vacant site, which is reasonable for close-packed solids, but inappropriate for bcc structures. Thus, one equates

$$E_v = E(\vartheta + \Omega) - E(\vartheta) \tag{A7.4.1}$$

and since $\Omega$ is $O(1/N)$ of the total volume $\vartheta$, one can Taylor expand in Eq. (A7.4.1) to find

$$E_v \simeq \Omega \frac{\partial E}{\partial \vartheta} \,. \tag{A7.4.2}$$

One can now calculate the 'departure from Joule's law', $\frac{\partial E}{\partial \vartheta}$, purely thermodynamically; after some modest manipulation (March, 1987), one finds

$$\frac{E_v}{k_B T_m} = \left\{ \frac{(\gamma - 1)(c_v/k_B)}{S(o)} \right\}^{1/2} . \tag{A7.4.3}$$

Here $\gamma = c_p/c_v$ is the ratio of the specific heats in the liquid metal just above the melting temperature $T_m$. $S(o)$ is the long wavelength limit (i.e. $k \to 0$) at the melting temperature $T_m$. $S(o)$ is related, via fluctuation theory, to the isothermal compressibility $k_T$ by the result

$$S(o) = \rho k_B T K_T \tag{A7.4.4}$$

where $\rho$ is the atomic number density. $S(k)$ is the liquid structure factor.

Rashid and March (1989; see also Alonso and March 1989) have considered Eq. (A7.4.3) in relation to a number of metals. The relation is useful, though not fully quantitative, as evidenced by the fact that the average value of $E_v/k_B T_m$ from experiment for the metals considered by the above workers is approximately 11, whereas Eq. (A7.4.3) predicts a number close to 8 for close-packed metals.

*Comparison with Statistical Mechanical Theory Appropriate to (say) Argon*

As mentioned earlier, Faber (1972) with metals specifically in mind, has expressed the vacancy formation energy $E_v$, for a pair force model and in the absence of relaxation, in terms of the pair potential, assumed to possess a Fourier transform as in the pseudopotential description of simple metals, and the liquid structure $S(k)$ referred to immediately above.

Bhatia and March (1984) have employed the alternative $r$-space description of Minchin *et al.* (1974), their formula for the vacancy formation energy $E_v$ being

$$E_v + p\Omega = -\frac{\rho}{2} \int g\phi \, d\mathbf{r} - \frac{\rho}{6} \int r \frac{\partial \phi}{\partial r} g \, d\mathbf{r} . \tag{A7.4.5}$$

Here $p$ is the pressure, $\Omega$ the atomic volume, $\rho$ the number density, $g(r)$ the atomic pair correlation function in the liquid just above $T_m(g(r) - 1$ is the Fourier transform of the liquid structure factor $S(k)$ reduced by unity) and $\phi(r)$ the (assumed) pair potential.

Bhatia and March (1984) were concerned specifically with condensed phases of rare gases, rather than metals discussed earlier in this Appendix. Nevertheless the results are related to those obtained above for metals and will therefore be discussed below.

If one invokes the virial equation for the pressure $p$, namely

$$p = \rho k_B T - \frac{\rho^2}{6} \int r \frac{\partial \phi}{\partial r} g(r)\, d\mathbf{r} \tag{A7.4.6}$$

one notes that putting $p = 0$ yields the second term on the right-hand side of Eq. (A7.4.6) for $E_v$ as $-k_B T_m$. Since $E_v$ is known empirically, it is apparent that the term involving $\partial\phi/\partial r$ makes only a modest contribution to $E_v$. Thus Eq. (A7.4.5) becomes

$$E_v = -\frac{\rho}{2} \int g\phi\, d(\mathbf{r}) - k_B T\,. \tag{A7.4.7}$$

For argon, Bhatia and March argued that since $g = 0$ within the core diameter $\sigma$ and the direct correlation function $c(r)$ of Ornstein and Zernike defined in terms of the so-called total correlation function $h(r) = g(r) - 1$ by

$$h(r) = c(r) + \rho \int h\left(\mathbf{r} - \mathbf{r}'\right) c(\mathbf{r}')\, d\mathbf{r}' \tag{A7.4.8}$$

is zero for hard spheres in the Percus-Yevick approximation outside the core (see, for example, Appendix 3.1 of March, 1990), then following the study of Woodhead-Gallaway, Gaskell and March (1968) where (in the so-called mean spherical approximation, MSA)

$$c(r) = -\frac{\phi_{lr}(r)}{k_B T}\,, \quad r > \sigma \tag{A7.4.9}$$

($\phi_{lr}$ denotes the long-range part of the pair potential), Eq. (A7.4.7) can be rewritten in the form:

$$2E_v = \left\{\rho k_B T \left[\int g(r)\, c(r)\, d\mathbf{r} - \frac{2}{\rho}\right]\right\}. \tag{A7.4.10}$$

One must again caution here that Eq. (A7.4.10) immediately above is appropriate for argon, but will not apply as it stands for the metals discussed earlier in this Appendix.

However, for the condensed rare gases, Eq. (A7.4.10) can be rewritten in terms to the total correlation function $h(r) = g(r) - 1$ as

$$\frac{E_v}{k_B T_m} = \frac{1}{2}\rho \int h(r)\, c(r)\, d\mathbf{r} - \frac{1}{2}\tilde{c}(q = 0) - 1 \tag{A7.4.11}$$

where $\tilde{c}(q)$ is the Fourier transform of $c(r)$ defined in Eq. (A7.4.8) above.

Utilizing now the convolution relation (A7.4.8) between $h(r)$ and $c(r)$ at the point $r = 0$ and the fact that in dense liquids $g(r = 0) = 0$, one can readily show that

$$-1 = c(r = 0) + \rho \int h(r)\, c(r)\, d\mathbf{r}\,. \tag{A7.4.12}$$

Using this relation to eliminate the integral term in Eq. (A7.4.11) leads directly to the Bhatia-March result

$$\frac{E_v}{k_B T_m} = \frac{1}{2}\left[c(r = 0) - \tilde{c}(q = 0) + 3\right]_{T_m}\,. \tag{A7.4.13}$$

In this admitted simple model then, the vacancy energy in units of the thermal energy $k_B T_m$ at melting, is determined entirely by the direct correlation function $c(r)$ of the liquid just above the freezing point. Because $c\,(r = 0)$ is a large negative number, and $\tilde{c}\,(q = 0)$ is approximately the same (e.g. in a hard sphere model they differ by unity, as shown by Bhatia and March (1984)), it is evident why the ratio $E_v/k_B T_m$ is a large number, found empirically to be $\sim 10$.

## A8.1. *An Embedded Atom Potential for hcp Metal Zr*

Goldstein and Jónsson (1995) have extended the embedded atom method (EAM), discussed elsewhere in the body of Chap. 8, to deal with the hcp metal Zr. The non-ideal $c/a$ ratio and the elastic responses have been incorporated in the fitting procedure which these authors employ. They assume simple functional forms for the pair interaction, atomic electron density and embedding function. They parametrize the functions by fitting to experimental values of (i) cohesive energy, (ii) equilibrium lattice constants, (iii) single crystal elastic constants and (iv) vacancy formation energy.

Furthermore an equation of state proposed by Rose *et al.* (1984) is employed to reproduce the pressure dependence of the cohesive energy. Taking into account the anisotropic elastic response of crystal Goldstein and Jónsson also reproduce dimer data and a high energy sputtering potential in order to extrapolate the range of utility of their potential into regions of extreme electron density-both high and low.

These workers then obtain fairly good accord with other experimentally observed properties. They have applied their potential to the calculation of stacking fault and self-interstitial formation energies and we summarize their main findings on these latter properties immediately below.

### A8.2. *Stacking Fault Energy*

A satisfactory potential must evidently stabilize the hcp structure of Zr with respect to other types of lattice. During the fitting procedure employed by Goldstein and Jónsson (1995), stability with respect to the fcc and bcc lattices was ensured.

As these workers point out, another test for the structural stability is to calculate the $I_2$ stacking fault energy. The fault occurs in the basal plane and corresponds to a translation in one close-packed plane to give the stacking ... ABABABCACAC .... The stacking fault energy must be positive and the theoretical estimates of Legrand (1984) suggest it should be large.

Goldstein and Jónsson (1995) have computed the $I_2$ stacking fault energy by employing a system of 1584 atoms consisting of 22 planes of 72 atoms. They combined two equivalent hcp crystals, translating one with respect to the other to give the desired stacking. They point out that, in effect, this creates two stacking faults as periodic boundaries are employed to model the infinite lattice. The size of the system was selected such that the two faults, separated by 10 planes, would not affect each other with the range of the interaction extending over some three layers. Goldstein and Jónsson obtained with their EAM potential a stacking fault energy of $\sim 120\ \mathrm{mJm^{-2}}$, which is considerably lower than the theoretical estimate ($> 300\ \mathrm{mJm^{-2}}$) of Legrand (1984). It is relevant here to note that Finnis-Sinclair potentials were generated by Igarashi *et al.* (1991) for Co, Zr, Ti, Ru, Hf, Zn, Mg and Be. These workers reported a very low value of the stacking fault energy with their potential namely $\sim 30\ \mathrm{mJm^{-2}}$ and they noted that it did not prove possible to obtain a reasonably high stacking fault energy without compromising the lattice stability under deformation. (See also footnote before Sec. 6.5.3).

### A8.3. *Self-Interstitial Formation Energies*

To test the potential further, under non-equilibrium conditions, Goldstein and Jónsson (1995) also computed the self-interstitial formation energies and their results are reproduced in Table A8.1 below.

The interstitial sites are as proposed by Johnson and Beeler (1981) and the calculations were carried out by Goldstein and Jónsson (1995) on 4 system sizes, from 1152 to 4800 atoms, in order to investigate possible size effects. The interstitial atom was introduced by these authors to the system was permitted to relax while the volume of the system was held constant. Though the calculated energies showed differences, Goldstein and Jónsson note that the effect of

Table A8.1. Self-interstitial formation energy (in eV) and $I_2$ stacking fault energies (in $mJm^{-2}$) for Zr (after Goldstein and Jónsson, 1995).

| Self-interstitial configuration | Goldstein-Jósson (1995) | Willaime-Massobrio (1991) | Oh-Jósson (1988) | Igarashi *et al.* (1991) |
|---|---|---|---|---|
| $B_c$ | 4.1 | $B_o$ | $B_o$ | 8.1 |
| $B_o$ | 4.6 | 4.3 | 4.7 | 9.2 |
| $B_c$ | 4.2 | $B_o$ | * | 8.8 |
| $C$ | 4.2 | 4.3 | 4.5 | 7.7 |
| $O$ | 4.6 | 4.5 | 4.6 | 7.5 |
| $T$ | 4.2 | $C$ | $C$ | 9.1 |
| $S$ | 4.2 | $C$ | 4.9 | 8.7 |
| $I_2$ stacking fault energy | 121 | * | * | 27 |

* above indicates no data is available.

system size is tiny ($< 1\%$): the values they report being for a system with 2048 lattice atoms. At the time of writing a comparison of the self-interstitial formation energies with experiment is not possible.

Thus, in Table A8.1, their results are compared with previously reported computations. For such a comparison, Goldstein and Jónsson make several points as follows:

(i) The potentials which accurately reproduce the non-ideal $c/a$ ratio find stable minima for each of the proposed sites. The potentials resulting in a near-ideal $c/a$ ratio predict the $B_c$, $T$ (Oh and Johnson, 1988; Willaime and Massobrio, 1990), $B_t$ and $S$ (Willaime and Massobrio, 1990) sites to be unstable: atoms introduced into these positions decay to the $C$ or $B_o$ sites.

(ii) The values of Igarashi *et al.* (1991) for the interstitial formation energies are roughly twice those of Goldstein and Jónsson (1995).

(iii) Goldstein and Jónsson find the $B_c$ configuration to be the most stable. Although the $C$, $T$ and $S$ configurations relax to the same energies, the sites remain unique with a separation of 0.59 Å between $C$ and $T$ and 0.49 Å between $C$ and $S$.

## A9.1. *Experimental Techniques on Positron Spectroscopy*

Positrons can have several states each of which gives a characteristic lifetime $\tau_i = 1/\lambda_i$. The probability of an annihilation at time $t$ is the sum of exponential decay components:

$$-\frac{dn(t)}{dt} = \sum_i I_i \lambda_i \exp[-\lambda_i t] \,,$$

$$\sum_i I_i = 1 \tag{A9.1.1}$$

with relative intensities $I_i$. This is normally analyzed by computers in order to extract lifetime values $\tau_i$ and relative intensities $I_i$ associated with the different components. The timing pulses are obtained by differential constant-fraction discrimination. The time delays between the start and stop signals are converted into amplitude pulse the heights of which are stored into a multichannel analyser. A typical time resolution is (200–250) ps.

Since there are only 2–3 decades of the exponential part of Eq. (A9.1.1) in an experimental lifetime spectum, typically only two lifetime components can be used to analyse the spectra in metals. The separation of two lifetimes is successful only if $\lambda_1/\lambda_2 > 1.5$.

The average lifetime given by

$$\tau_{av} = \int dt\, t\left(-\frac{dn(t)}{dt}\right) = \sum_i I_i \tau_i \tag{A9.1.2}$$

is a good and statistically accurate parameter. It can be used to label various states of the sample, but part of the underlying physical information connected with different annihilation modes is lost.

## A9.2. *Lineshape Parameters of Doppler-Broadening Spectroscopy*

For characterizing the 511 keV line, the parameter $S$ is defined as the ratio of the counts in the central region of the annhilation line of total number of the counts in the line. The wing parameter $W$ is the relative fraction of the counts in the wing regions of the line. The annihilations with valence electrons fall predominantly in the region of the $S$ parameter due to their low momentum. Therefore, the $S$ parameter reflects the behaviour of valence annihilation. On the other hand, the $W$ parameter reflects the behaviour of core annihilations due to the high momentum of core electrons.

The absolute values of the parameters are meaningless. The relative values like $S/S_{ref}$ and $W/W_{ref}$ are practically comparable between various experiments. When positrons are trapped, the lineshape is characteristic of the trapping defect. In a vacancy-type defect, the density of valence electrons is reduced. This leads to the narrowing of their momentum distribution which is seen as an increase in $S$. On the other hand, the core electrons are decreased in a vacancy-type defect. This leads to a decrease in the core annihilation parameter $W$.

# References


Abraham, F. F., Brodbeck, D., Rafey, R. A. and Rudge, W. E., *Phys. Rev. Lett.* **73**, 272 (1994).

Ackland, G. J., Tichy, G., Vitek, V. and Finnis, M. W., *Phil. Mag.* **A56**, 735 (1987).

Ackland, G. J. and Vitek, V., *Phys. Rev.* **B41**, 10324 (1990).

Ackland, G. J., Finnis, M. W. and Vitek, V., *J. Phys.* **F18**, L153 (1988).

Ackland, G. J. and Tichy, G., Vitek, V. and Finnis. M. W., *Phil. Mag.* **A56**, 15 (1987).

Aharony, A., Feder, J., eds. *Fractals in Physics, Essay in Honor of Mandelbrot, B. B.*, (North-Holland, 1990).

Alfred, L .C. R. and March, N. H., *Phil. Mag.* **2**, 985 (1957).

Alonso, J. A. and March, N. H., *Electrons in Metals and Alloys*, (Academic Press, London, 1989).

Alonso, J. A. and March, N. H., *Phys. Chem. Liquids* **20**, 235 (1989).

Alonso, J. A. and March, N. H., *Phys. Chem. Liquids* **20**, 235 (1989).

Alshits, V. I., in: *Elastic Strain Fields and Dislocation Mobility*, eds. Indenbom, V. L. and Lothe, J., (Elsevier Science Publishers B. V., 1992).

Ananthakrishna, G., Kubin, L. P. and Martin, G., *Non-linear Phenomena in Materials Science III*, (Trans. Tech. Publications, Switzerland, 1994).

Aoki, K. and Izumi, O., *Nippon Kinzoku Gakkaishi*, **41**, 170 (1977), (in Japanese).

Aoki, M. and Pettifor, D. G., *Mat. Sci. Engn.* **A176**, 19 (1994).

Arharov, B. E., Sym. UHC, *Phys. Met.* **16**, (1955) (in Russian).

Arharov, B. E., Sym. UHC, *Phys. Met.* **19**, 23 (1958) (in Russian).

Arkhipov, N. K., Klotsman, S. M., Polikarhova, M. P., Timopheev, A. N. and Shepatkovskii, O. P., *Fizika Metall. Metallovede*, **62**, 1181 (1986), (in Russian).

Arponen, J., Hautojärvi, P., Nieminen, R. and Pajanns, E., *J. Phys.* **F3**, 2092 (1973).

Asaro, R. J., *J. Phys.* **F5**, 2249 (1975).

Ashcrobt, N. W., *Phys. Lett.* **23**, 48 (1966).

Atkinson, W. and Caberera, W., *Phys. Rev.* **138**, A764 (1965).

Badura, K., Brossmann, U., Würschum, R. and Schaefer, H. E., Materials Sci. Forum, *Proceedings of ICPA-10*, eds. He,Y. J., Cao, B. S. and Jean, Y. C., Vols. 175–178, 1995, pp. 295, Tran. Tech. Publications, Switzerland.

Bak, P. and Tang, C., *Physica* **38**, 5 (1989).

Bak, P. and Creutz, M., in: *Fractals and Disordered Systems*, eds. Bunde, A. and Havlin, S., (Springer, Berlin, 1993).

Baker, J. A. and Coleman, P. G., *J. Phys.* **Cm1**, 39 (1989).
Bai, Y. L., Lu, C. S., Ke, F. J. and Xia, M. F., *Phys. Lett.* **A185**, 196 (1994).
Balluffi, R. W., in: "Interfacial Segregation," *Amer. Soc. Metals*, (1977).
Bardeen, J., *Phys. Rev.* **52**, 688 (1937).
Barnett, D. M. and Lothe, J., *Physica Norvegica* **7**, 13 (1973).
Basinski, Z. S. and Basinski, S. J., *Prog. Mater. Sci.* **36**, 89 (1992).
Bauchard, E., Lapasset, G. and Planes, J., *Europhys. Lett.* **13**, 73 (1990).
Beeler, Jr. J. R. and Esterling, *Metals* **19**, Ohio Park: American Society for Metals.
Beling, C. D., Simpson, R. I., Charlton, M., Javobeen, F. M. and Griffith, T. C., *Appl. Phys.* **A42**, 111 (1987).
Bergersen, B., Pajanne, E., Kubica, P., Stott, M. J. and Hodges, C. H., *Solid State Commun.* **15**, 1377 (1974).
Beringer, R., Montgomery, C. G., *Phys. Rev.* **61**, 222 (1942).
Berko, S., in: Positron Solid-State Physics, eds. Brandt, W. and Dupasquier, A., (North-Holland, Amsterdam, 1983) pp. 64–145.
Berko, S., Haghgoohe, M. and Mader, J. J., *Phys. Lett.* **A63**, 335 (1977).
Berko, S., Singh, D. J. and von Stetten, E. C., *J. Phys. Chem. Solids* **52**, 1485 (1991).
Bernardes, A. T. and Morevia, J. G., *Phys. Rev.* **B49**, 15035 (1995).
Bhatia, A. B. and March, N. H., *J. Chem. Phys.*, **80**, 2076 (1984).
Bhusham, B., *Tribology and Mechanics of Magnetic Storage Devices* (Springer, New York, 1990).
Bhusham, B. and Koinkar, V. N., *Appl. Phys. Lett.* **64**, 1653 (1994).
Bilby, B. A., Progr. *Solid Mech.* **1**, 331 (1960).
Bilby, B. A., Bullough, R. and Smith, E., *Proc. Roy. Soc.* **A231**, 263 (1955).
Bilby, B. A., Cottrell, A. H. and Swinden, K. H., *Proc. Roy. Soc.* **A272**, 304 (1963).
Bilby, B. A. and Eshelby, J. D., in: *Fracture*, eds. Liebowitz, H., Vol. 1, (Academic Press, New York, 1968) pp. 100-178.
Binnig, G., Quate, C. F.and Gerber, Ch., *Phys. Rev. Lett.* **56**, 930 (1986).
Blandin, A., Daniel, E. and Friedel, J., *Phil. Mag.* **4**, 180 (1959).
Blazej, M., Flores, F. and March, N. H., *Mol. Phys.*, (1995) (in press).
Blazej, M. and March, N. H., *Phys. Rev.*, **E48**, 1782 (1993).
Bollman, W., *Crystal Defects and Crystalline Interfaces*, (Springer Berlin, 1970).
Bordoni, D. G., *Rci. Sci.* **19**, 851 (1949).
Born, M. and Green, H. S., *Proc. Roy. Soc.* **A188**, 10 (1946).
Born, M. and Huang, K, *Dynamical Theory of Crystal Lattices* (Clarendon, Oxford, 1956).
Boronski, E. and Nieminen, R. M., *Phys. Rev.* **B34**, 3820 (1986).
Bowden, F. P. and Taber, D., *The Friction and Lubrication of Solids* (University Press, Oxford, 1950).
Bowden, F. P. and Taber, D., *Friction and Lubrication* (Methuen, London, 1979).
Bowen, D. K., Christion, J. W. and Taylor, G., *Can. J. Phys.* **45**, 903 (1967).
Boyer, L. L. and Hardy, J. R., *Phil. Mag.* **24**, 647 (1971).
Bradley, A. J. and Taylor, A., *Proc. Roy. Soc.* **A159**, 56 (1937).

Brandon, D. G., Ralph, B., Ranganathan, S. and Wald, M. S., *Acta Metall.* **12**, 813 (1964).
Brandt, W. and Dupasquier, A. (eds.), *Positron Solid-State Physics* (North Holland, Amsterdam, 1983).
Brandt, W. and Paulin, R., *Phys. Rev.* **B5**, 2430 (1972).
Brede, M. and Haasen, P., *Acta Metall.* **36**, 2003 (1988).
Brown, L. M., in: *Dislocation Modelling of Physical systems*: eds. Ashby, M. F., Bullough, R., Hartley, C. S. and Hirth, J. P., (Pergaman, Oxford, 1981) p. 51.
Brown, R. C., Warster, J. March, N. H., Berrin, R. C. and Bullough, R., *Phil. Mag.* **23**, 555 (1971).
Brusa, R. S., Duarte Naia, M., Margoni, D. and Zecca, A., in: Positron Annihilation, *Proceedings of ICPA-10*, eds. He, Y. J., Cao, B. S. and Jean, Y. C., or Materials Science Forum, Vols. 175–178 (Trans. Tech. Publications, Switzerland, 1995) pp. 655.
Budd, H. F. and Vannimenus, J., *Phys. Rev. Lett.* **31**, 1218 and 1430 (1973).
Bullough, R. and Perrin, R. C., in: *Dislocation Dynamics*, eds. Rosenfield, A. R., Hahn, G. T., Bement Jr. A. L. and Jaffee, R. I. (McGraw-Hill, 1968) p. 175.
Bullough, R. and Pemin, R. C., *Proc. Roy. Soc.* **A305**, 541 (1968).
Burke, E. C. and Hibbard, Jr. W. R., *Trans. AIME* **194**, 295 (1952).
Burns, S. J., *Scr. Metall.* **20**, 1489 (1986).
Buttner, F. H., Funk, E. R. and Udin, H., *TAIME* **194**, 40 (1952).
Callaway, J., *Energy Band Theory* (Academic Press, New York, 1964).
Cao, B. H. and Zhang, H. T., *Acta Physica Sinica* **35**, 750 (1986), (in Chinese).
Carbotte, J. P. and Kahana, S., *Phys. Rev.* **A139**, 213 (1965).
Carlsson, A. E., in: *Solid State Physics*, eds. Ehrenreich, H., Seitz, F. and Turnbull, D., **43**, 1 (1990).
Castaing, J., Veyssiere, P., Kubin, L. P. and Rabier, J., *Phil. Mag.* **A44**, 1407 (1981).
Ceferley, D. M. and Alder, B. J., *Phys. Rev. Lett.* **45**, 566 (1980).
Celinski, Z., Heinrich, B., Cochran, J. F., Myrtle, K. and Arrott, A. S., in: *Science and Technology of Nanostructured Magnetic Materials*, eds. Hadjipanayis, G. C. and Prinz, G. A., (Plenum, New York, 1991).
Celli, V. and Flytzanis, N., *J. Appl. Phys.* **41**, 4443 (1970).
Ceperley, D. M. and Alder, B. J., *Phys. Rev. Lett.* **45**, 566 (1980).
Chambers, L. G., *Integral Equations* (Inter. Textbook Co., Ltd., 1976) p. 133.
Chelikowsky, J. R., *Phys. Rev.* **B19**, 686 (1979).
Chen, N. K. and Pond, P. B., *J. Met.* **4**, 1085 (1952).
Chen, Y. F., Wang, K. X. and Wang, Z. M., Kexue Tongbao, *Sci. Bull.* **21**, 974 (1980), (in Chinese).
Cheng, C., Needs, R. J., Heine, V. and Churcher, N., *Europhys. Lett.* **3**, 475 (1987).
Cheung, K., Yip, S. and Argon, A. S., *J. Appl. Phys.* **69**, 2088 (1991).
Chicois, J., Fougeres, R., Guichon, G., Hamel, A. and Vincent, A., *Acta Met.* **34**, 2157 (1986).
Ching, E. S. C., Langer, J. S. and Nakamish, H., *Phys. Rev. Lett.* **76**, 1087 (1996).
Christian, J. W., *Metall. Trans.* **A14**, 1237 (1983).

Christian, J. W. and Crocker, A. G., in: *Dislocations in Solids*, eds. Nabarro, F. R. N., Vol. 3 (North-Holland, Amsterdam, 1980) p. 165.
Coble, R. L., *J. Appl. Phys.* **34**, 1679 (1963).
Cohen, M. L. and Heine, V., in: *Solid State Phys.* **24**, 37 (1970).
Conrad, H. and Sprecher, A. F., in: *Dislocation in Solids*, Ed. Nabarro, F. R. N., Vol. 8 (North-Holland, Amsterdam, 1989).
Cooper, M. J., *Phil. Mag.* **8**, 811 (1963).
Corbel, C., Puska, M. and Nieminen, R. M., *Phys. Rev.* **B34**, 3820 (1986).
Corless, G. K. and March, N. H., *Phil. Mag.* **6**, 1285 (1961).
Cottrell, A. H., *An Introduction to Metallurgy* (Eduard Arnold, Bath, 1976) p. 40.
Cottrell, A. H., *Dislocations and Plastic Flow of Crystals* (Oxford University Press, London 1953 — in Chinese, trans. by Kê,T.S., 1960).
Cottrell, A. H., *The Mechanical Properties of Matter* (John Wiley & Sons, Inc. London, 1964).
Cottrell, A. H., *Theory of Crystal Dislocations*, (Documents on Modern Physics), (Gordon and Breach, 1964).
Cox, J. J., Horne, G. T. and Mehl, R. F., *Trans. ASM.* **49**, 116 (1957).
Crais, R. A., *Phys. Rev.* **B6**, 1134 (1972).
Cui Guowen, *Surfaces and Interfaces* (Tsinghua University Press, Beijing, (in Chinese), 1990).
Cyrot-Lackmann, F., *Adv. Phys.* **16**, 393 (1967).
Czacher, A., in *Inelastic Scattering of Neutrons*, Vol. I (IAEA, Vienna, 1965) p. 181.
Dacorogna, M., Ashkenazi, J. and Peter, M., *Phys. Rev.* **B26**, 1527 (1982).
Daniuk, S., Sob, M. and Rubaszek, A., *Phys. Rev.* **B43**, 2580 (1991).
Darken, L. S. and Gurry, R. W., *Physical Chemistry of Metals* (McGraw-Hill, New York, 1963).
Dauskardt, R. H., Haubensak, F and Ritchie, R. O, *Acta Metall. Mater.* **38**, 143 (1990).
Daw, M. S., in: *Atomistic Simulation of Materials*, eds. Vitek, V. and Srolovitz, D. J., (Plenum, New York, 1989).
de Arcangelis, L., Hansen, A., Herrmann, H. J. and Roux, S., *Phys. Rev.* **B40**, 877 (1989).
de Heer, W. A., Knight, W. D., Chou, M. Y. and Cohen, M. L., in: *Solid State Physics*, eds. Ehrenreich, H. and Turnbull, Vol. 40 (Academic Press, Orlando, 1987) p. 93.
de Wit, R., *Int. J. Engng. Sci.* **19**, 1475 (1981).
Dehjar, E. Y., *Problems on Metal Physics and Physical Metallurgy* (1957, in Russian).
Deng, K. M., He, Z. X., Lung, C. W., Wang, K. L. and Wang, J. L., *Int. J. Engn. Sci.* **29**, 79 (1991).
Deng, W., Xiong, L. Y., Lung, C. W., Wang, S. H. and Guo, J. T., Ph.D. thesis, 1994.
Deng, W., Xiong, L. Y., Lung, C. W., Wang, S. H. and Guo, J. T., *J. Mater. Sci. Lett.* **14**, 291 (1995).
Deng, W., Xiong, L. Y., Lung, C. W., Wang, S. H. and Guo, J. T., *ibid*, in press.

Deng, W., Xiong, L. Y., Lung, C. W., Wang, S. H. and Guo, J. T., Materials Sci. Forum, *Proceedings of ICPA-10*, eds. He, Y. J., Cao, B. S. and Jean, Y. C., Vol. 175–178 (Trans. Tech. Publications, Switzerland, 1995) p. 339.

Depersio, J. and Escaig, B., *Phys. Stat. Sol.* (a), **40**, 393 (1977).

Dlubek, G., Brummer, O., Meyendorf, N., Hautojärvi,P., Vehanen, A. and Yli-Kauppila, J., *J. Phys.* **F9**, 1961 (1979).

Dong, L. K., Zhang, H. Y. and Lung, C. W., *Int. J. Solids Structures*, **25**(7), 707 (1989).

Dorn, J. E., in: *Dislocation Dynamics*, eds. Rosenfield, A. R., Hahn, G. T., Bemeyt, A. L. and Jaffe, R. I. (McGraw Hill, New York, 1968) p. 27.

Douglass, R. W., Crier, C. A. and Jaffee, R. L., Rept. Office of Naval Research, Battel Memorial Institute, 1961.

Duan, Y. S. and Duan, Z. P., *Int. J. Engn. Sci.* **24**, 513 (1986).

Duan, Y. S. and Zhang, S. L., *Int. J. Engn. Sci.* **28**, 689 (1990).

Duan, Y. S. and Zhang, S. L., *ibid.*, (1991).

Ducastelle, F. and Cyrot-Lackmann, F., *J. Phys. Chem. Solids* **31**, 1295 (1970).

Duesbery, M. S. and Foxall, R. A., *Phil. Mag.* **A20**, 719 (1969).

Duesbery, M. S. and Richardson, G. Y., *CRC Critical Review in Solid State and Materials Science* **17**, 1 (1991).

Duesbery, M. S., in: *Dislocations in Solids*, ed. Nabarro, F. R. N., Vol. 8 (North Holland, Amsterdam, 1989).

Dugdale, D. S., *J. Mech. Phys. Solids* **8**, 100 (1960).

Dupasquier, A. and Ottaviani, G., in: *Positron Spectroscopy of Solids*, eds. Dupasquier, A. and Mills, jr. A. P. (IOS Press, Tokyo, 1995).

Dupasquier, A. and Somoza, A., Positron Annihilation, in: *Proc. ICPA-10*, 1995, or Materials Science Forum, Vols. 175–178 (Trans. Tech. Publ., Switzerland, 1995) p. 35.

Duxbury and Li, in: Disorder and Fracture, eds. Charmet, J. C., Roux, S. and Guyon, E., NATO ASI Series, Series B, Physics, Vol. 235, Chapter 8, pp. 141, (Plenum Press, New York).

Duxbury, P. M. and Li, Y. S., in: *Proceedings of the SIAM Conference on Random Media and Composites*, (Leesbury, Virginia, 1988).

Duxbury, P. M. and Leath, P. L., *J. Phys.* **A20**, L411 (1987).

Duxbury, P. M., Leath, P. L. and Beale, *J. Phys. Rev.* **B36**, 367 (1987).

Edmondson, B., *Proc. Roy. Soc.* **264**, 176 (1961).

Eldrup, M., Mogensen, O. E. and Evans, J. H., *J. Phys.* **F6**, 499 (1976).

Engoy, T., Maloy, K. J., Hansen, A., and Roux, S., *Phys. Rev. Lett.* **73**, 834 (1994).

Eshelby, J. D., in: *Dislocations in Solids*, ed. Nabarro, F. R. N., Vol. 1 (North-Holland, Amsterdam, 1979) p. 167.

Eshelby, J. D., in: *Solid State Physics*, eds. Seitz, F. and Turnbull, D., Vol. 3 (Academic Press, New York, 1956) pp. 79–144.

Eshelby, J. D., *Proc. Roy. Soc.* **A241**, 376 (1957).

Eshelby, J. D., *Proc. Roy. Soc.* **A252**, 561 (1959).

Eshelby, J. D., Read, W. T. and Shochley, W., *Acta Metlall.* **1**, 251 (1953).

Fantozzi, G., Esnout, C., Benoit,W. and Ritchie, I. G., *Prog. Mat. Sci.* (Pergamon, Oxford), **27**, 311 (1982).
Farmer, J. D., in *Dimensions and Entropies in Chaotic Systems*, ed. Mayer-Kress, E. (Springer, Berlin, 1986) p. 54.
Fast, L., Wills, J. M., Johonsson, B. and Eriksson, O., *Phys. Rev.* **B51**, 17431 (1995).
Feder, J. *Fractals* (Plenum, New York, 1988).
Fen, D., *Metal Physics*, Vol. I (China Scientific Publlo., Beijing, 1987 in Chinese).
Fenn, Jr. R. W., Hibbard, Jr. W. R. and Lepper, Jr. H. A., *Trans. AIME* **188**, 175 (1950).
Ference, T. G. and Balluffi, R. W., *Scripta Metall* **22**, 1929 (1988).
Fineberg, J., Gross, S. P, Marder, M. and Suinney, H. L., *Phys. Rev. Lett.* **67**, 457 (1991).
Finnis, M. W., Sinclair, J. E., *Phil. Mag.* **A50**, 45 (1984).
Finnis, M. W., Sinclair, J. E., *ibid* **53**, 161 (1986).
Finnis, M. W., Paxton, A. T., Pettifor, D. G., Sutton, A. P. and Ohta, Y., *Phil. Mag.* **A58**, 143 (1988).
Flinn, P. A., *Trans. AIME*, **218**, 145 (1960).
Flores, F. and March, N. H., *J. Phys. Chem. Solids* **42**, 439 (1981).
Fonda, R. W., Yan, M. and Luzzi, D. E., *Phil. Mag. Lett.* **71**, 221 (1995).
Fong, C. Y., Walter, J. P. and Cohen, M. L., *Phys. Rev.* **B11**, 2759 (1975).
Frank, F. C., in: *Symposium on the Plastic Deformation of Crystalline Solids*, 1950, pp. 150, Office of Naval Research, Pitts burgh.
Frenkel, J. and Kontorova, T., *Phys. Z. Sowj.* **13**, 1 (1938).
Frenkel, Y. E., *Introduction to Theory of Metals* (Chinese Scientific Pub. Co., Beijing, 1957 — in Chinese).
Friedel, J., in: The Physics of Metals, ed. Ziman, J. M. (Cambridge University Press, Cambridge, 1969) p. 340.
Friedel, J., in *Dislocations* (Pergamon Press, Paris, 1964 — in English).
Friedel, J., in: *Dislocations in Solids*, ed. Nabarro, F. R. N., Vol. 1 (North Holland, Amsterdam, 1979) p. 1–32.
Friedel, J., *Ann. Phys.* (Paris), **1**, 257 (1976).
Fumi, F., *Phil. Mag.* **46**, 1007 (1955).
Gairola, B. K. D., in: *Continuum Models of Discrete System*, eds. Brulin, O. and Hsich, R. K. T. (North-Holland, Amsterdam, 1981) p. 55.
Galebiewska-Lasota, A. A., *Int. J. Engng. Sci.* **17**, 329 (1979).
Gao, F. and Zhang, H. T., *Acta Physica Sinica* **37**, 1315 (1988), (in Chinese).
Gehlen, P. C., Beeler, Jr. J. R. and Jaffee, R. I., in: *Interatomic Potentials and Simulation of Lattice Defects* (Plenum, New York, 1972).
Genould, P., Singh, A. K., Manuel, A. A., Jarlborg, T., Walter, E., Peter, M. and Weller, M., *J. Phys.* **F18**, 1933 (1988).
Gifkins, R. C. and Snowden, K. U., Trans. *Metall. Soc. AIME* **239**, 910 (1967).
Gilgien, L., Galli, G., Gygi, F. and Car, R., *Phys. Rev. Lett.* **72**, 3214 (1994).
Goldstein, A. S. and Jónsson, H., *Phil. Mag.* **B71**, 1041 (1995).
Gong, B. and Lai, Z. H., *Engn. Fract. Mech.* **44**, 991 (1993).

Gong, B., Lai, Z. H., Ninomi, M. and Kobayashi, T., *Acta Met. Sinica* (English Edition), Series A, **6**, 121(1993).

Goodman, D. A., Bennett, L. H. and Watson, R. E., *Scripta Metall.* **17**, 91 (1983).

Gou Qingquan, *Kexue Tongbao* **6**, 22 (1962), (in Chinese).

Gou Qingquan, Su Weinhue, Dai Shouyu, *J. Nature Science of Jilin University* **1**, 229 (1962), (in Chinese).

Granato, A. V., *Phys. Rev. Lett.* **27**, 660 (1971).

Granato, A. V., in *Internal Friction and Ultrasonic* Attenuation in Solids Proc. ICIFUAS-9, ed. Kê, T. S. (Ge Tingsui), July 17–20, 1989, Beijing, China (Int. Acad. Publishers, Pergamon Press, 1990) p. 3.

Greenberg, B. A., Katsnelson, M. I., Koreshkov, V. G., Oselskii, Yu. N., Peschanskikh, G. V., Trefilov, A. V., Shamanaev, Yu. F. and Yakovenkova, L. I., *Phys. Stat. Sol.* (b), **158**, 441 (1990).

Greenough, A. P., *Phil. Mag.* **43**, 1075 (1952).

Gremaud, G., *J. de Physique*, **42**, C5-1114 (1981).

Griffith A. A., *Proc. Int. Congr. Appl. Nech* **55**, (1924).

Grosskreutz, J. C. and Mughrabi, H., in: *Constitutive Equations in Plasticity*, ed. Argon, A .S. (MIT Press. Cambridge, Massachusetts, 1975).

Gschneidner K. A., Theory of Alloy Phase Formation, ed. Bennett *The Metallurgical Society of the AIME* (Warendale, 1980) p. 1.

Gullikson, E. M. and Mill, jr, A. P., *Phys. Rev. Lett.* **57**, 376 (1986).

Gundersen, S. A. and Lothe, J., Springer Series on Wave Phenomena, eds. Parker, D. F. and Maugin, G. A., Vol. 7 (Springer, Berlin), 1988).

Gupta, R. P. and Siegel, R. W., *Phys. Rev. Lett.* **39**, 1212 (1977).

Gupta, R. P. and Siegel, R. W., *Phys. Rev.* **B22**, 4572 (1980).

Gupta, R. P. and Siegel, R. W., *J. Phys.* **F10**, L7 (1980).

Ha, K. F. and Wang, X. H., *Acta Met. Sinica* (English Letters), **8**, 1 (1995).

Ha, K. F., Xu, Y. B., Wang, X. H. and An, Z. Z., *Acta Metall. Mater.* **38**, 1643 (1990).

Ha, K. F., Yang, C. and Bao, J. S., *Scripta Met. Mat.* **30**, 1065 (1994).

Hafner, J. and Heine, V, *J. Phys.* **F13**, 2479 (1983).

Haken, H., *Advanced Synergetics*, (Springer,Berlin, 1983) (also available in Chinese).

Haken, H., *Synergetics, An Introduction*, 3rd ed. (Springer, Bernn, 1983) (also available in Chinese).

Hansen, A., Hinrichen, E. L., Maloy K. J., and Roux, S., *Phys. Rev. Lett.* **71**, 205C (1993).

Hansen, H. E., Eldrup, M., Linderoth, S., Nielson, B. and Petersen, K., in: Positron Annihilation, eds: Coleman, A., Skarma, B. and Diana, C. (North-Holland, Amsterdam, 1982) p. 432.

Hansen, H. E, Nieminen, R. M. and Puska, M. J., *J. Phys.* **F14**, 1299 (1984).

Harper, J. and Dorn, J. E., *Acta Metall.* **5**, 654 (1957).

Hastings, H. M. and Sugihara, G., *Fractals: A User's guide for the natural sciences* (Oxford University Press, Oxford, 1993).

Hautojärvi, P. (ed.), *Positrons in Solids, Topics in Current Physics*, **12**, (1979).

Hautojärvi, P., *Solid State Commun.* **11**, 1049 (1972).
Hautojärvi, P. and Corbel, C., in: *Positron Spectroscopy of Solids*, eds. Dupasquia, A, and Mills, jr. A. P. (IOS press, Amsterdam, Oxford, 1995).
Hautojärvi, P., Huomo, H., Puska, M. and Vehanen, A., *Phys. Rev.* **B7**, 4326 (1985).
Hautojärvi, P., Johansson, J., Vehanen, A., Yli-Kauppila, J. and Moser, P., *Phys. Rev. Lett.* **44**, 1326 (1980).
Havlin, S. and Ben-Avraham, D., *Adv. Phys.*, **36**, 695 (1987).
He Qin, Xiong Lianyue and Lung, C. W., *Kexue Tongbao* **17**(11), 500 (1966) (in Chinese).
Hede, B. B. J. and Carbotte, J. P., *J. Phys. Chem. Solids* **33**, 727 (1972).
Hedin, L. and Lundqvist, S., in: *Solid State Phys.* eds. Ehrenreich, H., Seitz,F. and Turnbull, D. **23**, 2 (1969).
Heine, V., *Comments on Condensed Matter Physics* **16**, 379 (1994).
Heine, V. and Weaire, D., in: *Solid State Phys.* eds. Ehrenreich, H., Seitz, F. and Turnbull, D., **24**, 249 (1970).
Heinrichs, J., *Phys. Rev.* **B32**, 4232 (1985).
Heinz, A. and Neumann, P., *Acta Metall.* **38**, 1933 (1990).
Herman, F. and March, N. H., *Solid State Commun.* **50**, 725 (1984).
Herring, C., *J. Appl. Phys.* **21**, 437 (1950).
Herring, C., *Phys. Rev.* **57**, 1169 (1940).
Herrmann H. J., *Physica* **38**, 192 (1989).
Herrmann, H. J., *Physica A* **163**, 359 (1990).
Herrmann, H. J. and Roux, S. (eds.), *Statistical Models for the Fracture of Disordered Media, Random Materials and Processes*, Vol. 2 (North-Holland, Amsterdam, 1990).
Herrmann, H. J. and Roux, S., in: *Statistical Models for the Fracture of Disordered Media*, eds. Herrmann, H. J. and Roux, S., Vol. 2, (Elsevier Science Publishers B. V., 1990).
Herrmann, H. J. and de Arcangelis, L., in: *Disorder and Fracture*, eds. Charmet, C., Roux,S. and Guyon, E. (Plenum Press, 1990) pp. 149–163.
Hidalgo, C, De Diego, N. and Moser, P., *Appl. Phys.* **A40**, 25 (1986).
Hieber, H., Mordike, B. L. and Haessen, P., *Platinum Metall. Rev.* **8**, 102 (1967).
Hinrichaen, E. L., Hansen, A. and Roux, S., *Europhys. Lett.* **8**, 1 (1989).
Hirsch, P. B., *J. Phys.* **40**, C6 (1979), France.
Hirsch, P. B., Proc. *5th Int. Conf. Crystallography* (Cambridge University, Pres, 1960) p. 139.
Hirsch, P. B., *Progress in Materials* **36**, 63–88 (1992).
Hirth, J. P., in: Elastic *Strain Fields and Dislocation Mobility*, eds. Indenbom, V. L. and Lothe, J., MPCMS, Vol. 31 (Elsevier, 1992) p. 237.
Hirth, J. P. and Lothe, J., *Theory of Dislocations*, 2nd ed. (John Wiley & Sons. Inc., New York, 1982).
Hoagland, R. G., Daw, M. S., Foiles, S. M. and Baskes, M. I., *J. Mater. Res.* **5**, 313 (1990).
Hodges, C. H., *Phys. Rev. Lett.* **25**, 284 (1970).

Holas, A. and March, N. H., *Phys. Rev.* (1995), (to appear).
Holzwarth, V. and Etzmann, U, *Phil. Mag. Lett.* **70**, 75 (1994).
Hornbogen, E., *International Materials Reviews* **34**, 277 (1989).
Hosson, J. Th. M. de, *Int. J. Quant. Chem.* **XVIII**, 575 (1980).
Huang, Z. H., Tian, J. F. and Wang, Z. G., *Materials Science and Engineering* **A118**, 19 (1989).
Huang, W., Mura, T., *J. Appl. Phys.* **41**, 5175 (1970).
Hume-Rothery, W., Mabbolt G. W. and Channel-Evans, K. M., *Phil. Trans. Roy. Soc. London* **A233**, 1 (1934).
Hume-Rothery, W., Smallman, R. E. and Haworth, C. W., *Structure of Metals and Alloys* (Institute of Metals, London, 1969).
Hutchison, M. M. and Louat, N., *Acta Met.* **10**, 255 (1962).
Hutchison, T. S. and McBride, S. L., *Canad. J. Phys.* **50**, 906 (1972).
Ichimaru, S., *Rev. Mod. Phys.* **54**, 1017 (1982).
Igarashi, M., Kantha and Vitek, V., *Phil. Mag.* **B63**, 603 (1991).
Indenbom V. L. and Lothe J. (eds), Elastic Strain Fields and Dislocation Mobility; *Modern Problems in Condensed Matter Sciences* (*MPCMS*), Vol. 31, (Series eds.), Agranovich, V. M. and Maradudin, A. A. (North-Holland, 1992).
Indenbom, V. L., in: Elastic Strain Fields and Dislocation Mobility, eds. Indenbom, V. I. and Lothe, J., *MPCMS*, Vol. 31 (North-Holland, 1992) p. 1.
Indenbom, V. L. and Chernov, V. M., in: *Elastic Strain Fields and Dislocation Mobility*, eds. Indenbom, V. I. and Lothe, J. (Elesvier Science Publishers B. V., 1992).
Indenbom, V. L., Petukhov, B. V. and Lothe, J., in: *Elastic Strain Fields and Dislocation Mobility*, eds. Indenbom, V. L. and Lothe, J. (Elesvier Science Publishers B. V., 1992) p. 489.
Ivanov *et al.*, *Phil. Mag.* **B69**, 1185 (1994).
Jaeger, H. M., Nagel, S. B. and Behroner, R. P., *Physics Today* **32**, (1996).
Jena, P., Ponnambalam, M. J. and Manninen, M., *Phys. Rev.* **B24**, 2884 (1981).
Jensen, K. O. and Walker, A. B., *J. Phys.: Condens. Matter.* **2**, 9757 (1990).
Jiang, B. and Lung, C. W., ICTP Preprint of the International Centre for Theoretical Physics, Trieste, Italy, 1995, (IC/95/209).
Jiang. J and Lung, C. W., *J. Mater. Sci. Technol.* **12**, 69 (1996).
Jiang, J., Zhou, X. Z., Zhu, J. and Lung. C. W., in: *Positron Annihilation Proc. ICPA-10*, or Mater. Sci. Forum, Vols. 175–178, 1995, pp. 395.
Jiang, J., Xiong, L. Y., Lung, C. W., Ji, G. K., Liu, N. Q. and Wang, S. Y., in: Positron Annihilation, *Proc. ICPA-6*, 1982, pp. 496.
Jiang, J, Xiong, L. Y., Zhu, J. and Lung, C., W., (to be submitted for publication).
Jillson, D. C., *Trans. AIME* **188**, 1129 (1950).
Johnson, M. D. and March, N. H., *Phys. Lett.* **3**, 313 (1963).
Johnson, R. A., *Phys. Rev.* **B6**, 2094 (1972).
Johnson, R. A., *Phys. Rev.* **B37**, 3924 (1988).

Johnson, R. A., in *Computer Simulations in Materials Science*, eds. Arsenault, R. J., Beeler, Jr, J. R. and Estering, D. M. (American Society for Metals, Ohio Park, 1987).

Johnson, R. A. and Beeler, J. R., *Interatomic Potentials and Crystalline Defects*, ed. Lee, J. K. (The Metallurgical Society Of AIME, New York, 1981).

Jones, W. and March, N. H., *Theoretical Solid State Physics*, Vols. I and II (Dover, New York, 1985).

Kadic. A. and Edelen, D. G., *A Gauge Theory of Dislocations and Disclinations* (Springer, Berlin, 1983).

Kahana, S., *Phys. Rev.* **117**, 123 (1960).

Kahana, S., *Phys. Rev.* **129**, 1622 (1963).

Kanzaki, H., *J. Phys. Chem. Solids* **2**, 24 (1957).

Kanzaki, H., *J. Phys. Chem. Solids* **2**, 107 (1957).

Kear, B. H. and Wilsdorf, H. G. F., *Trans. AIME* **224**, 382 (1962).

Kê, T. S., Advances in Science of China, *Physics* Vol. 3, (Science Press, Beijing, 1990b) pp. 1–113.

Kê, T. S., *J. Appl. Phys.* **19**, 295 (1948).

Kê, T. S., *J. Appl. Phys.* **20**, 274 (1949)

Kê, T. S., *J. Appl. Phys.* **20**, 1226 (1949).

Kê, T. S., *Phys. Rev.* **71**, 533 (1947).

Kê, T. S., *Phys. Rev.* **72**, 41 (1947).

Kê, T. S., *Proceedings of the 9th International Fraiction and Ultrasonic Attenuation in Solids*, 1989, Beijing, China. (Kê, T. S. *et al.*), (International Academic Publishers, Pergamon Press, Beijing, 1990a) p. 113.

Kelly, A., *Strong Solids*, 2nd edition, (Oxford: Clarendon Press, 1973).

King, A. M. and Smith, D. A., *Acta Cryst.* **36**, 335 (1980).

Kirkwood, J. G., *J. Chem. Phys.* **3**, 300 (1935).

Kléman, M., in: *Dislocations in Solids* Vol. 5, ed. Nabarro, F. R. N. (North-Holland, Amsterdam, 1980) p. 243.

Kluin, J. E. and Hehenkamp, Th., *Phys. Rev.* **B44**, 11597 (1991).

Kobayashi, S. and Ohr, S. M., *Phil. Mag.* **A42**, 763 (1980).

Kohn, W., *Phys. Rev. Lett.* **2**, 393 (1959).

Kohn, W. and Yaniv, A., *Phys. Rev.* **B20**, 4948 (1979).

Kojima, H., and Suzuki, T., *Phys. Rev. Lett.* **21**, 896 (1968).

Kondo, K., *Proc. 2nd Japan Nat. Congr. Appl. Mech.* (Science Council of Japan, 1952) p. 41.

Kong, Q. P., see Kê, T. S., 1990a, pp. 419.

Kosevich, A. M., *Dislocations in Solids*, ed. Nabarro, F. R. N. Vol. 1, (North-Holland, Amsterdam, 1979) p. 33.

Kosugi, T. and Kino, T., *J. Phys. Soc. Japan* **58**, 4269 (1989).

Kosugi, T. and Kino, T., *Mater. Sci. Eng.* **A164**, 368 (1993).

Kraft, T., Marcus, P. M., Methfessel, M. and Scheffler, M., *Phys. Rev.* **B48**, 5886 (1993).

Krasko, G. L., *Int. J. Refract Hard Met.* (UK), **12**, 251 (1993-4).

Kreuzer, H. J., Lowy, D. N. and Gortel, Z. W., *Solid State Commun.* **35**, 381 (1980).
Krim, J., Solina, D. H. and Chiarello, R., *Phys. Rev. Lett.* **66**, 181 (1991).
Kirm, J, and Widom, A., *Phys. Rev.* **B38**, 12184 (1988).
Kröner, E., *Arch. Rat. Mech. Anal.* **4**, 273 (1960).
Kröner, E., On Gauge Theory in Defect Mechanics, Proc. 6th Symp. on Trends in Application of Pure Mathematics to Mechanics, *Lecture Notes in Physics*, 1986, pp. 249.
Kubica, P. and Stewart, A. T., *Phys. Rev. Lett.* **34**, 852 (1975).
Kubin, L. P., *Rev. Deform. Behav. Mater* **4**, 181 (1982).
Kunin, I. A., in: The Mechanics of Dislocations, *Proc. Int. Symp.* 1986, pp. 69.
Kusmiss, J. H., Stewart, A. T., *Adv. Phys.* **16**, 471 (1967).
Lagerlof, K. P. D., Heuer, A. H., Castaing, J., Riviere, J. P. and Mitchell,T. E., *J. Am. Cer. Soc.* **77**, 385 (1994).
Lai, Z. H., Ma, C. X. and Conrad, H., *Scripta Met. Mat.* **27**, 527 (1992).
Lai, Z. H., Conrad, H., Chao, Y. S., Wang, S. Q and Sun, J., *Scripta Metall.* **23**, 305 (1989).
Landau, L. D. and Lifshitz, E. M., *Field Theory* (Peking, 1961 in Chinese) pp. 90–94.
Landman, U., *Nature* **374**, 607 (1995).
Landman, U., Barrett, R. N., Cheng, H. P., Cleveland, C. L. and Luedtke, W. D., *Computations for the Nano-Scale*, eds. Blochi, P. E., Joachina, C. and Fisher, A. J. (Kluwer, Dordrecht, 1993) pp. 75–113.
Langer, J. S., *Phys. Rev. Lett.* **70**, 3592 (1993).
Lannoo, M. and Allan, G., *J. Phys. Chem. Solids* **32**, 637 (1971).
Laub, W., Oswald, A., Muschik, T., Gust, W. and Fournelle, R. A., in: *Solid-State Phase Transformations*, eds. Johnson, W. C., Howe, J. M., Laughlin, D. E. and Soffa, W. A. (Warrendale, Pennsylvania, 1994) p. 1115.
Lavrent'yer, F. F. and Salita, O. P., *Phys. Met. Met.* (USSR) **48**, 108 (1979).
Lawley, A., van den Sype. J. and Madin, R., *J. Inst. Metals* **91**, 23 (1962-63).
Lee, D., *Acta Metall.* **17**, 1057 (1969).
Legrand, P. B., *Phil. Mag.* **B49**, 171 (1984).
Leibfried,G., in: *Dislocations and Mechanical Properties of Crystals* eds. Fisher, J. C., Johnston, W. G., Thomson, R. and Vreeland, Jr. (Wiley, New York, 1957) p. 495.
Li, J. C. M., in: *Dislocation Modelling of Physical Systems*, eds. Hartley, C. S. *et al.* (Oxford: Pergamon, 1981) p. 498.
Li, S. S.; *Acta Mech. Sinica*, **18** 350 (1986), (in Chinese).
Li, S. S. and Lin, G. H., *Acta Phys. Sinica* **31**, 38 (1982), (in Chinese).
Liang, S. J. and Pope, D. P., *Acta Metall.* **25**, 485 (1977).
Liebowitz, H., *Fracture* Vols. I–VII (Academic Press, New York, 1984).
Lifshitz, E. M., *Soc. Phys. JETP* **2**, 73 (1956).
Lihaqiaor, B. A. and Hayinov, P. Y., *Introduction to Theory of Disclinations*, (Translated by Ding, D. H. and Zhou, R. S. in Chinese Wuhan University Press, 1989) p. 20.
Lin, I. H. and Thomson, R., *Scripta Metall.* **17**, 1301 (1983).

Lin, Z. R., Chokshi, A. H. and Langdon, Y. G., *J. Mater. Sci.* **23**, 2712 (1988).
Lindhard, J., Kgl. *Danske Mat-fys. Medd.* **28**, 8 (1954).
Ling, D. D. and Gelatt, C. D., *Phys. Rev.* **B22**, 557 (1980).
Liu, C. T., Lee, E. H. and McKameg, C. G., *Scripta Metall.* **23**, 1875 (1989).
Liu, C. T., McKamey, C. G. and Lee, E. H., *ibid* **24**, 385 (1990).
Liu, C. T. and George, E. P., *ibid* **24**, 1285 (1990).
Liu, C. T., *ibid* **27**, 25 (1992).
Liu, S. H., in *Lecture Notes of ICTP Working Party on Electrochemistry*, 1990, Trieste, Italy.
Liu, S., Xiong, L. Y. and Lung, C. W., *Phys. Stat. Sol.* (a), **99**, 97 (1987).
Llanes, L. and Laird, C., *Mater. Sci. Engn.* **A157**, 21 (1992).
Long, Q. Y., Deng, J., Jiang, J and Lung, C. W., Submitted for publication.
Long, Q. Y., Li, S. Q., and Lung, C. W., J. Phys. D: *Appl. Phys.* **24**, 602 (1991).
Long, Q. Y., Fu, R., Zhang, T. Y. and Lung, C. W., *J. Phys. D: Appl. Phys.* **22**, 991 (1989).
Lothe, J., in: Elastic Strain Fields and Dislocation Mobility, eds. Indenbom, V. I. and Lothe, *J., MPCMS* Vol. 31 (North-Holland, 1992) p. 175.